MASS SPECTROMETRY IN
SCIENCE AND TECHNOLOGY

Mass Spectrometry in Science and Technology

F. A. White

JOHN WILEY & SONS, INC., NEW YORK · LONDON · SYDNEY

Library of Congress Catalog Card Number: 68-19783
Printed in the United States of America

Dedicated to A. O. C. NIER

Preface

Mass spectrometry has entered a new era. It has emerged from the university environment where it was conceived and had its embryonic growth. Today, a variety of mass spectrometers stand in the major industrial laboratories throughout the world. Their appearance is neither fortuitous nor a technical exhibit of corporate affluence. Rather, these instruments represent a capital investment without which many facets of basic and applied science cannot properly be examined.

The discovery of new phenomena, the development of new materials and devices, and the improvement of system reliability have become increasingly dependent on the access scientists and engineers have to highly detailed information. This information is forthcoming only through the medium of modern instrumentation—which is no longer a mere adjunct of science but an integral part of science itself. To paraphrase the apt remark of Alfred North Whitehead, if we are on a higher imaginative level today, it is probably not because we have finer imaginations but because we possess better instruments. In the physical sciences two major classes of analytical tools have allowed us to address detailed questions to nature and receive answers of great significance. The first category is derived from the ingenious utilization of the electromagnetic spectrum (e.g., optical, infrared, and microwave spectroscopy, x-ray diffraction and fluorescence, and nuclear and paramagnetic resonance). A second class makes use of the "particles" of modern physics; the electron microscope and mass spectrometer are prime examples.

Despite the potential wealth of data that a mass spectrometer would seem to provide, mass spectrometry advanced at only a modest pace until the last decade. Before World War II the mass spectrometer remained a rather specialized tool in the hands of atomic and nuclear physicists and a small number of chemists and geologists. Spectrometers displayed the isotopic structure of the elements, measured the binding energies of nuclei, identified the fissionable isotope of uranium (U^{235}), and made possible the discovery of a large number of new nuclides arising from radiation-induced transmutations. It was not employed to identify chemical compounds on a large scale, although its potentialities were early recognized by those in the petroleum industry. After World War II a phenomenal growth occurred in the application of isotopes—

primarily those that were radioactive. Their general availability and the reasonably simple instrumentation required for their use made them attractive candidates for thousands of tracer applications. Radiotagging became almost fashionable, and it gave rise to the birth of many industrial enterprises. I would be the last to discount the tremendous impact that radiotracers have made on modern science and technology, but the time has now arrived to consider the many fields in which the detection of nonradioactive atoms and stable isotopes can be important.

Thus the principal objective of this book is to make a serious appraisal of the mass spectrometer's present and potential relevance to a large number of professional fields. Few instruments contribute to a broader research spectrum than do the modern mass spectrometers which can furnish data relating to the charge, mass, and kinetic energy of ions, atoms, and molecules. Nevertheless, the magnitude of the contribution of these instruments to research, engineering, or product monitoring will be largely commensurate with the creativity and resourcefulness of those who apply them. Therefore in my opinion the best measure of the success of this book will be the extent to which engineers and scientists in diverse fields are stimulated to consider whether spectrometers can assist them in their own areas of endeavor.

This type of undertaking presents a single author with a challenging but difficult assignment. It is one that also calls for a more general treatment of the subject matter than is found in the usual technical monograph. I am hopeful, however, that the intended trade-off of analytical detail for a broader perspective will be welcomed by the nonspecialist. Further, although in no sense is this work a textbook, an effort has been made to present mass spectrometry in a manner that will be helpful to students having interdisciplinary interests.

Predicting the growth rate of any specific technical area is almost impossible, and the same comment can be made for the use of mass spectrometric instruments. Who in an earlier age could have guessed the need for microscopes? A century ago there were only about 50 optical microscopes in America. By 1936 a single manufacturer in the United States produced its Serial No. 250,000, and the current *annual* production rate (including student models) is probably close to one-half this figure. Commercial electron microscopy did not have its advent until 1940; by 1953, 500 instruments were being utilized by industrial, government, medical, and university laboratories. Electron microscopes now number in the thousands. Similarly, it seems difficult to overestimate the impact of mass spectrometric instrumentation during the next few decades.

To begin with, nature has been generous in providing two-thirds of the elements with more than a single isotope. Roughly 300 naturally occurring isotopes are distributed among the 92 chemical species. This variety provides an unambiguous fingerprint for many elements and it has allowed the spectrometer to detect the presence of trace-element impurities many orders of magnitude below those that can be detected by conventional microchemical methods. The increasing importance of impurity-sensitive devices and materials (e.g., semiconductors and superconductors) has underscored the importance of mass spectral assays, and mass spectrometers have already monitored impurities in the parts-per-billion range. In favorable cases chemical assignment on the basis of the detection of only a thousand atoms ($\sim 10^{-19}$ gram) may be technically feasible in the future. Many papers have recently appeared relating to species identification of atoms and molecules in the upper atmosphere, but they can be viewed as special cases of a much larger class of measurements. Studies relating to plasmas, molecular dissociation, charge exchange, sputtering, and residual gas analysis demand detailed analyses without which components or large engineering systems can be neither properly designed nor appraised. In the metallurgy field the mass spectrometer can help to answer questions relating to high-temperature diffusion and impurity concentrations at grain boundaries. The same statement can be made for the difficult areas of surface chemistry and surface physics. Other problems involving mass transfer phenomena may range from laboratory tests of the wear characteristics of gears and bearings to the rapidly expanding field of environmental science. Tracer studies in biological and medical research can undoubtedly be enhanced by employing advanced techniques in mass spectral assay.

Thus mass spectrometry stands at the crossroads of many technical disciplines. It marks the point of intersection of nuclear and solid-state physics, chemistry and materials research, plasma physics and electrical engineering, and atmospheric science and studies of the earth's mantle. Prerequisite to specific measurements in many of these fields will also be the availability of stable isotopes that have been separated or enriched to a high degree of isotopic purity. In the same sense that a crystallographer requires a single or well-ordered crystal the modern mass analyst often requires atoms of a single mass number. Polycrystalline structures and naturally occurring polyisotopic compositions rarely represent the optimum materials for research. Their entropy is too high and their potential is too low. Even in highly practical situations the availability of separated isotopes greatly enhances the sensitivity of the mass spectrometric approach. The oxygen content in metals and semiconductors, uranium concentrations in lakes and rivers, trace nutrients in plant life, or build-up of noxious gases in a space capsule—all, in principle, become more amenable to measurement if separated isotopes can be introduced at some stage of the analysis.

Isotopic tagging or coding of products also offers intriguing possibilities that may be limited more by economic considerations than by technical feasibility. At the very least, I would expect mass spectrometry to play an important role in process monitoring, quality control, and the establishment of standards.

It is a pleasure to acknowledge both direct and indirect contributions that have led to this book. Professor A. O. C. Nier of the University of Minnesota has been a continual source of encouragement for many years, and more than any other scientist he has been responsible for the growth of mass spectrometry throughout the United States. Among the many others with whom I have been privileged to communicate over an extended period are Dr. A. J. Ahearn of the U.S. National Bureau of Standards, Dr. A. E. Cameron of the Oak Ridge National Laboratory, Dr. W. D. Davis of the General Electric Research and Development Center, Dr. W. M. Gibson of the Bell Laboratories, W. M. Hickam of the Westinghouse Research and Development Center, Professor M. E. Inghram of the University of Chicago, and Dr. R. E. Honig of the RCA Laboratories. Former General Electric colleagues at the Knolls Atomic Power Laboratory who shared many early years of experience include T. L. Collins, L. A. Dietz, J. L. Mewherter, F. M. Rourke, and J. C. Sheffield, and to this staff I shall always be indebted. More recently, graduate students at the Rensselaer Polytechnic Institute have been a source of inspiration and assistance. I should like to express my appreciation to John Wiley staff members for extending many courtesies to me, and I am grateful to the following publishers for permission to use previously reported materials: Academic Press, American Institute of Physics, *American Scientist,* Cambridge University Press, Elsevier Publishing Company, Institute of Physics and the Physical Society (London), McGraw-Hill Book Company, National Academy of Sciences, National Aeronautics and Space Administration, North-Holland Publishing Company, Pergamon Press Ltd., Prentice-Hall, Inc., and Reinhold Publishing Corporation. A special acknowledgement is also in order not only to those scientists whose specific work is identified in these pages but also to the many others whose contributions have made mass spectrometry what it is today.

Miss Judy Stemp provided secretarial assistance and Mathew Trzepacz assumed responsibility for many of the graphics. Their help has been invaluable in the preparation of this book.

FREDERICK A. WHITE

Rensselaer Polytechnic Institute
Troy, New York
January, 1968

Contents

Chapter 1

History

The English scientist Francis William Aston is often cited for his pioneering studies that set the stage for modern mass spectrometry. Yet Aston himself credits the imagination and insight of Sir William Crookes with the first modern concept of the isotopic structure of the chemical elements. In an address presented to the Chemical Section of the British Association at Birmingham in 1886 Crookes announced [1]:

"I conceive, therefore, that when we say the atomic weight of, for instance, calcium is 40, we really express the fact that, while the majority of calcium atoms have an actual atomic weight of 40, there are not a few which are represented by 39 or 41, a less number by 38 or 42, and so on. Is it not possible, or even feasible, that these heavier and lighter atoms may have been in some cases subsequently sorted out by a process resembling chemical fractionation? This sorting out may have taken place in part while atomic matter was condensing from the primal state of intense ignition, but also it may have been partly effected in geological ages by successive solutions and reprecipitations of the various earths. This may seem an audacious speculation, but I do not think it beyond the power of chemistry to test its feasibility."

EARLY THEORIES

With this concept in mind, Crookes pursued his remarkable researches on the rare earths. Commenting on the optical spectrum of yttrium, he said "Here, then, is a so called element whose spectrum does not emanate equally from all its atoms; but some atoms furnish some, other atoms others, of the lines and blends of the compound spectrum of the element. Hence the *atoms of this element differ probably in weight,* and certainly in the internal motions they undergo." Suggesting that the same might be the case for elements generally, he coined the term "meta-elements" to explain the phenomenon of fractional atomic weights. But this theory of "meta-elements" was abandoned when more refined chemical methods showed that the spectra observed by Crookes resulted from an actual mixture of rare earth elements, each having a characteristic spectrum and atomic weight. Nevertheless, the isotopic structure of the elements had been conceived.

By the turn of the century the measured values of "atomic weights" became more accurate and consistent. It was clear that many atomic weights were very close to integers, when represented by a mass scale based on a value of 16 for oxygen. The correlation could hardly be fortuitous, but how did the many elements having weights not even close to integral numbers fit into the scheme of things? Strangely enough, the revival of Crookes' idea of "meta-elements" came from a new and exciting field being developed by Wilhelm Roentgen, Pierre and Marie Curie, and others. In Aston's own words: "Apart from speculative considerations [of Crookes], the theory of isotopes had its birth in the gigantic forward wave of human knowledge inaugurated by the discovery of radioactivity" [2].

Can we then give a single name or date that marks the discovery of the isotopic composition of matter and the advent of mass spectrometry? No! Contributions came from laboratories in England, France, Switzerland, Germany, and the United States during the feverish burst of activity that was the prelude to the 1913 postulation, by Niels Bohr, of the nuclear atomic model.

In the preceding century important basic discoveries were (a) the law of chemical combining weights—John Dalton, 1808; (b) the distinction of atoms and molecules—Amedeo Avogadro, 1811; (c) the unit weight hypothesis—the physician and biochemist William Prout, 1815, and (d) the periodic classification of the elements—Dmitri Mendeleev, 1868. It is interesting to note that this latter milestone was achieved in the same decade when an issue of *Scientific American* contained the statement: "the formula for water, according to recent discoveries, must soon be changed from HO to H_2O."

Then an instrumental observation was made by Eugene Goldstein, a German physicist who was exploring electrical discharges at low pressure. In a rudimentary precursor to a cathode ray tube he noticed streamers of light that appeared to be emanating from holes in a perforated cathode disk. He surmised that these luminous streamers must be associated with some type of rays or particles that travelled in a direction opposite to the usual cathode rays. He chose to call these rays "Canalstrahlen," because of their canal or capillary-like appearance in the tube. In 1898, using a similar apparatus, W. Wien showed that these rays could be deflected in a sufficiently strong magnetic field. Thus the stage was set for the later development of evacuated tubes, strong magnetic fields, and general methods for deflecting atomic and molecular charged particles.

RADIOACTIVITY

As stated above, the first clear evidence for the polyisotopic composition of matter came through research with the radioactive elements. Isotopes per se were generally unsuspected and unsought. In atomic physics the excitement was associated with the discovery of radioactivity of uranium by Henri Becquerel (1896), the identification of the alpha particle as a helium nucleus by Lord Rutherford (1903), and the mass-energy equivalence proposed by Albert Einstein (1905). But many chemists were also spending long hours examining the heavy elements with new tools. In 1906 B. Boltwood [3] at Yale discovered a new element that he termed *ionium,* which he identified as having the properties of thorium. Subsequent work conducted in Germany suggested that ionium and thorium were chemically inseparable. In 1907 mesothorium was discovered by Otto Hahn, and this element was found by W. Marckwald and F. Soddy [4] to be chemically indistinguishable from radium.

These and other studies seemed to point up similar chemical identities for the radioactive species, and certain regularities in radioactive decay chains were noticed by other workers. Collectively these caused Frederick Soddy (an early co-worker—and golf partner—of Rutherford) to issue a memorable report on radioactivity in 1910. He concluded [5]:

"These regularities may prove to be the beginning of some embracing generalisation, which will throw light, not only on radioactive processes, but on elements in general and the Periodic Law. Of course, the evidence of chemical identity is not of equal weight for all the preceding cases, but the complete identity of ionium, thorium and radiothorium, of radium and mesothorium 1, of lead and radium D, may be considered thoroughly established . . . . *The recognition that elements of different atomic weights may possess identical properties seems destined to have its most important application in the region of inactive elements, where the absence of a second radioactive nature makes it impossible for chemical identity to be individually detected.* Chemical homogeneity is no longer a guarantee that any supposed element is not a mixture of several of different atomic weights, or that any atomic weight is not merely a mean number. The constancy of atomic weight, whatever the source of the material, is not a complete proof of homogeneity, for, as in the radio elements, genetic relationships might have resulted in an initial constancy of proportion between the several individuals, which no subsequent natural or artificial chemical process would be able to disturb. If this is the case, the absence of simple numerical relationships between the atomic weights becomes a matter of course rather than one of surprise."

ISOTOPES DEFINED

The brilliant exposition by Professor Soddy provided an answer to the many anomolies in atomic weights. Newton had once remarked that "nature is pleased with simplicity" and Soddy's proposal was attractive to fellow scientists on this basis. His thesis was, of course, substantially the same as that of Sir William Crookes, who had earlier postulated the existence of "meta-elements." But this time the postulate was supported by the first threads of experimental evidence. Subsequently Soddy coined a word of Greek derivation, "isotope," (iso, equal; tope, place) that was quickly to find its way into the scientific literature: "The same algebraic sum of the positive and negative charges in the nucleus when the arithmetical sum is different gives what I call *isotopes* or isotopic elements because they occupy the same place in the periodic table. They are chemically identical, and save only as regards the relatively few physical properties which depend upon atomic mass directly, physically identical also."

SPECTROSCOPIC EVIDENCE OF ISOTOPES

New experimental evidence in support of the isotopic structure of matter was soon forthcoming. It appeared on the photographic plates of both optical and x-ray spectroscopes, and finally in the detection of faint lines in the camera of a mass spectrograph.

The photographic darkroom of Ernest Rutherford's laboratory was the scene of the search for new optical spectral lines; in 1912 Russell and Rossi [6] decided to compare the emission spectrum of pure thorium with that of a mixture of thorium and ionium. The presence of a large fraction of ionium was observed by its alpha-particle activity. Not a single additional spectral line for the thorium-ionium mixture could be found when it was compared with pure thorium. The possibility that ionium had no arc spectrum in the observed optical spectrum seemed highly improbable. The only remaining explanation supported the chemical identity of the samples and the isotope postulate, namely, that the cloud of valence electrons might be the same, but the two nuclei could have different weights.

What seemed convincing from an examination of the optical region of the electromagnetic spectrum was substantiated when Rutherford and his colleagues examined the x-ray region. The prelude to their specific experiment was the brilliant work by Moseley, who showed that many metallic elements could be excited to yield a characteristic x-ray spectrum—unique for each element and independent of the compound or physical state of the specimen. Rutherford chose to look at a member of the uranium family, Radium-B and its products. The characteristic L radiation corresponded exactly to the expected wavelength for ordinary lead, and subsequent studies made the x-ray assignment of lead conclusive.

THE ISOTOPES OF LEAD

By about 1915 the collective results of many investigators seemed to indicate that the final decay product of all heavy elements was lead. It was also known that the emission of discrete alpha and beta particles would suggest an integral change in atomic number (nuclear charge) and atomic weight. Thus thorium subjected to 6 alpha decays should result in a net change of 24 units (i.e., from 232 to 208). But lead from non-radioactive sources was known to be approximately 207.2—far from an integral value. A plausible explanation was advanced by Professor Soddy, who suggested that the atomic weight of any lead sample might be related to the source—and thus its relative concentration of uranium or thorium. He reported a value of greater than 207.7 for thorite mined in Ceylon. In 1916 T. W. Richards [7] of Harvard analyzed lead from uranium-rich ore from Norway that yielded a value as low as 206.1. The results were conclusive. Several decay series of these heavy elements were responsible for the multiplicity of daughter products. The final stable nuclei that remained were those of lead, but the atomic weights were indeed quantized: they were Pb^{206}, Pb^{207}, and Pb^{208}.

It is also of historical interest to note that this element, widely distributed in nature, was one of the first known to the ancients. The Babylonians used it in engraving, and extensive use of lead was made by the Romans for writing tablets, water pipes, and coins. (Large deposits were found in the 18th century in North America, and today the tri-state area of Missouri, Kansas, and Oklahoma is one of the world's chief sources of the metal.) Oddly enough, this element—the stable endproduct of radioactive decay—is also one of the most extensively used materials for radiation shielding in today's nuclear-power industry.

J. J. THOMSON'S POSITIVE RAY ANALYZER

At the same time that radiochemists were finding an explanation for the variations in the atomic weight of lead, the prototype of the modern mass spectrograph was being assembled at the Cavendish Laboratory in England. J. J. Thomson, stimulated by Crookes' earlier work, embarked on a more detailed investigation into the properties of "positive rays." The essential features of his apparatus are contained in Figure 1.1. Positively charged atoms formed in a low-pressure discharge tube were accelerated and collimated by a fine-bore brass tube of several centimeters in length ("the finer the bore, the more accurate results obtained"). These charged atoms or ions then proceeded into a low-pressure region along an axis between two electrostatic plates and the pole pieces of an electromagnet. The ions were stopped by a fluorescent screen that allowed direct observation of the final focal point of any particular ion trajectory.

Thus with an apparatus having several elements common to a television tube, Thomson demonstrated that a heavy charged particle could be analyzed with respect to its mass and charge. Consider, for example, a small deflection produced by an electric field strength E_y on a particle of mass m, charge e moving in the x direction with a velocity v. Simple kinematics show that if the angle of deflection is small, a displacement should be observed owing to the interaction of the particle and the field corresponding to

$$y = kE_y \frac{e}{mv^2}. \tag{1.1}$$

Similarly, if the electric field is removed, and a magnetic field of intensity B is applied so as to cause a deflection along the z coordinate, the displacement will be related by an expression

$$z = k' \frac{Be}{mv}. \tag{1.2}$$

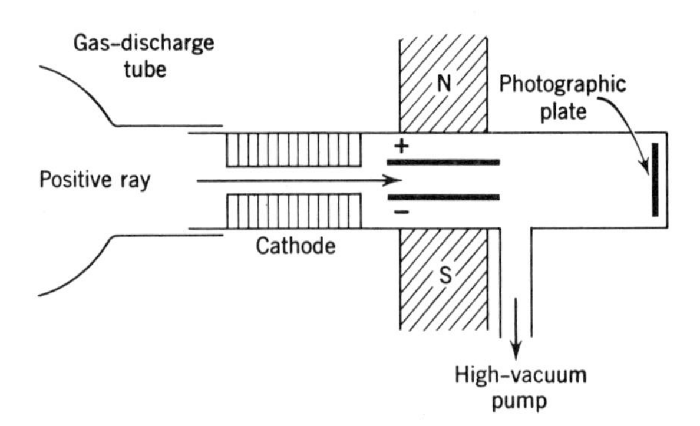

Fig. 1.1 Thomson's positive ray analyzer.

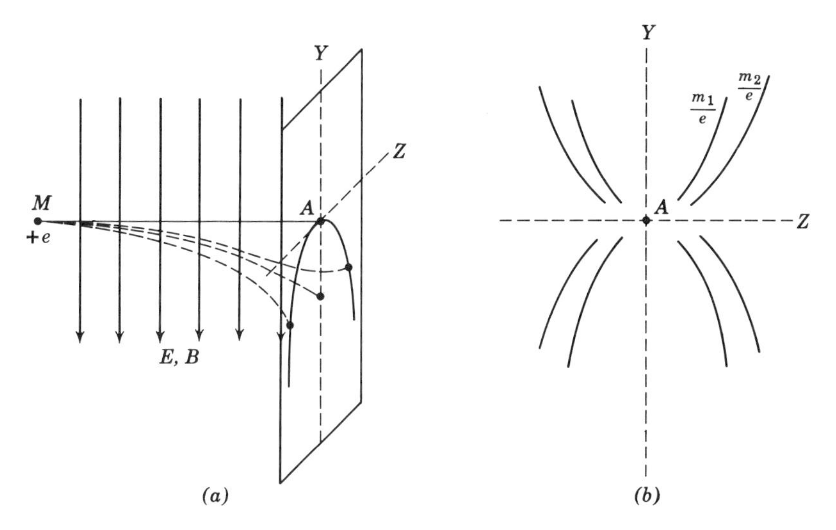

Fig. 1.2 Parabolic trajectories of charged particles.

The constants k and k' depend only upon the dimensions and geometry of the specific apparatus. For the case when both electric and magnetic fields are superposed, the ion will impinge upon the fluorescent screen at a point (y,z), where z/y will be proportional to the ion velocity, and z^2/y is proportional to the mass to charge ratio m/e. A parabola results for ions of a single mass but which have multiple velocities. Ions having different m/e values produce different parabolas, as indicated in Figure 1.2.

Thomson's earlier work in 1879 had demonstrated that the electron was a constituent of all atoms, and in 1903 the charge of the electron had been measured by cloud-chamber analysis. Thus Thomson could confidently interpret the positive ray trajectories that he observed on the fluorescent screen. Using photographic plates to supplement his fluorescent detector, he saw clearly that the recorded ion streams were discrete, sharply-defined families of particles and not mere blurs. This work constituted an important experimental proof that the individual atoms of the same element had approximately the same mass. It also marked the advent of a completely new field of spectroscopy, the spectroscopy of *mass*. Accurately forecasting the future, Thomson said "I feel sure that there are many problems in chemistry which could be solved with far greater ease by this than by any other method. The method is surprisingly sensitive, more so even than that of spectrum analysis [optical], requires an infinitesimal amount of material and does not require this to be specially purified. The technique is not difficult if appliances for producing high vacuum are available."

THE ISOTOPES OF NEON

Continued refinements of the parabolic ray analyzer soon led to an important new discovery. By the summer of 1912 Thomson could resolve and clearly distinguish parabolas corresponding to atomic mass differences of less than 10%. Several gases were subjected to analysis, and in November a small purified sample of neon gas was examined—this at a time when "there was probably less than one gramme in existance."

In a subsequent address to the Royal Institution in January of 1913 Thomson said [8]:

"I now turn to the photograph of the lighter constituents; here we find the lines of helium, of neon (very strong), of argon, and in addition there is a line corresponding to an atomic weight 22, which cannot be identified with the line due to any known gas. I thought at first that this line, since its atomic weight is one-half that of CO_2, must be due to a carbonic acid molecule with a double charge of electricity, and on some of the plates a faint line at 44 could be detected. On passing the gas slowly through tubes immersed in liquid air the line at 44 completely disappeared, while the brightness of the one at 22 was not affected."

Analysis of subsequent samples of neon in later investigations also indicated a spectral line at mass position 21 as well as at mass 20 and 22. The evidence for the existence of the isotopic structure of matter was thus confirmed, completely independent of any correlation with nuclides appearing as daughter products of radioactive decay.

ASTON'S MASS SPECTROGRAPH

As soon as the existence of isotopes had been clearly established, several workers initiated definite programs to improve the resolution and sensitivity of apparatus that would resolve atomic masses. Two names will always stand out in connection with such early work. The first is that of F. W. Aston, a contemporary of Thomson at the Cavendish Laboratory at Cambridge. The second investigator is A. J. Dempster, Professor of Physics at the Ryerson Physical Laboratory at the University of Chicago.

Aston's work was monumental—both with respect to the improved instrumentation that he developed and for his systematic assay of many elements to determine their isotopic structure. Aston's new instrument was conceived from an optical analogue. Instead of the superposed electric and magnetic fields of Thomson, he proposed that electric and magnetic fields act successively upon a positive ray, or ion, in a fashion similar to the refraction of a light beam by two prisms. Consider an unresolved ion beam passing through collimating slits S_1 and S_2 so that it enters the electric field established by the parallel plate electrodes P_1 and P_2 (see Figure 1.3). The positive ions will undergo a clockwise deflection dependent upon their kinetic energies (and charge), and a portion of the beam will pass through the aperture at S_3. The ions then pass through a magnetic field region in a counterclockwise fashion, and are brought to a focus according to their mass, on a line OX, whose specific angle with respect to the incident beam depends on several parameters. Aston showed that, for small angles, this arrangement of opposing electric and magnetic deflections could bring a "chromatic" mass spectrum to a focus. Use of a photographic plate then permitted a convenient calibration of a mass scale, with the relative isotopic abundance being crudely measured by the intensity of line blackening.

Physical characteristics of this mass spectrograph included a 20,000–50,000-V discharge tube as an ion source, a 5-cm length brass electrode separated by 2.8-mm glass spacers, and a magnet with a pole face diameter of 8 cm. Electrostatic potentials and the magnetic field could be programmed to focus ions from hydrogen to singly ionized atoms of mercury.

With this instrument Aston surveyed a number of gases and metallic elements to detect whether these elements were polyisotopic; he also pioneered the establishment of a precision mass scale that would allow positive identification of molecules and subsequent precision measure-

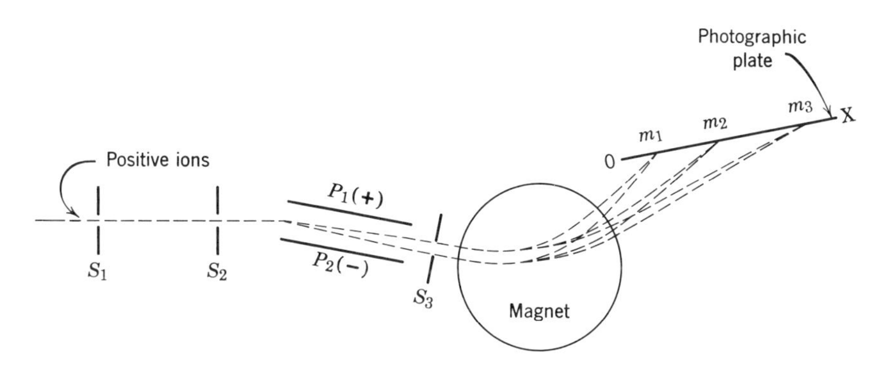

Fig. 1.3 Aston's mass spectrograph.

ments of nuclear binding energies. By 1922 he had suggested that a third isotope of neon might exist, and discovered that multiple isotopes existed in nature for chlorine, bromine, krypton, xenon, mercury, silicon, potassium, and rubidium. By this same date Dempster had also reported his observations on lithium, magnesium, and zinc.

The impact of these early studies can hardly be overemphasized. It led to the ultimate discovery that *most* rather than a few of the elements were isotopic, which justified a new "whole number rule" that had always been so appealing to both philosopher and physicist. It also generated the intensive efforts toward improved instrumentation that allowed hypotheses of nuclear structure to be validated or rejected. Furthermore, these developments came just a half century after a nineteenth-century scientist (J. S. Stas) had said: "I have arrived at the absolute conviction, the complete certainty, so far as it is possible for a human being to attain certainty in such matters, that the law of Prout [integer atomic weights] is nothing but an illusion, a mere speculation definitely contradicted by experience."

SPECTROGRAPHS OF DEMPSTER, BAINBRIDGE, MATTAUCH, AND NIER

In 1918 in the United States, Professor Dempster reported the first 180° mass spectroscope, an instrument which was to serve as the prototype for many subsequent analyzers. Although Aston's instrument possessed both velocity and mass focusing properties, it was restricted to small collimated beams. The Dempster design possessed direction focusing properties (i.e., it focused ion beams possessing a large angular divergence).

The basic design of Dempster was adopted by Harvard's Bainbridge, who added a velocity filter to limit the initial energy distribution of ions entering the magnets. This refinement greatly improved the mass resolving power and the precision of relative mass measurements. It was succeeded by an ingenious combination of electrostatic lens and magnet that combined both direction and velocity focusing properties. This latter design, reported by Mattauch and Herzog in 1934, is one of the most important types available today. Among the early spectrographs, the 60° magnetic sector type designed by Nier also stands high on the list of important instruments.

FIRST COMMERCIAL DEVELOPMENTS

During the 1930's the first "schools" of mass spectrometry were established at Harvard, the University of Chicago, and the University of

Minnesota by Professors Bainbridge, Dempster, and Nier, respectively. The discovery (with an optical spectrograph) of deuterium in 1931 by Professor Urey at Columbia was also of historical significance, as it provided a new means of isotopic labelling in biological systems and also stimulated isotopic research generally. The end of the decade then saw the beginning of industrial research interest in mass spectrometry. At the Westinghouse Research Laboratory exploratory studies were initiated by J. A. Hipple—relating to gaseous discharges and basic electrical phenomena. At the Consolidated Engineering Corporation in Pasadena, the personal interest of a major stockholder, Herbert Hoover, Jr., was in a large measure responsible for the early contributions of the firm's instrumentation effort for geological groups and the petroleum industry [9]. Initial interest focused on the determination of very small quantities of hydrocarbon impurities in soil gases for the purpose of oil prospecting; subsequently attention was directed to the problems of analysis in gasoline refining streams. Thus, in 1943, Consolidated sold its first commercial mass spectrometer to the Atlantic Refining Company.

The advent of World War II marked the beginning of some additional uses for mass spectrometry as the United States decided that an all-out effort should be made to produce U^{235} for atomic bombs. Professor Nier's 60°, 6-in. radius instrument was then used as the prototype after which dozens of instruments were patterned, and these were supplied to the Oak Ridge plant by the General Electric Company. The function of these spectrometers was to monitor the production of uranium and provide accurate ratios of U^{235} to U^{238}. The synthetic rubber program of the war also provided a stimulus to mass spectrometry, but the work of the Manhattan Project was a prime generating factor during the early years of the 1940–1950 decade.

During the 1950–1960 decade, the vast potential applications for mass spectrometric measurements became more evident. New instruments were made available by Consolidated Electrodynamics, Associated Electrical Industries Ltd. (England), the Bendix Corporation, and Nuclide Analysis Associates. A large number of specialized instruments were also built by scientists in universities and government laboratories. In March of 1953 a meeting of historical importance took place in Pittsburgh with the founding of a professional mass spectrometer group, Committee E-14, sponsored by the American Society of Testing Materials. It was not a small meeting, but it contrasted sharply in scope and representation to the Fourteenth Annual Conference on Mass Spectrometry in Dallas, May 1966. In attendance at this latter meeting were staff members from well over 200 industrial, university, government, and independent research institutions, representing 39 states and 15 foreign countries. In 1967, publication plans were announced for an international journal relating exclusively to mass spectrometry and ion physics. Mass spectrometry had indeed entered a new era.

REFERENCES

[1] W. Crookes, *Nature,* **34,** 423 (1886).
[2] F. W. Aston, *Mass Spectra and Isotopes,* Arnold, London, 1933, p. 8.
[3] B. B. Boltwood, *Amer. J. Sci.,* **22,** 537 (1906).
[4] F. W. Aston, *Isotopes,* Arnold, London, 1922, p. 7.
[5] *Ibid.,* p. 8.
[6] A. S. Russell and R. Rossi, *Proc. Roy. Soc.,* **87A,** 479 (1912).
[7] T. W. Richards and C. Wadsworth, *J. Amer. Chem. Soc.,* **38,** 2613 (1916).
[8] F. W. Aston, *Isotopes,* Arnold, London, 1922, p. 34.
[9] F. A. White, "American Industrial Research Laboratories," Public Affairs Press, Washington, D.C. (1961), p. 146.

Chapter 2

Types of Mass Spectrometers

The number of mass spectrometers that have been proposed, designed, and successfully constructed is exceedingly large. Despite the proliferation of specific instruments, however, most mass spectrometers can be included in one of the following classes:

1. Single magnetic analyzers.
2. Double-focusing spectrometers.
3. Multiple magnet systems.
4. The cycloidal mass spectrometer.
5. Cyclotron resonance types.
6. The time-of-flight spectrometer.
7. Quadrupole and rf mass filters.
8. Special types.

Nearly all of the above types are available in commercial versions with advanced engineering features that are suitable for precision analysis or basic research.

SINGLE MAGNETIC ANALYZERS

The 180° Sector

The semicircular, homogeneous field design was first reported by Dempster [1]; and it is still widely employed either as a single magnetic analyzer, or in conjunction with electrostatic lenses. Typical ion trajectories for the 180° magnet arrangement are shown in Figure 2.1, assuming a highly collimated ion beam.

Let singly charged ions be produced and accelerated through a potential V, thus acquiring a kinetic energy eV. If this accelerating potential is large compared with the initial energy distribution of the ions, we can assume that all ions enter the magnetic field with a discrete velocity which is given by

$$eV = \tfrac{1}{2}mv^2, \tag{2.1}$$

where m is the mass of the ion, e is the electronic charge, and v is the terminal velocity of the ion after acceleration. For positive ions having a charge state $n(n = 1, 2, 3 \ldots)$, where n denotes the number of electrons stripped from a neutral atom, the more general expression results:

$$neV = \tfrac{1}{2}mv^2. \tag{2.2}$$

If the magnetic field B is perpendicular to the velocity vector of the ions, the ions will be deflected into a circular orbit, resulting from the balancing of centrifugal and centripital forces:

$$Bnev = \frac{mv^2}{R}, \tag{2.3}$$

where R is the radius of curvature. Eliminating v from the above two equations yields the general expression relating the radius of curvature of an ion in a homogeneous magnetic field to the several other parameters:

$$R = \frac{1}{B}\left(\frac{2mV}{ne}\right)^{1/2}. \tag{2.4}$$

If the magnetic field strength B is expressed in gauss, m is given in atomic mass units, V is the accelerating potential in volts, and n is the multiplicity of electronic charge, then the radius of curvature, in centimeters, is approximated by the relation

$$R = \frac{144}{B}\left(\frac{mV}{n}\right)^{1/2}. \tag{2.5}$$

The above suggests that for ions accelerated through the same potential, those of larger mass number will have a larger radius of curvature. It will also be noted that an ambiguity remains with respect to the mass number and charge state. A doubly charged magnesium ion of mass 24 will have approximately the same radius of curvature as will a C^{12} atom that is only singly ionized. Therefore, unless the magnetic analyzer can resolve the small momentum difference corresponding to these two species, only a single peak will be observed at the mass-equals-12 spectral position. The equation can also be differentiated to yield an expression of the general form

$$\frac{2\Delta R}{R} = \frac{\Delta m}{m} + \frac{\Delta V}{V} - \frac{2\Delta B}{B}. \tag{2.6}$$

For a homogeneous magnetic field and ions having a negligible energy spread, the last two terms will vanish and the expression reduces to

$$\frac{\Delta m}{m} R = 2\Delta R. \tag{2.7}$$

The quantity ($\Delta m/m\ R$) gives a measure of the mass dispersion, or the separation of resolved ions along the focal plane. Thus, in Figure 2.1, if R_1 and R_2 are the radii of curvature of masses m_1 and m_2, respectively, the mass dispersion of these two isotopes well be

$$\left(\frac{m_1 - m_2}{m_1}\right) R_1 = 2(R_1 - R_2). \tag{2.8}$$

For isotopes that differ by only one mass number ($\Delta m = 1$), the mass dispersion, D, along the focal plane will have a magnitude

$$D = c\frac{R}{m} \tag{2.9}$$

where the constant c will depend upon the magnetic sector angle.

Thus the distance along the 180° focal plane between adjacent isotopes will increase directly with the radius of curvature, and it will be inversely proportional to the mass number. Hence analyzers with small radii of curvature and having a reasonable mass separation for light gases may be inadequate for the analysis of metals of high atomic mass number.

Although the 180° analyzer possesses good directional focusing properties, there are definite limitations on the sharpness of the line image, even if the ions are monoenergetic and the magnetic field is perfectly

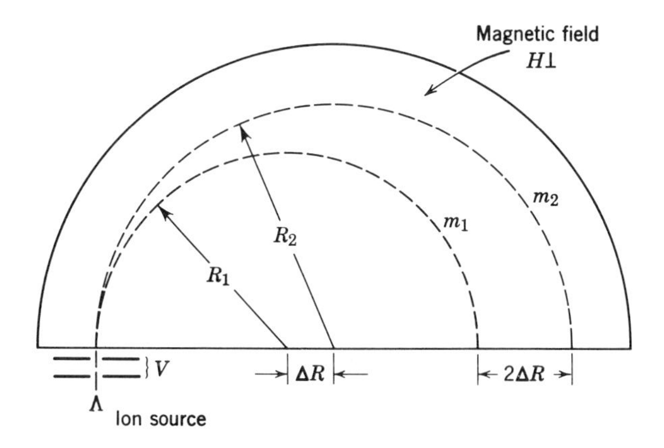

Fig. 2.1 Ion paths in a homogeneous 180° magnetic sector.

homogeneous. Consider Figure 2.2, in which ions leaving a source have a half-angle of divergence, α, and rays are drawn with equal radii from centers of circular orbits from points 0, 0′, and 0″. When α is large, the central ray will not coincide with the peripheral rays, and at the 180° boundary the image width will be of magnitude α^2R. (At the 360° boundary, i.e., initial source position, this aberration will vanish, but so will all mass dispersion).

In most mass spectrometers, however, α is restricted to small angles so that the image in the focal plane remains reasonably sharp. Even with large angles of divergence, special methods may be used to reduce the α^2R line image. From Figure 2.2 it would appear that better focusing might result if one could decrease slightly the effective radius of curvature of the central ion trajectory relative to the paraxial rays. Such a perturbation can, in fact, be accomplished by the introduction of a thin magnetic shim [2], appropriately contoured and indicated schematically by the dotted outline. Other schemes involve providing special pole-piece contours at the magnet boundaries. For high-resolution work,

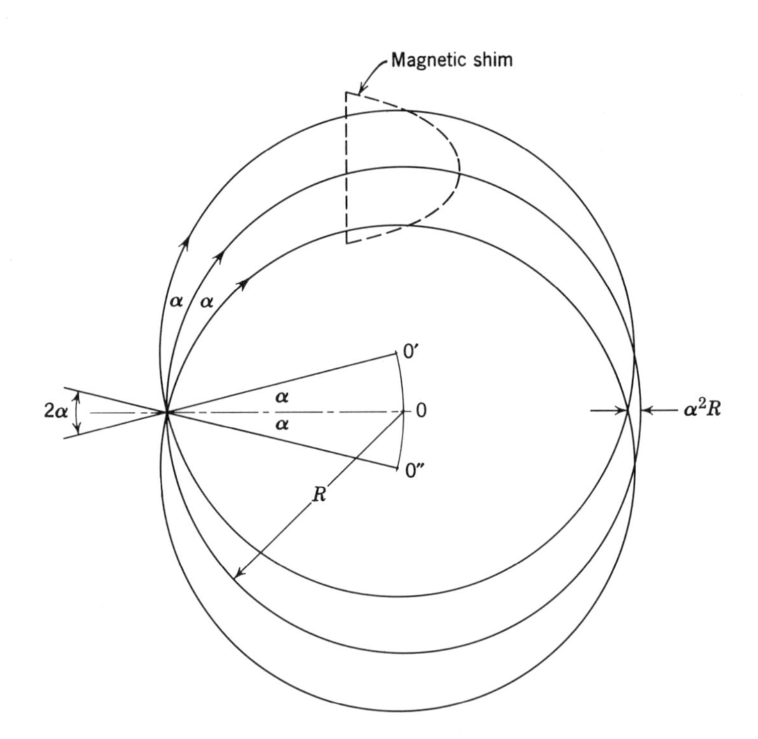

Fig. 2.2 Angular aberration in the 180° sector.

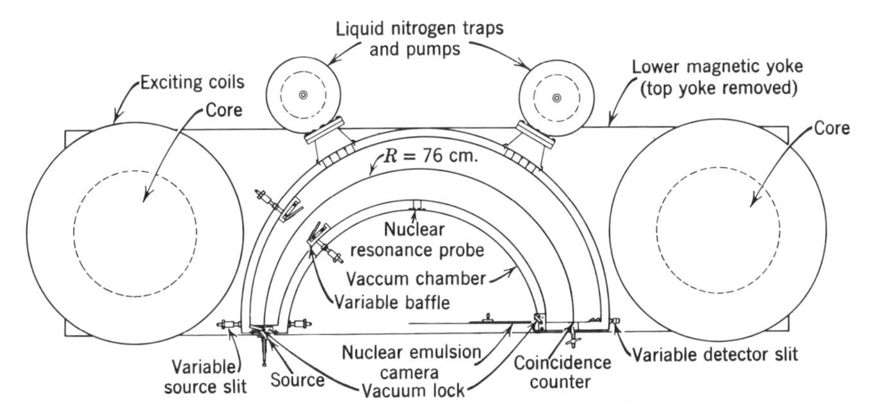

Fig. 2.3 Mass spectrometer and alpha-particle analyzer.

however, combinations of electrostatic and magnetic analyzers provide a better general solution.

There are several desirable features to the 180° geometry. The ion trajectories lie completely within the analyzing field—thus there are no field boundary effects and the theoretical resolution of an instrument can usually be approximated in practice. Ions of all masses are focused along the 180° boundary so that the simultaneous read-out of mass spectra can be conveniently recorded by photographic plates. The vacuum chamber may be separate from the magnet, or the top and bottom pole pieces can function as an integral part of the analyzing vacuum housing. If the magnet is of large dimensions, the 180° analyzer may also serve as a wide-angle alpha-particle analyzer. A 76-cm, 180° radius-of-curvature spectrometer used both as a high-resolution alpha-particle spectrometer and a mass analyzer has been reported by White et al. [3], (Figure 2.3), and in this instance the upper and lower magnetic yokes served as the top and bottom surfaces of the vacuum analyzing chamber.

The 60° Sector

The 60° magnetic sector was introduced by Nier [4] in the late 1930's. Many instruments built for the Oak Ridge National Laboratory during World War II were patterned after Nier's design, and the 60° sector has been among the more popular configurations in hundreds of commercial spectrometers. Figure 2.4 indicates the general magnet configuration and ion trajectory. The wedge-shaped magnet provides direction focusing for divergent ions; the source, detector slit, and apex of the magnetic

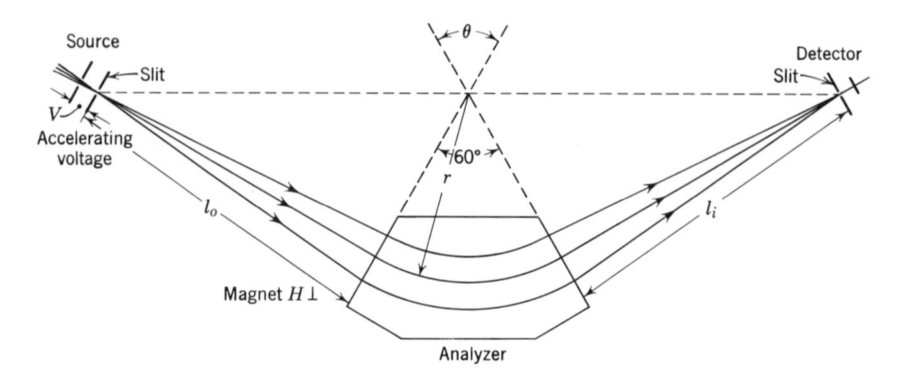

Fig. 2.4 The 60° magnetic analyzer.

sector are on a common line. Both symmetric and asymmetric geometries have been investigated, but the symmetric (equal object-image distances from the magnetic boundary) type is the one usually constructed. A mathematical treatment for the general case of an ion entering and leaving a magnetic sector at right angles has been developed by Herzog. [5].

If l_o is the object distance (source slit to magnetic boundary), l_i is the image distance, r is the effective radius of curvature of the ion trajectory, and θ is the angle of the magnetic sector, then the directional focus equation has the form

$$(l_{om} - g_m)(l_{im} - g_m) = f_m{}^2, \tag{2.10}$$

where $f_m = r/\sin\theta$ and $g_m = f_m \cos\theta$.

Here f_m is the focal length of the system, as in the optical case. In the symmetric situation, $l_o = l_i$ so that

$$l_{om} = l_{im} = f_m + g_m \tag{2.11}$$

or

$$l_{om} = l_{im} = r(\csc\theta + \cot\theta). \tag{2.12}$$

For an angle θ of 60°, $l_o = l_i = r\sqrt{3}$. The ion source and detector slits are thus placed at a distance of $1.732r$ from the magnetic boundaries. This sector angle has proved to be a convenient one for instruments the radii of which have ranged from 5 to 40 cm. There is adequate space to accommodate an ion source housing and appropriate differential pumping. Also, the magnetic shielding of electron multiplier detectors is simplified. For large radii of curvature instruments, the fringing field may be negligible so that magnetic shielding may be unnecessary.

For equal radii of curvature, the magnets, either permanent or electromagnetic, also require considerably less material than do the 180° type.

The 90° Sector

The symmetric 90° magnetic sector represents another extremely useful configuration, and it is also available in many commercial forms. From (2.12) we note that $l_o = l_i = r(\csc 90° + \cot 90°) = r$. The path length of an ion is thus intermediate between the 180° and 60° instruments having a comparable mass resolution. Thus the probability that an ion will undergo scattering by residual gas molecules in the vacuum analyzing chamber is less than in the 60° case, but greater than for the 180° deflection.

Figure 2.5 shows a typical configuration for a 90° instrument and indicates a practical correction which must be made for most sector magnets for proper ion focusing. The general ion optics equations presume a sharply defined magnetic boundary from which object and image

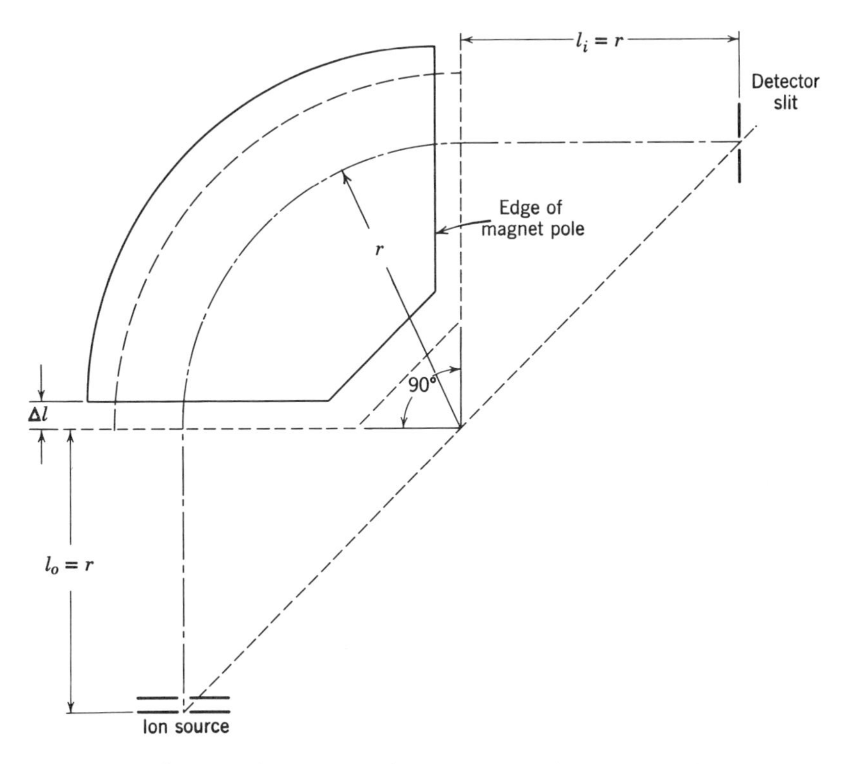

Fig. 2.5 A 90° magnetic sector displaced from its theoretical position to correct for the fringing field.

distances can be computed. Of course, no such well-defined boundary exists. In practice it is convenient to assume that the magnetic fringing field extends about one gap width beyond the physical edge of the magnetic sector [6]. This assumption appears to be quite valid for cases where the gap is small compared to the magnet radius. Thus a magnet should be displaced a distance, Δl, from the position indicated by the dotted line, which represents the idealized situation. In large analytical instruments, however, a final "best focus" position is often achieved experimentally, with the magnet being displaced in small increments with respect to the analyzing tube.

There is no reason why sector magnets of any angle cannot be constructed, but the 180°, 60°, and 90° have become somewhat standard. It is also unlikely that very small angular sectors ($< 10°$) would yield comparable performance, as the inhomogeneous fringing field would begin to have dimensions comparable to that of the primary homogeneous field region.

DOUBLE-FOCUSING SPECTROMETERS

In the single magnetic analyzer there is no provision for focusing ions differing in initial energy. If ions are emitted from a thermal or surface ionization source the ion beam may have a negligible energy spread, but this is not true in the general case. The effect of an ion energy spread may be decreased by using a very high accelerating potential—thus decreasing the $\Delta V/V$ contribution to the image width (2.6)—but this is not an attractive solution. The most useful method of providing energy (or velocity) focusing is to employ a radial electrostatic field. By radial electrostatic field is meant a field generated by a voltage difference applied across the plates of a cylindrical condenser. If ions are deflected through such a field, there exists a simple relationship between the radius of curvature r_e, the potential V through which ions have been accelerated, and the electrostatic field strength $\mathcal{E}$ existing between the plates of the cylindrical lens. If r_e is in centimeters, V is in volts, and $\mathcal{E}$ is in volts per centimeter, then

$$r_e = \frac{2V}{\mathcal{E}}. \tag{2.13}$$

This is a mass independent relationship that holds for electrons, protons, or ions. It is important in mass spectrometry because it permits the selection of ions having a fairly discrete energy. Thus, in Figure **2.6**, if narrow collimating slits are used at the ends of a cylindrical condenser, ions having a kinetic energy $e(V + \Delta V)$ and $e(V - \Delta V)$ will

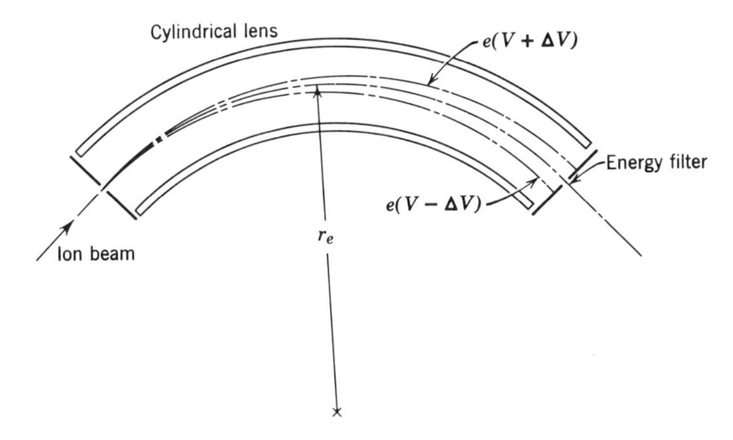

Fig. 2.6 Electrostatic lens providing energy filtering of an ion beam.

be filtered out from the ion beam, allowing the transmission of only a narrow energy group. Hence this analyzer can furnish a monoenergetic ion group to a subsequent magnetic analyzer, and yield a highly resolved mass spectrum.

A radial electrostatic field can also be used to accomplish another objective. General equations have been developed by Mattauch and Herzog [7], [8], who analyzed (a) the problem of electrostatic directional focusing and (b) the relationship of combined electrostatic and magnetic fields. The latter analysis has shown that by using a suitable combination of electrostatic and magnetic fields, one may obtain directional focusing for an ion beam of a given mass-to-charge ratio, even if it is heterogeneous in energy. The term *double focusing* is thus ascribed to those situations in which angular aberrations and velocity aberrations effectively cancel.

Letting l_{oe} and l_{ie} denote the object and image distances, r_e the mean radius of curvature of the electrostatic analyzer, ϕ the sector angle, and f_e the focal length of the lens, a relationship has been found to be valid that is similar to the magnetic case, namely,

$$g_e = \frac{r_e \cot(\sqrt{2}\,\phi)}{2} = f_e \cos(\sqrt{2}\,\phi),$$

and

$$f_e = \frac{r_e}{\sqrt{2}\sin(\sqrt{2}\,\phi)}, \tag{2.14}$$

for

$$l_{oe} = l_{ie},$$

$$l_{oe} - g_e = f_e,$$

$$l_{oe} = l_{ie} = g_e + f_e = f_e \cos(\sqrt{2}\,\phi) + f_e$$

$$= \frac{r_e}{\sqrt{2}} \frac{1 + \cos(\sqrt{2}\,\phi)}{\sin(\sqrt{2}\,\phi)} = \frac{1}{\sqrt{2}} \csc(\sqrt{2}\,\phi) + \cot(\sqrt{2}\,\phi). \quad (2.15)$$

Two of the most widely used electrostatic sector angles in mass spectrometry have been $\phi = 31° 50'$ and $\phi = 90°$. The respective object or image distances from the electrostatic boundary are then found to be $(r_e + r_e/\sqrt{2}) = 1.707r_e$, and $0.35r_e$.

The particular combination of electrostatic and magnetic field parameters that provide double focusing is given by satisfying (2.10), (2.14), and the following expression simultaneously:

$$r_e(1 - \cos\sqrt{2}\,\phi) + \sqrt{2}\,l_{ie} \sin\sqrt{2}\,\phi$$
$$= \pm\{r_m(1 - \cos\theta) + l_{om}[\sin\theta + \tan\delta\,(1 - \cos\theta)]\}. \quad (2.16)$$

The angle δ is the angle between the velocity vector of the ion and the normal to the magnetic field boundary. The positive sign is employed when the ion is deflected in the same sense in both electrostatic and magnetic fields; the negative sign is used when the ion undergoes successive deflections in opposing directions.

Two of the best examples of double-focusing spectrometers are the consecutive electrostatic-magnetic field configurations of Mattauch and Herzog [9], [10], and the general type developed by Nier and his collaborators.

The Mattauch-Herzog design has the remarkable property of possessing double focusing for all masses. Figure 2.7 is a schematic diagram of the system that comprises a $\pi/4\,\sqrt{2}$ (31° 50′) electrostatic sector followed by a 90° magnetic analyzer.

The ion source is located at the principal focus of the electrostatic lens so that ions emerge from the energy analyzer in a parallel beam. This is consistent with (2.14), which indicates that if $l_{oe} = r_e/\sqrt{2}$, $l_{ie} = \infty$. The ion beam thus enters the magnetic sector as a bundle of parallel rays, which is equivalent of having $l_{om} = \infty$. With this restriction and for $\theta_m = 90°$, $l_{im} = 0$—as in (2.10). Hence ions of all masses will be focused along a line which is coincident with the second magnetic field boundary. Because there is no intermediate focal point, the distance Δ

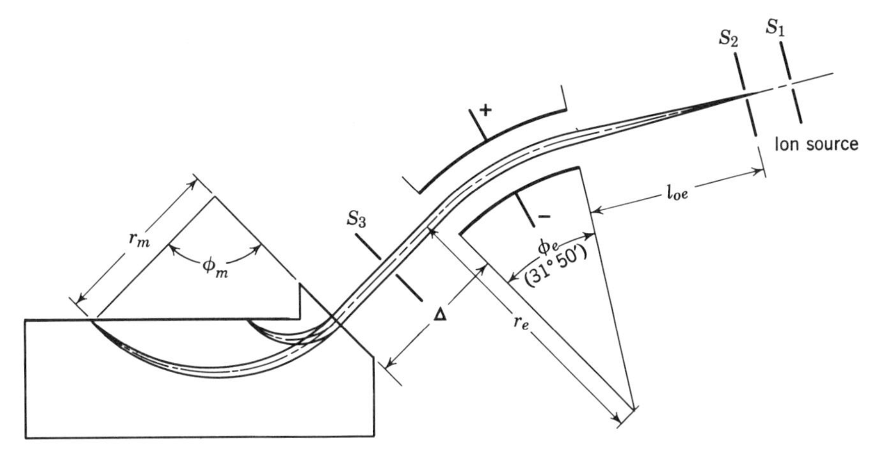

Fig. 2.7 Double-focusing system of Mattauch and Herzog.

between the electrostatic and magnetic field boundaries is indeterminate and may be chosen at will.

The range of m/e values which can be simultaneously focused on a photographic plate will depend on the range values of r_m allowed by the magnet configuration. In some commercial instruments a mass range extending from boron to uranium can be monitored with a single setting of the magnetic field.

A collimating slit of variable aperture between the electrostatic and magnetic analyzer permits acceptance into the magnet of an ion beam having either a small or large energy distribution. The combined mass and energy focusing of this system is excellent, providing a reasonably sharp mass spectrum, even if the spread in ion energies is an appreciable fraction ($\sim 5\%$) of the total accelerating potential. The resolution and mass dispersion depend respectively on the selection of values of r_e and r_m. The first instrument of this type had an $r_e = 28$ cm and a maximum $r_m = 24$ cm. A modified Mattauch-Herzog instrument, having different angular parameters, has $r_e = r_m = 254$ cm. This very large spectrometer, built by Stephens et al. [11], incorporates many special features and utilizes a spherical electrostatic sector rather than the conventional cylindrical geometry. Many other versions of this basic spectrometer have been built and successfully used for mass measurements, isotopic abundance measurements, and general analytical work. Mass spectrometric doublets can be measured with precision and the instrument can be used with virtually any ion source. Further, because many masses can be simultaneously monitored and integrated, wide fluctuations in ion beam intensity do not seriously interfere with the accurate recording of isotopic abundance ratios.

The high-resolution spectrometers built by Nier and co-workers at the University of Minnesota are improved versions of the somewhat similar tandem system devised by Dempster [12] and by Bainbridge and Jordan [13]. The Dempster instrument comprised a 90° electrostatic lens followed by a 180° magnetic sector. The latter instrument included a $\sqrt{2}\pi/2$ (127° 17′) electrostatic lens followed by a 60° magnet. Both systems employed symmetric arrangements of the components.

The two high-resolution instruments built by Nier and his associates consist of a 90° electrostatic lens followed by a 60° magnetic sector. The first instrument (Nier and Roberts) had the parameters $r_e = 18.87$ cm and $r_m = 15.24$ cm. [14] An enlarged version of this tandem arrangement was designed by Quisenberry, Scolman, and Nier with $r_e = 50.31$ cm and $r_m = 40.64$ cm. [15] A schematic diagram is shown in Figure 2.8. It will be noted that whereas the electrostatic lens is symmetric with respect to object and image slits, S_1 and S_3, the magnet is asymetrically located with respect to S_3 and S_4. This arrangement provides not only first-order energy focusing. but both first and second order directional focusing, thereby allowing wide limits on the half-angle of divergence (α) that can be employed without angular aberration. These two instruments have yielded mass measurements of the highest precision for an exceedingly large number of elements and isotopes.

Important features of the above instruments include (a) highly stabilized accelerating potentials and magnetic fields, (b) a method of "peak-matching" ions of nearly equal masses by bringing them sequentially into focus along identical ion trajectories, and (c) a complete electrical detection system replacing the photographic plate. As shown in Figure 2.8, the magnetic sector contains an auxiliary mass spectrometer whose output, from a pair of ion beam collectors, is connected to a differential amplifier, and linked to the voltage supplies of the ion source and electrostatic analyzer. If the accelerating potential or the magnetic field changes for any reason, the differential amplifier senses this change in the auxiliary tube, and compensation is provided for in a feed-back loop. The sequential focusing of two ions of a mass-doublet, having small mass differences, has proved to be quite successful. A voltage scan corresponding to a small fraction of the mass scale is related directly to a small change in resistance ΔR of a voltage divider. By measuring the precise resistance change $\Delta R/R$, it can be shown that ions differing by a small fractional mass, ΔM, are related by $\Delta R/R = k\ (\Delta M/M)$ (see Chapter

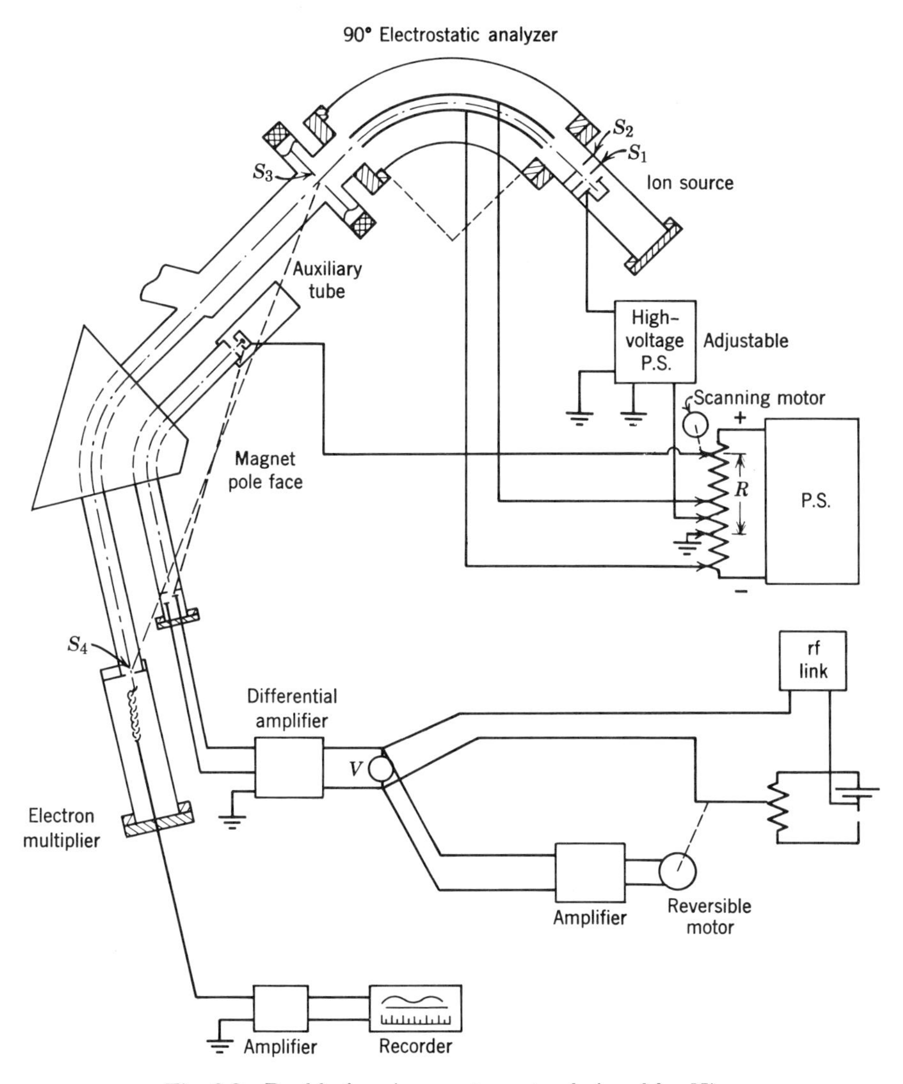

Fig. 2.8 Double-focusing spectrometer designed by Nier.

5). The output currents from two mass peaks are then alternately presented on an oscilloscope and when two peaks are exactly superposed or "matched," the ΔR or ΔM values can be recorded. Electronic recording in this manner circumvents several sources of error associated with the recording of ion spectra with photographic plates.

MULTIPLE MAGNET SYSTEMS

An increasingly important class of spectrometers used for research in nuclear physics comprise instruments having more than one magnetic analyzer (see Chapter 5). The first such tandem magnet system was reported by Inghram and Hayden [16] in 1954. In this instrument ions were mass analyzed in the first magnet, given an additional acceleration in a region between the two magnets, and reanalyzed in a magnetic sector which deflected the ion beam in a direction opposite to that of the first analyzer. The general orientation of the magnets has led to this scheme's being called an "S" tandem analyzer. Subsequently White and Collins [17] reported on the construction of a two-stage magnetic analyzer comprising two 90° magnets with 30.5-cm-radius of curvature arranged in a "C" type configuration so that deflection of the ion beam proceeded in the same sense. No additional acceleration was given to the ions in the region between the two magnets, this being a "drift" tube only. In this latter instrument (Figure 2.9) nuclear magnetic resonance regulation (NMR) was also introduced to provide both short- and long-term stabilization of the tandem magnet system [18]. The construction of these multiple magnet systems was carried out to achieve something quite different from the ultrahigh-resolution, double-focusing spectrometers. One technical objective was to measure isotopes existing in very small relative abundances.

In conventional mass spectrometers comprising a single magnet, isotopic abundance ratios for adjacent mass numbers (in the mass range of approximately 100) can rarely be measured for ratios greater than $10^4/1$. Even if resolution is high, peak-to-valley ratios are limited by phenomena such as small-angle scattering from residual gas molecules in the vacuum analyzer, reflection from metal surfaces, imperfections in ion optics, and charge exchange. However, if an ion beam is constrained to pass through two identical momentum analyzers, allowing but a single mass number to enter into the second magnet, considerable discrimination can be achieved with respect to spurious ions. Providing the collimating slit that allows the ion beam to enter the second magnet is small compared to the mass dispersion, the second magnet will reanalyze the spectrum with a mass discrimination approximating the first. In other words, if adjacent isotopes can be filtered through a single momentum analyzer so that $10^4/1$ ratios can be observed, two magnets—operating as independent filters—might be expected to provide a $10^8/1$ discrimination from adjacent isotopes.

As examples of the multiple magnet system for both the "S" and "C" configurations, two instruments will be cited that have equal radii

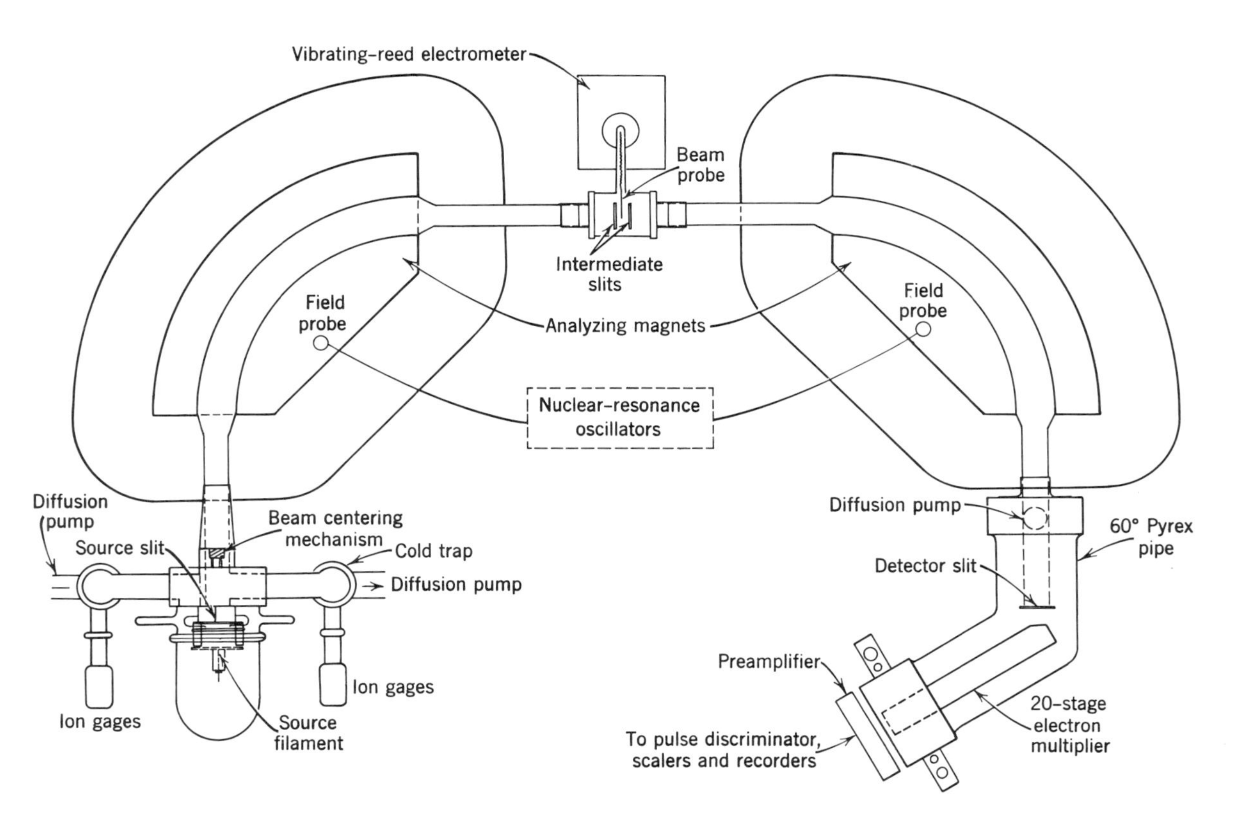

Fig. 2.9 Two-stage spectrometer with "C"-shaped geometry.

of curvature and several other common features. An assessment of their relative advantages is, therefore, somewhat easy to make. One spectrometer is a so-called three-stage mass spectrometer, reported by White, Sheffield, and Rourke [19], which employs two 90°, 50.8-cm. radius-of-curvature magnets and a single 90° electrostatic lens of equal radius of curvature. Figure 2.10 is a schematic diagram of the trajectory and the general configuration of the magnets and exciting coils. Ions are accelerated from a surface ionization source, analyzed in the first 90° magnet, and form an image at an intermediate slit (of variable aperture)

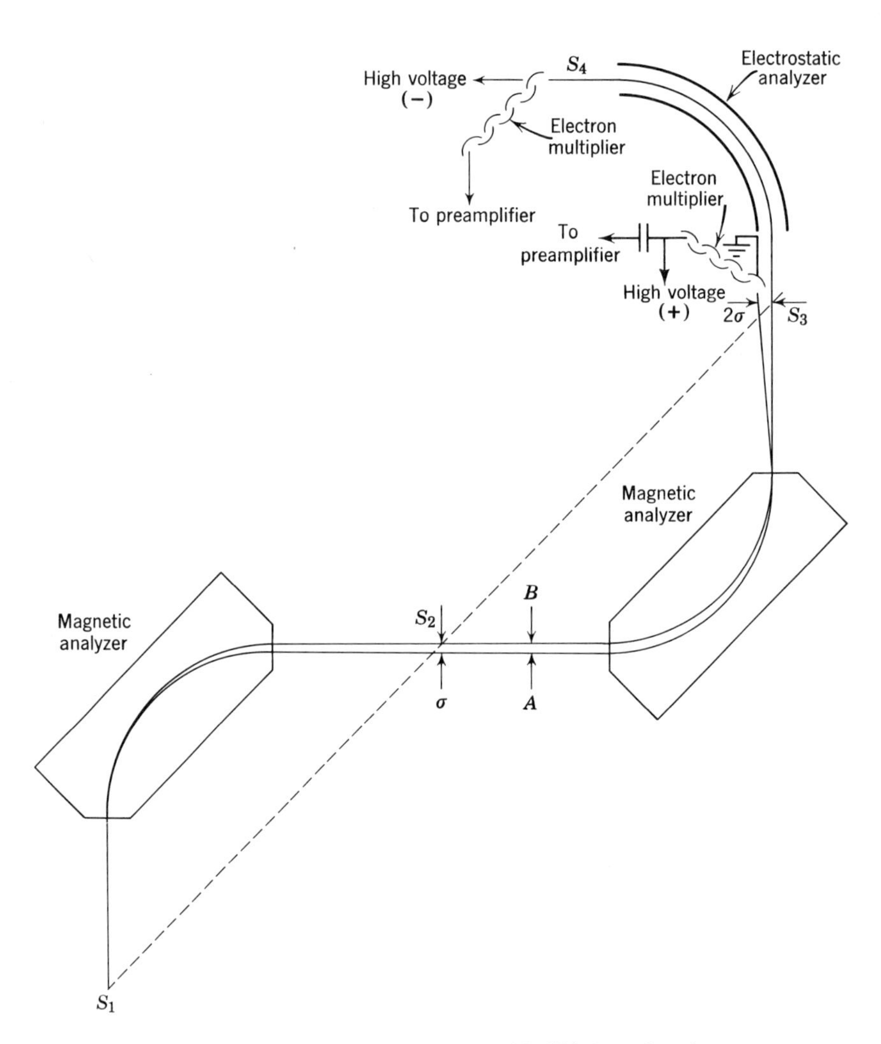

Fig. 2.10 Three-stage spectrometer with "S"-shaped trajectory.

between the two magnets. The second magnet (plus the 90° electrostatic lens) then serves as a symmetric double-focusing system that essentially reanalyzes all ions which have already been mass resolved. It is clear from the geometry that if the intermediate slit is opened to permit the passage of two ion groups (i.e., adjacent masses A and B) the dispersion at the focal point of the second magnet S_3, is twice that at the intermediate slit S_2. With the large radius of curvature, this arrangement permitted the use of two electron multipliers, as indicated, so that simultaneous pulse counting of ions could be made of two mass groups. (Note that it is necessary to operate the first dynode of the electron multiplier at S_3 at ground potential to prevent beam defocusing.) Measurements at S_2, S_3, and S_4 verified the assumption that not only did the second magnet yield an improvement in abundance sensitivity but also that an even lower background ion current might be achieved at the end of the electrostatic section. At mass numbers 100 and greater, abundance sensitivities of 10^7 to 10^8 were achieved, and a measurement of the Na^{23} ion beam to that observed at the 23.5 mass position was greater than $10^{10}/1$.

A "C"-shaped instrument, recently completed at the Rensselaer Polytechnic Institute, has been reported by White and Forman [20] and is shown in the schematic diagram Figure 2.11 and the photo of Figure 2.12. This four-stage spectrometer essentially comprises two complete double-focusing systems in tandem. All components are symmetrically arranged and all analyzing elements have a mean radius of curvature of 50.8 cm. Adjustable slits are placed at all intermediate focal points, S_1, S_2, S_3, S_4 and at the focal detector position S_5. The magnet design in this latter instrument is somewhat more compact and economical with respect to the use of iron, but both have exciting coils that require only low-current power supplies. In each case the electromagnets are mounted on roller bearings that provide for convenient magnet positioning in order to optimize the ion optics. Provision has been made at S_3 for an auxiliary port so that a sputter ion source can be added; this particular magnetic configuration also lends itself to the detection of a neutral ion beam. Such a neutral beam can be generated by having a primary ion beam traverse a thin foil (see Chapter 13). With such a provision, ratios of neutral and multiple-charged ions can be monitored by a proper programming of the second double-focusing system.

The abundance sensitivity of this instrument is comparable to the "S"-shaped configuration at comparable slit apertures and pressures, with an abundance sensitivity of $3 \times 10^8/1$ being measured in the mass range of 150. The "S"-shaped geometry is clearly preferred when high resolution and abundance sensitivity are prerequisite. This geometry is also required if a multiplicity of masses is to be simultaneously recorded at a final detector. However, for the measurement of isotopic ratios a full mass unit apart, the "C"-shaped trajectory appears to give comparable performance. Of course, in this latter case, the intermediate slit aperture must be appreciably less than the dispersion, otherwise resolution will vanish (see Figure **2.2**). But with this restriction the "C" geometry appears to give improved peak shapes; furthermore, stability requirements for the magnets and source voltages are less stringent.

Scanning of the mass spectrum can be accomplished by varying either the accelerating potential (and electrostatic lens potentials) or the magnetic fields, but the former is easier to achieve. Magnets can be cohered either by using nuclear magnetic resonance, or by connecting the coils of both magnets in series and using current regulated power supplies.

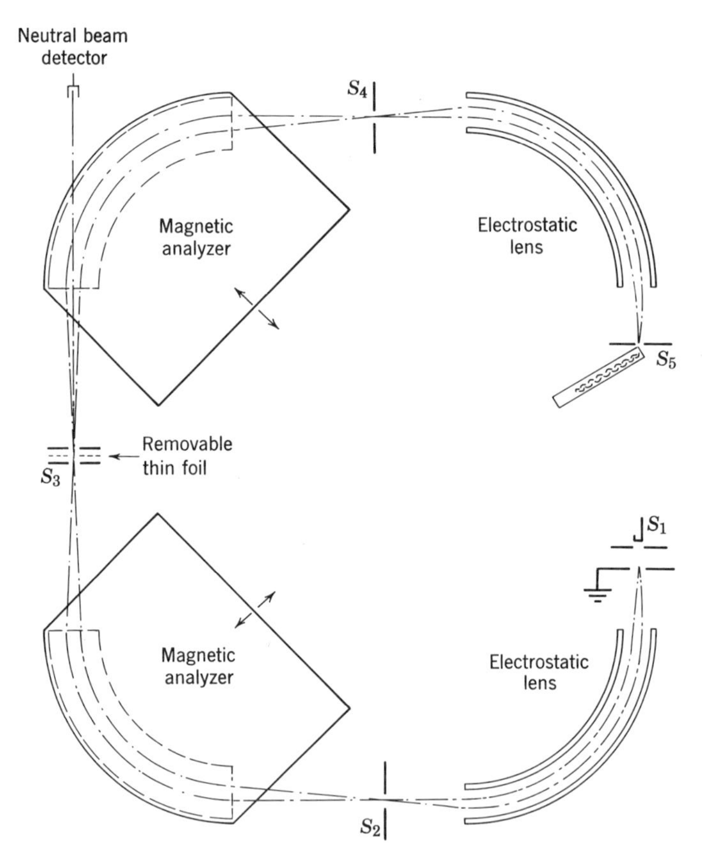

Fig. 2.11 Four-stage mass spectrometer schematic of "C"-shaped geometry of magnets and electrostatic lenses.

Fig. 2.12 Four-stage mass spectrometer.

Commercial versions of tandem magnet systems are available, both as analytical mass spectrometers and as partial pressure gas analyzers with small radii of curvature.

THE CYCLOIDAL MASS SPECTROMETER

Although the tandem arrangement of electrostatic and magnetic fields has many advantages, a somewhat simpler and more compact geometry will satisfy both energy and momentum focusing conditions. In a scheme reported in 1938, Bleakney and Hipple [21] showed that if a homogeneous electrostatic field is superposed upon a uniform magnetic field, perfect double focusing can be achieved in a plane normal to the magnetic field. This crossed-field arrangement and the resulting cycloidal trajectory of ions is indicated in Figure 2.13. Here, a is the radius of the generating circle of the cycloid, b is the distance from its center, and θ is the generating angle. The rectilinear coordinates in terms of these parameters are then given by the simultaneous equations

$$x = a\theta - b \sin \theta$$

$$y = a - b \cos \theta. \tag{2.17}$$

It will be noted that the distance along the x axis for a complete cycle will be $d = 2\pi a$, and it can be shown that the radius of the generating circle will be related to the electrostatic field strength $\mathcal{E}$, the magnetic field strength B, and the mass to charge ratio of the ion by

$$a = (\mathcal{E}/B^2)(m/ne). \tag{2.18}$$

Hence

$$d = (2\pi\mathcal{E}/B^2)(m/ne). \tag{2.19}$$

The remarkable property of the cycloidal system is that within wide limits the distance d is invariant in a given (m/ne) ion group. Regardless of the initial velocity distribution of an ion beam (ΔV), or its divergence angle from normal entry—providing only that its path is in a plane perpendicular to the magnetic field vector, all ions of a given mass-to-charge ratio will have a common object-image distance. A schematic diagram of a typical mass spectrometer system is shown in Figure **2.14**.

The resolution of such a spectrometer in the case of photographic detection will be

$$\frac{m}{\Delta m} = \frac{d}{\Delta d} = \frac{d}{S_o}, \tag{2.20}$$

where S_o is the object slit width.

Scanning of a mass spectrum can be achieved either by (a) variation of the magnetic field strength with a fixed ion-accelerating potential and crossed electrostatic field, or (b) maintaining B invariant and changing the magnitude of the electrostatic field $\mathcal{E}$.

A cycloidal instrument of large dimensions (object-image distance $\simeq$15 cm) has been reported by Voorhies, Robinson, Hall, Brubaker, and Berry [22], in which the mass triplets C_2H_4, N_2, and CO were completely resolved. This instrument employed an electromagnet that provided a magnetic field intensity of 10,000 gauss across a gap of approximately 3 cm. Most modern commercial versions, however, em-

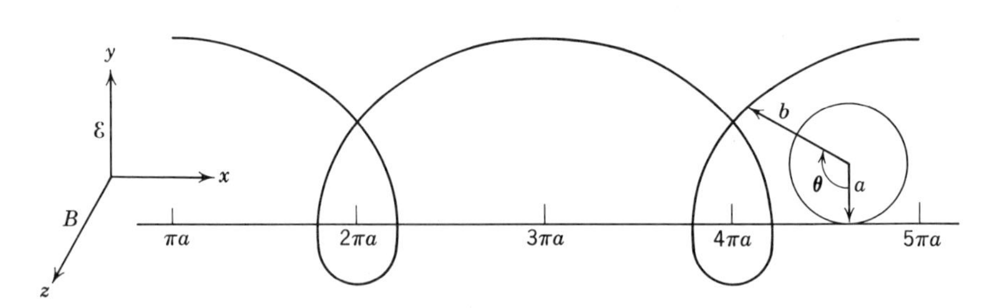

Fig. 2.13 Cycloidal path of an ion in a crossed E-B field.

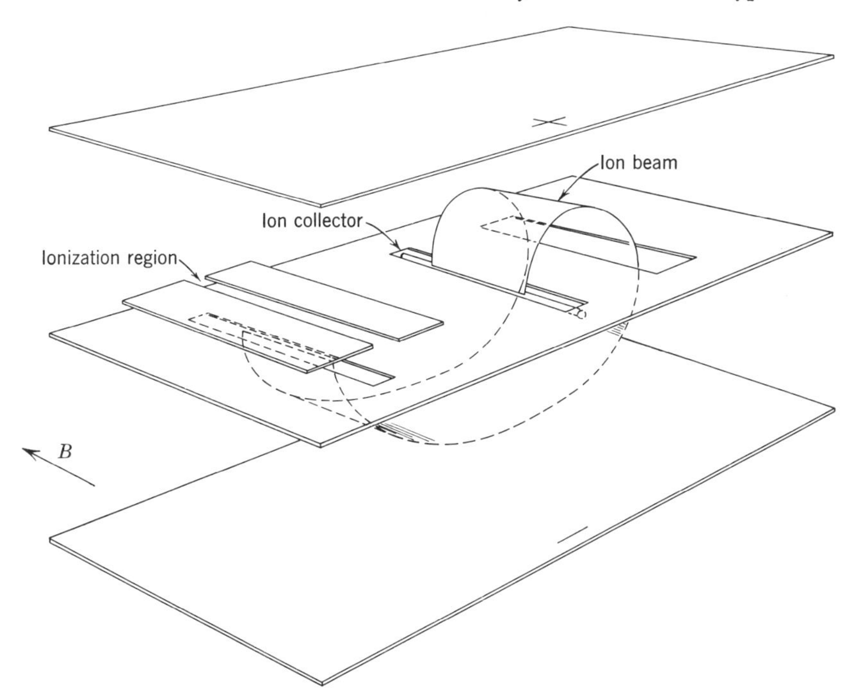

Fig. 2.14 The cycloid-type mass spectrometer.

ploy small permanent magnets that provide fields of about 4500 gauss. These spectrometers are effectively used as residual gas analyzers, and they are available with automatic and continuous voltage-scan control units.

CYCLOTRON RESONANCE TYPES

A very large number of spectrometers have been devised based on the cyclotron principle that was first used by Lawrence and Livingston [23] to accelerate ions to extremely high energies. Lawrence demonstrated that charged particles would undergo a spiral path if an appropriate radio frequency voltage was applied to semicircular "dees" and if the ions were also constrained by a magnetic field. For a singly charged ion, the angular frequency ω_c of the rf field was shown to have the relation

$$\omega_c = \frac{e}{m} B. \tag{2.21}$$

Sommer, Thomas, and Hipple [24] were among the first to apply the cyclotron principle to mass spectrometry, and they termed their device the *omegatron.*

Figure 2.15 is a schematic diagram of the instrument. One difference between this device and a cyclotron is that charged particles are not accelerated across a narrow gap between "dees," but rather they are continuously subjected to acceleration in an extended homogeneous radiofrequency field. If the frequency of the rf field, $E = E_o \sin \omega_0 t$, is equal to the cyclotron frequency ω_c of an ion, the ion will undergo a spiral trajectory until it is intercepted by the collector. For the resonant case, the ion radius will increase linearly with time according to the relation

$$r = \frac{E_o t}{2B}. \tag{2.22}$$

If r_o is the distance from the point where ions are generated, and we define resolution by the relation $m/\Delta m = \omega_c/\Delta\omega$, then it can be shown that the resolution of the omegatron is [25]

$$\frac{m}{\Delta m} = \frac{\omega_c r_o B}{2E_o} = \frac{r_o B^2 e}{2mE_o}. \tag{2.23}$$

It is clear from (2.23) that we should expect the highest resolution of this device to be for ions of low mass number. Resolution is indeed excellent for light gaseous ions—with a reported value of 35,000 for the mass

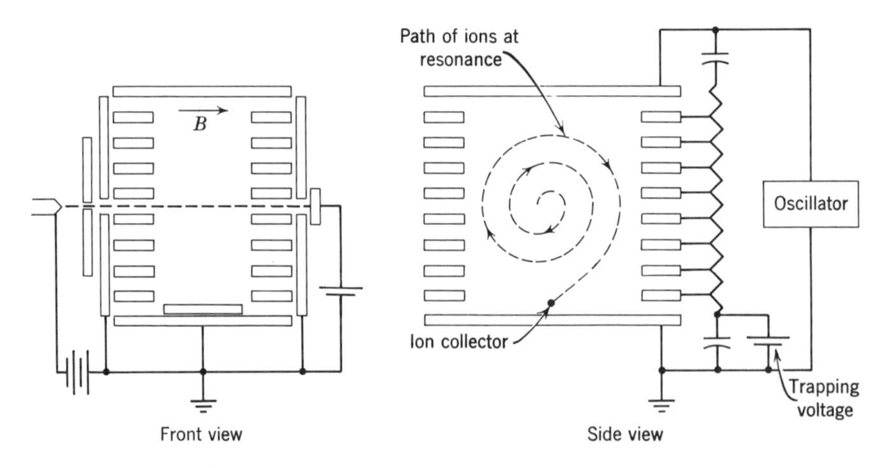

Fig. 2.15 Schematic diagram of the omegatron.

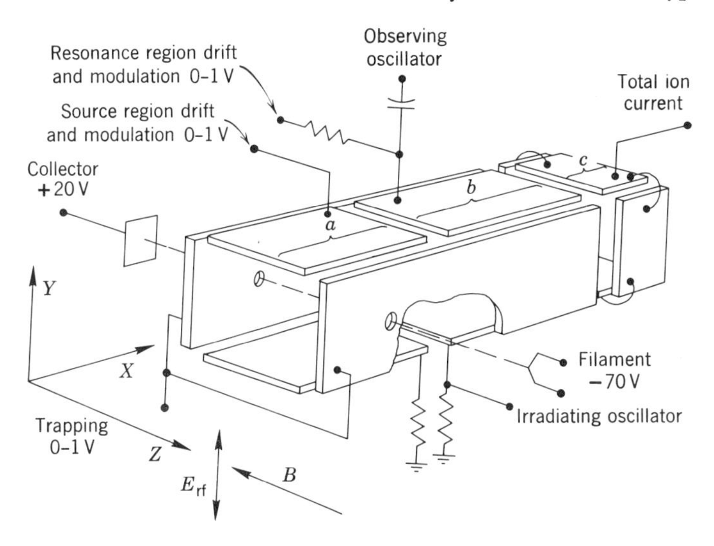

Fig. 2.16 Syrotron mass spectrometer (Varian Associates).

doublet ($H_2^+ - D^+$). Because of its simple and compact construction, the omegatron is also suitable as a residual gas analyzer.

A new version of the cyclotron resonance instrument (Figure 2.16) has recently been reported (a "Syrotron" Mass Spectrometer) by Varian Associates. It has two important differences from the original design. First, the region in which ions are produced by electron impact is separated from the radiofrequency field region. The ions are then "drifted" into the crossed magnetic and oscillating electric field by the action of a static electric field applied at right angles to the magnetic field. The net result is that the ions form a well-defined sheet moving into the analyzing region. This technique eliminates the difficulties of space charge of the electrons, which can distort the resonance condition. The second distinguishing feature of the Syrotron relates to the method of ion current detection. In the classical omegatron the expanding spiral of resonant ions emerges through a final slit and is measured by an electron multiplier or electrometer. The Syrotron method of ion detection is based on the absorption of radiofrequency energy, which gives rise to expanding spiral motions (i.e., the work done in accelerating the ions). Energy absorbed by the ions changes the Q of the resonant circuit, and the ΔQ can be observed by a phase-sensitive detector. Resonant ion currents have been reported as low as 10^{-16} A, and a resolution approaching 1000 has been claimed in the mass region of 200 amu.

Another cyclotron-type spectrometer that has undergone extensive development is the so-called *mass synchrometer,* designed by Lincoln Smith [26]. The basic features of this device are shown in Figure 2.17. Ions are accelerated through a source slit S_1, pass through a semicircular orbit, and arrive at a multiple slit assembly S_2, S_3, S_4. Slits S_2 and S_4 are grounded while S_3 is connected to a pulse generator. The period of revolution of a singly charged ion is

$$T = \frac{2\pi m}{eB}. \tag{2.24}$$

If a pulse is applied to the modulator plate S_3, the rf fields in the modulator will cause the velocity of the ion to be modulated in a sinus-

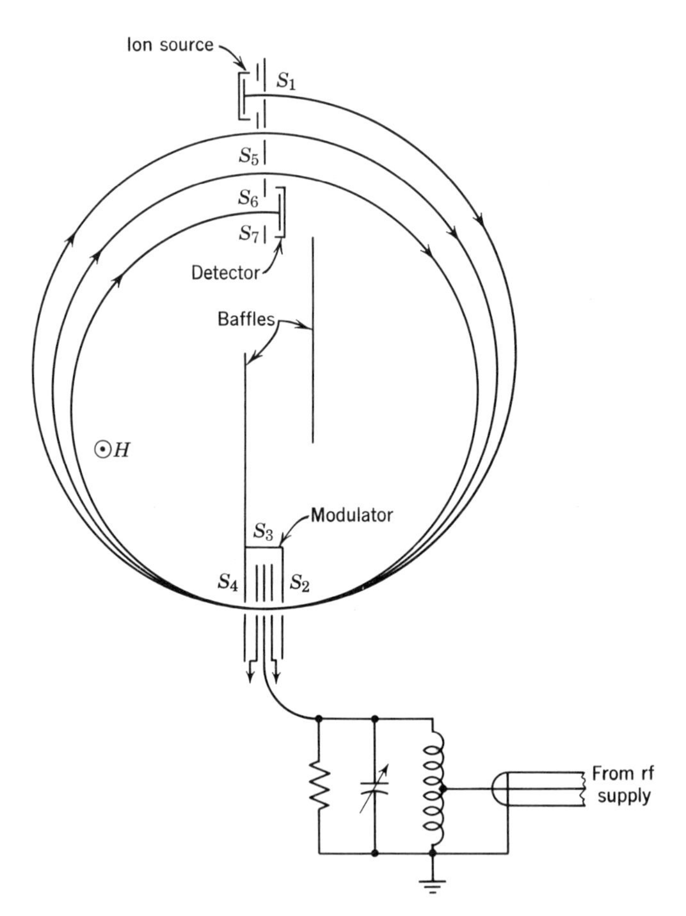

Fig. 2.17 Mass synchrometer designed by L. Smith.

oidal manner. If the voltage is of proper phase and amplitude, the ion group will be decelerated so that the ion trajectory will have a decreased radius and pass through slit S_5. Since from (2.24) we know that the time for the complete revolution of an ion is independent of its velocity, the ions can be kept in phase and undergo a further deceleration in the modulator. This will cause the ions to undergo a further shrinking in the diameter of their orbit, and they will eventually be intercepted by the ion detector. This mass synchometer has been exclusively used for the mass measurements, and very high resolution has been achieved. The correlation between mass and frequency is given by the simple relation

$$\frac{\Delta m}{m} = \frac{\Delta f}{f}. \tag{2.25}$$

The frequency f, and small frequency differences Δf can be measured with high precision, thus permitting exceedingly small mass differences to be observed.

THE TIME-OF-FLIGHT MASS SPECTROMETER

One of the first pulsed mass spectrometers was proposed by Stevens [27] in 1946; shortly thereafter an "ion velocitron" was described by Cameron and Eggers [28]. A prototype of the modern time-of-flight spectrometer was subsequently developed by Wiley and McLaren [29].

The essential components of this type of analyzer include (a) an electron gun for the production of ions, (b) a grid system for accelerating ions to uniform velocities in a pulsed mode, (c) an ion drift region, and (d) a sensitive fast-response ion detector. Suitable electronic circuitry is also required for translating the time-dependent arrival of ions of different velocities into a time base that is related to mass number.

A schematic diagram of a recent Bendix time-of-flight instrument [30] is shown in Figure 2.18. Ions are formed and accelerated in a source region in a manner to keep the ion injection time small compared to their transit time in the drift region. After traversing the drift space of 170 cm, they impinge upon a high-gain electron multiplier. The multiplier and associated output circuits register the arrival of single positive ions. Ions of a given energy traverse the drift region with a velocity, v_i, given by

$$v_i = \left(\frac{2neV}{m_i}\right)^{1/2}. \tag{2.26}$$

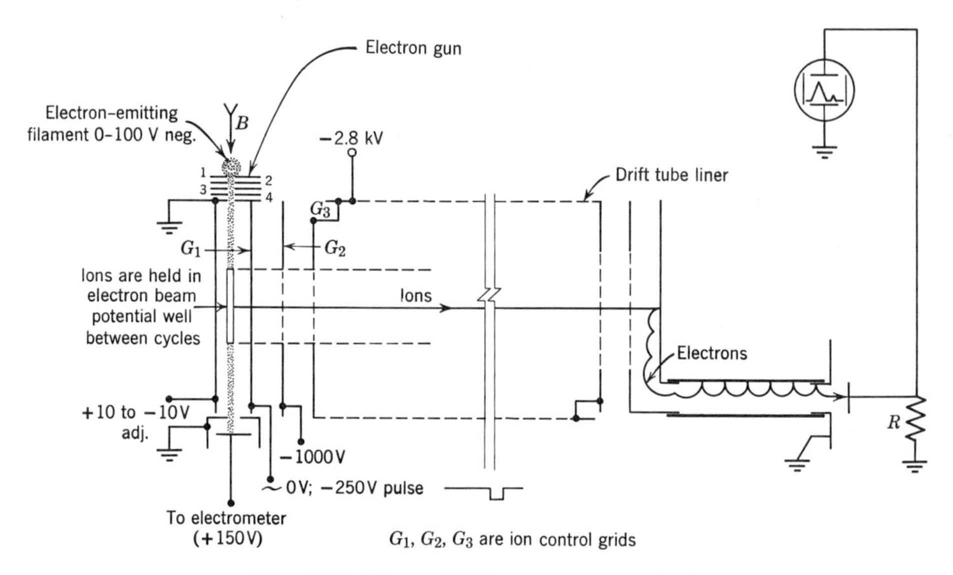

Fig. 2.18 Time-of-flight spectrometer [30].

where V is the potential through which ions of charge ne and masses m_i have been accelerated. If the drift-region length is L, the difference in transit time (Δt) for two ions of masses m_1 and m_2 will be

$$\Delta t = \frac{L(\sqrt{m_1} - \sqrt{m_2})}{\sqrt{2neV}}. \tag{2.27}$$

It is clear from the above that the time resolution will increase with increased drift tube length, and will decrease with increasing accelerating potentials.

A prime consideration with the time-of-flight instrument is the matter of ion injection. What is desired is the restriction of an "ion bunch" to the drift region that is neither spatially dispersed nor inhomogeneous in energy. If ions leave the source region at the improper time or with a spread in energy the resolution will substantially decrease and a background or "noise" spectrum will result. The problem becomes serious when the accelerating potential, in practice, is limited to about 3000 V and a reinjection of ions occurs each 100 μs. (The cycle of operation is triggered by a stable oscillator.)

Two methods of providing fairly discrete "ion bunches" have proved successful: (a) the pulsed and (b) the "continuous" ionization mode. In the pulsed mode the ionizing electron beam is pulsed, biased off, and the ions formed are initially accelerated by the drawout grid. A second grid provides the additional accelerating potential. This two-grid system has been shown to minimize both initial spatial dispersion and inhomogeneities in ion energy. By immediately applying a drawing-out pulse to the source region after ion formation, the two-field grid system allows slow ions to catch up with those of higher velocity—thus substantially "bunching" an ion group in time.

The "continuum" mode of operation is continuous only in a single sense. The *electron* beam is left on for a very large fraction of a cycle (e.g., 50–100 μs). During this interval the electrons are continually generating ions and it is assumed that a "potential well" is formed between the backing plate and the first drawing-out grid, G_1. This well serves as a trap for lower energy ions, but more energetic ions escape and are attracted to the backing plate by a slight bias. Under proper operating conditions, the subsequent pulsing of the G_1 grid results in a spectrum of improved resolution. An additional advantage of the "continuous" mode is the increase in sensitivity (10 to 100) owing to the longer duty cycle that results from having ions immediately available. This factor is an important one because in obtaining very fast mass spectral scans one is often limited with respect to the total number of recorded ions.

The time-of-flight instrument occupies a unique place in mass spectrometry as it provides a simple, rapid measurement of the abundance of various isotopes or elements comprising a sample. In practice, 10,000 to 100,000 spectra can be scanned per second. With the aid of suitable electronic circuitry it is thus possible to monitor reaction rates and to investigate reaction profiles of only 100-μs duration. The recently increased drift tube length of 170 cm has contributed to improved mass resolution, and the very high scanning rates have made possible improved ion statistics. Although rarely desired, a scan can also be made covering from 0 to 900 amu in 1.5 sec. In order to increase the dynamic range of the instrument a special technique has been developed. For example, to prevent multiplier saturation when very large ion peaks are present in the presence of smaller adjacent peaks, appropriate "gating" pulsing can be applied to the multiplier. Thus it is possible to suppress mass 40 without interferring with the recording of masses 39 or 41. This type of refinement has extended the practical range of sensitivity in identifying gas chromatograph effluent by orders of magnitudes.

QUADRUPOLE AND RF MASS FILTERS

A number of completely nonmagnetic spectrometers have made their appearance in recent years. Representative of this important class of mass filters that are finding wide application are (a) the quadrupole spectrometer devised by Paul and Steinwedel [31] and (b) the rf mass analyzer of Bennett [32]. Neither scheme purports to give high resolution, but both approaches have been sufficiently useful to warrant commercial development.

Figure 2.19 is a sketch of the quadrupole type mass filter. A cross section view (Figure 2.19*a*) indicates an idealized hyperbolic field configuration. The group of four cylindrical rod-shaped electrodes (Figure

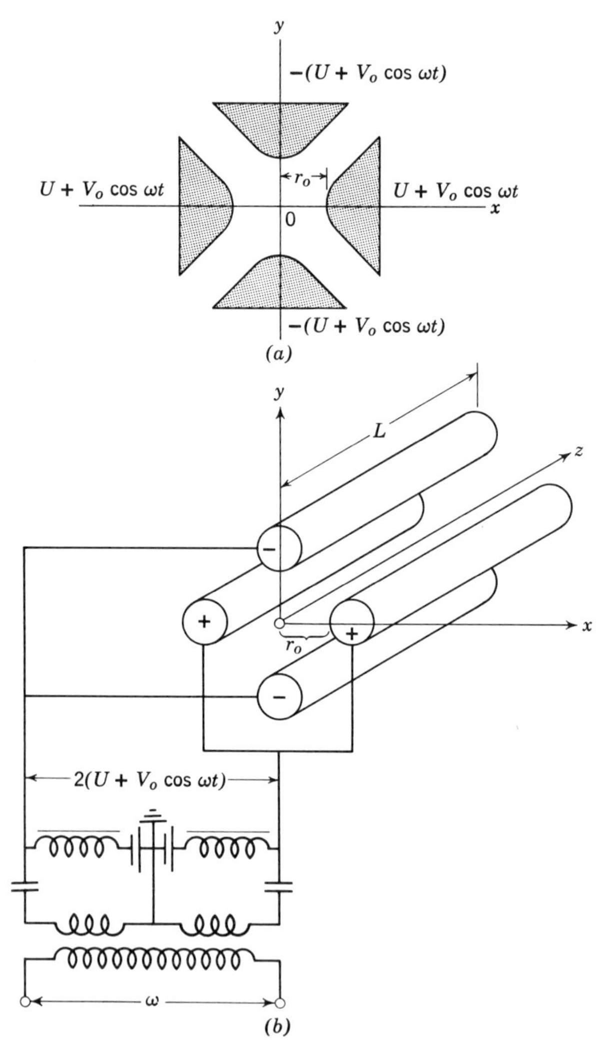

Fig. 2.19 The quadrupole mass filter.

2.19*b*) is representative of the commercial geometrical arrangement. The basic concept of this device is to provide a potential field distribution, periodic in time and symmetric with respect to the axis, which will transmit a selected mass group and cause ions of improper mass to be deflected away from the axis. In particular this mass selection scheme employs a combination of dc potentials plus a radiofrequency potential so that the transit time of the ions is long compared to the rf period. By proper programming of potentials and frequency, an ion of desired mass can be made to pass through the system while unwanted masses will undergo an oscillating trajectory of increasing amplitude so that they will ultimately be collected on one of the electrodes.

If the voltage applied to the quadrupoles consists of a fixed voltage U and the radiofrequency component $V_0 \cos \omega t$, the potential Φ may be expressed in terms of the rectilinear coordinates and the constant r_0, which is a measure of the electrode spacing. The potential will have the form [33]

$$\Phi = \frac{(U + V_0 \cos \omega t)(x^2 - y^2)}{r_0^2}, \tag{2.28}$$

so that the motion of an ion having a charge to mass ratio ne/m and moving in the z direction can be represented by the equations:

$$\ddot{x} + \frac{2ne}{mr_0^2}(U + V_0 \cos \omega t)x = 0, \tag{2.29}$$

$$\ddot{y} - \frac{2ne}{mr_0^2}(U + V_0 \cos \omega t)y = 0,$$

$$\ddot{z} = 0;$$

by letting

$$\omega t = 2\zeta; \qquad \frac{8neU}{mr_0^2\omega^2} = a; \qquad \frac{4neV}{mr_0^2\omega^2} = q \tag{2.30}$$

the orthogonal set of equations in x and y become:

$$\frac{d^2x}{d\zeta^2} + (a + 2q \cos 2\zeta)x = 0, \tag{2.31}$$

$$\frac{d^2y}{d\zeta^2} - (a + 2q \cos 2\zeta)y = 0.$$

Taken together, these relations represent a oscillating system in which the restoring force is periodic in time. The general solution of these equations takes an exponential form in which there exist finite limits on the range of values of a and q, if the amplitude of an ion oscillation is to remain bounded. Hence it is customary to construct a stability diagram which reveals the range of a/q values that are consistent with a real solution.

In principle, for a given potential field configuration it is only the ionic charge that determines whether an ion trajectory is stable or unstable. The initial position, and velocity in the z direction does not enter into the theoretical considerations for a stable ion trajectory. In practice, however, there are very real limits on the angular divergence of an entering ion beam, and the x, y, and z components of its initial velocity. As might be expected in such a system, there is also a convenient trade-off that can be made for transmission and resolution.

If the particles injected into the quadrupole are positive ions, the potential lines established by the dc field alone will have some focusing effect toward the z axis and a slightly divergent ion constrained to the x-z plane would undergo simple harmonic oscillations. The addition of the rf field will then superpose a high-frequency oscillation. The dc voltages in the y-electrodes tend to cause some defocusing, but the superposition of an rf voltage (which is usually high compared to the dc potentials) has the effect of a focusing action toward the z axis. This results from the fact that the mean trajectory of an ion with time (from the z axis) is determined by the *difference* in the opposing accelerations induced by the dc and rf potentials. A qualitative representation of an ion in the x-z plane is sketched in Figure 2.20.

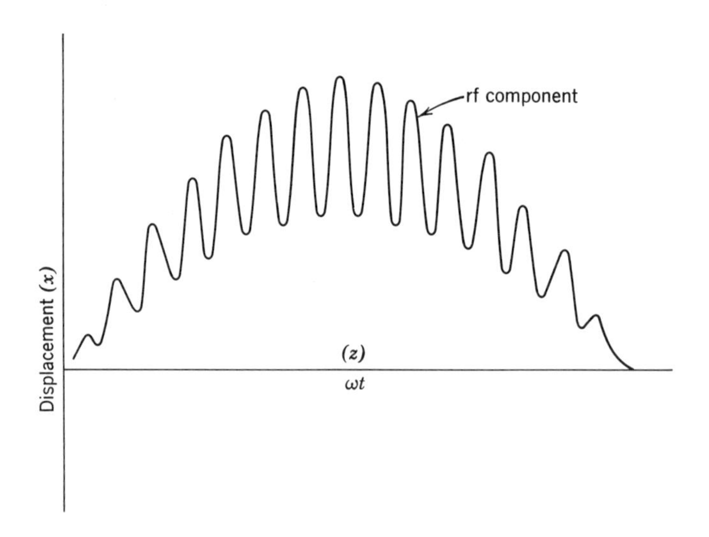

Fig. 2.20 Trajectory of an ion in the x-z plane of a quadrupole.

Table 2.1

Characteristics of a Quadrupole Mass Spectrometer [34]

Beam voltage	10–30 V
Diameter of entrance aperture	1.0 mm
Quadrupole spacing (r_0)	3.5 cm
Field length (L)	582 cm
Electrode construction (each)	60 wires in bundle
dc voltage for (U)	658 V*
rf voltage for (V_0)	3924 V*
Frequency ($\omega/2\pi$)	471 kHz
rf periods ion experiences	>600
rf power	290 W
Detector	17-stage electron multiplier
Mass scanning method	Fixed ω, simultaneous change of V_0 and U
Maximum resolution achieved	16,000

* For ions of 200 amu.

The modern version of the quadrupole spectrometer is a versatile instrument of high transmission and sensitivity. When equipped with an electron multiplier detector, it has been able to detect partial pressures in the 10^{-13} torr range. Commercial types are equipped with variable mass scanning sweeps (1, 10, and 100 μs/amu) so that rapidly changing concentrations of gases can be monitored on a continuing basis.

Typical parameters for a quadrupole instrument are shown in Table 2.1.

The Bennett radiofrequency mass spectrometer predates the quadrupole type, its development arising from the investigation of negative ions at the National Bureau of Standards shortly after World War II. This instrument operates in a manner somewhat analogous to a linear accelerator. A simplified schematic drawing of a "three-stage" tube is shown in Figure 2.21.

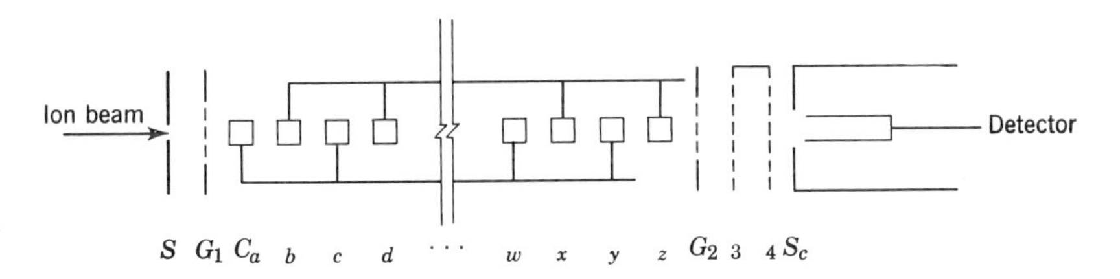

Fig. 2.21 The radiofrequency spectrometer devised by Bennett.

The tube consists of three groups of rf modulating grids, two field-free drift regions, a retarding electrode, and a detector. An ion beam is injected into the tube and passes successively through the rf grid system. If the ions are initially homogeneous in energy, the rf grid system will selectively accelerate or decelerate ions of different mass because ions of different masses will have different initial velocities. The total incremental velocity that will be added to an ion will, of course, depend on the phase of the rf grid cycle.

The maximum velocity increment will occur for unique values of ion velocity and phase angle. As opposed to the quadrupole mass filter, the rf-modulating voltage is usually small compared to the initial ion-accelerating voltage. The final retarding potential is selected to allow ions to proceed to the detector only if they have received the maximum incremental energy. A mass scan can then be achieved by appropriately (a) varying the initial ion beam energy, (b) tuning the rf oscillator, and (c) applying proper bias to the retarding electrode.

The principle of mass selection can be illustrated by considering only the first grid system in Figure **2.21**, in which the intergrid distance $S = ab = ba$ and grids a, b, and c are connected to the rf supply of angular frequency, ω. [35] At time $t = 0$, let the ions enter the rf field:

$$E_{a,b} = E \sin (\omega t + \theta). \tag{2.32}$$

At this time the adjacent grid will have the field strength

$$E_{b,c} = -E \sin (\omega t + \theta). \tag{2.33}$$

We have already stated that the incremental energy increase will be small compared with the initial kinetic energy of the incident ion beam. Let this increment then be

$$\Delta W = \Delta \left(\frac{mv^2}{2}\right) = v\,\Delta(mv) = v \int F\,dt. \tag{2.34}$$

When

$$F_{a,b} = eE_{a,b} = eE \sin (\omega t + \theta),$$
$$F_{b,c} = eE_{b,c} = -eE \sin (\omega t + \theta);$$

but S/v is the transit time between the grids ab, and bc; therefore

$$\Delta W = v \left[\int_0^{S/v} eE \sin (\omega t + \theta)\,dt - \int_{S/v}^{2S/v} eE \sin (\omega t + \theta)\,dt, \right. \tag{2.35}$$

$$\Delta\omega = \frac{evE}{\omega} \left[\cos \theta - 2 \cos \left(\frac{S\omega}{v} + \theta\right) + \cos \left(\frac{2S\omega}{v} + \theta\right). \right. \tag{2.36}$$

This energy increment has a maximum for the condition that

$$\frac{S\omega}{v} + \theta = \pi, \tag{2.37}$$

which simply states that ions which traverse the grid b at the instant of field reversal experience the maximum energy increase. If the frequency, ω, is varied, it has been shown that $\theta = 46° 26'$. The corresponding transit angle between the grids is then:

$$\frac{S\omega}{v} = 133° 34'. \tag{2.38}$$

If the velocity of the ions is expressed in terms of the kinetic energy of the ions

Then

$$eV_o = \tfrac{1}{2}mv^2 \tag{2.39}$$

$$m = \frac{0.266 \times 10^{12} V_0}{S^2 f^2}. \tag{2.40}$$

This is the optimum relationship of the parameters for a singly charged ion, where m is expressed in atomic mass units, V_o is in volts, S is in centimeters, and $f(\omega/2\pi)$ is the frequency of the rf field.

It permits the design of the system which will give the most effective mass filtering, and by appropriately adjusting the drift spaces, additional filtering will cause further discrimination from unwanted ions and provide increased resolution.

A somewhat analogous axial rf type, but which comprises 21 equally spaced modulator grids, has been reported by Redhead and Crowell [36]. Both types have found extensive applications where limitations of weight or size preclude the use of magnetic systems.

SPECIAL TYPES

The Cascade Analyzer

Two recently designed spectrometers will be mentioned because they represent a substantial departure from most of the previously discussed types. The first is a so-called cascade mass spectrometer in which "primary" and "secondary" ion beams can be simultaneously resolved within a single magnet. The second is basically the analytical counterpart to the electron beam probe, in that it can provide a chemical analysis of a sample within a very small region but with the distinction that ions rather than x-rays provide the means of impurity identification. Both are important tools for the investigation of surface phenomena.

The cascade spectrometer includes two successive 180° ion paths and differs from other two-stage spectrometers in that a large single magnet (rather than tandem magnets) includes both trajectories. One such instrument, having two 12-in. radius-of-curvature trajectories, was reported by White, Sheffield, and Rourke [37]. Included in this instrument is a semiconductor magnetic multiplier detector that is able to scan the entire focal plane of the second trajectory, thus providing high sensitivity for observing reflected or sputtered ions. More recently I have designed a similar magnet [38] at the Rensselaer Polytechnic Institute that includes the former cascade arrangement, but which also provides a multiplicity of ion paths. This magnetic analyser is shown in Figure 2.22; four distinct ion trajectories can be identified. The first consists of the double-focusing Mattauch-Herzog trajectory with source location at S_1. The second is the "cascade" path with a "primary" source at S_2 or S_2' and a secondary source at S_3. If S_3 is employed as a primary source of ions, and these ions are accelerated and caused to pass through a thin foil or gas cell, a third path exists. In this case positive and negative ions can be simultaneously monitored by the clockwise and counterclockwise paths that terminate at C' and S_2, respectively. Finally, S_4 can be used as a source for the independent 80° magnetic analyzing sector. It will also be noted that the magnetic boundary from C to C' serves as the common focal plane of ion

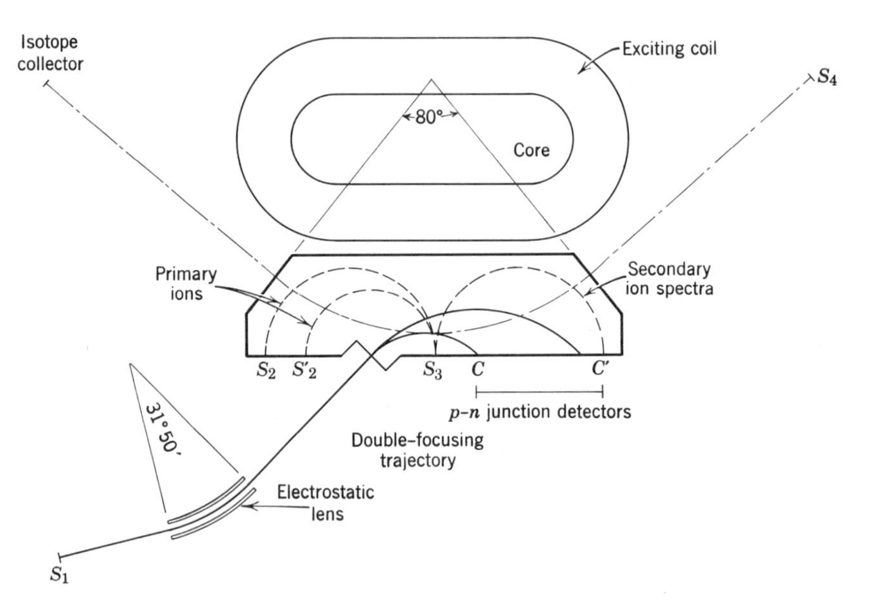

Fig. 2.22 Cascade analyzing magnet that provides for several ion trajectories.

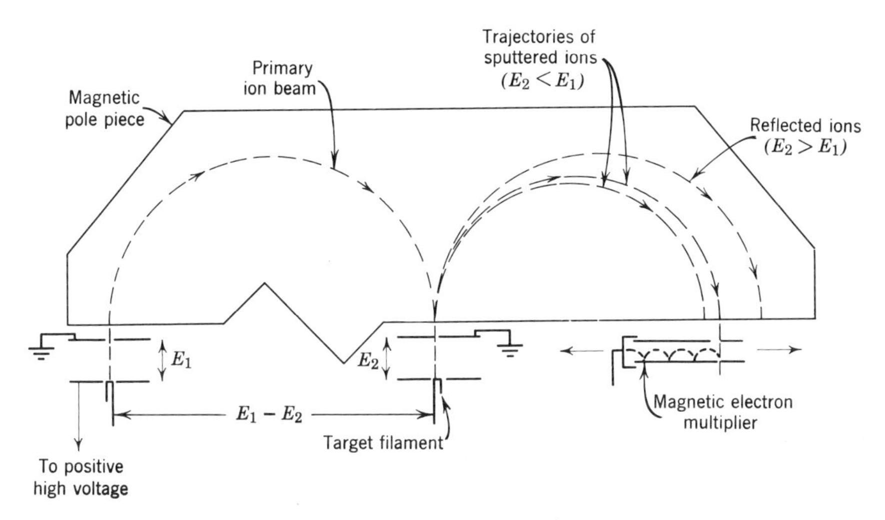

Fig. 2.23 Cascade ion paths for reflected and sputtered ions.

detection for the Mattauch-Herzog and cascade trajectories. The intent is to use an array of reverse biased *p-n* junctions for the simultaneous detection of a number of isotopes (see Chapter 4).

A discussion of this magnet will be restricted to the cascade ion path of Figure 2.23. There are three situations of specific interest [39]: (a) the case in which the primary accelerating voltage is less than the potential at S_3 (i.e., $E_2 > E_1$); (b) the case in which $E_2 < E_1$; and (c) pulsing of the primary ion source. If $E_2 > E_1$ and both potentials are positive with respect to ground, the ions that have been mass selected in the first trajectory will be decelerated in the second ion gun. The ions will have their velocity vector reduced to zero at some distance between the target filament and the grounded electrode; they will subsequently be reaccelerated and emerge with their original energy E_1. These electrostatically reflected ions will then have a radius of curvature that is equal to that of the primary ion beam. This scheme is somewhat equivalent to having two magnets in tandem and an improvement in "abundance sensitivity" of a factor of 100 has been noted. Such a scheme might also be useful for obtaining very pure isotopes of very small sample size by collecting them after they have been reanalyzed in the second reflected path.

The situation in which $E_2 < E_1$ will result in the bombardment of the target or specimen filament. The specific primary ion group that has been momentum analyzed may (a) be reflected from the surface, (b) penetrate into the filament with a mean range dependent upon the energy of the ions and the stopping power of the target material, (c) emerge after multiple collisions, or (d) cause sputtering of surface atoms of the filament. If the filament is heated to some temperature (T), primary particles can diffuse to the surface and a fraction of these will be amenable to analysis as ions.

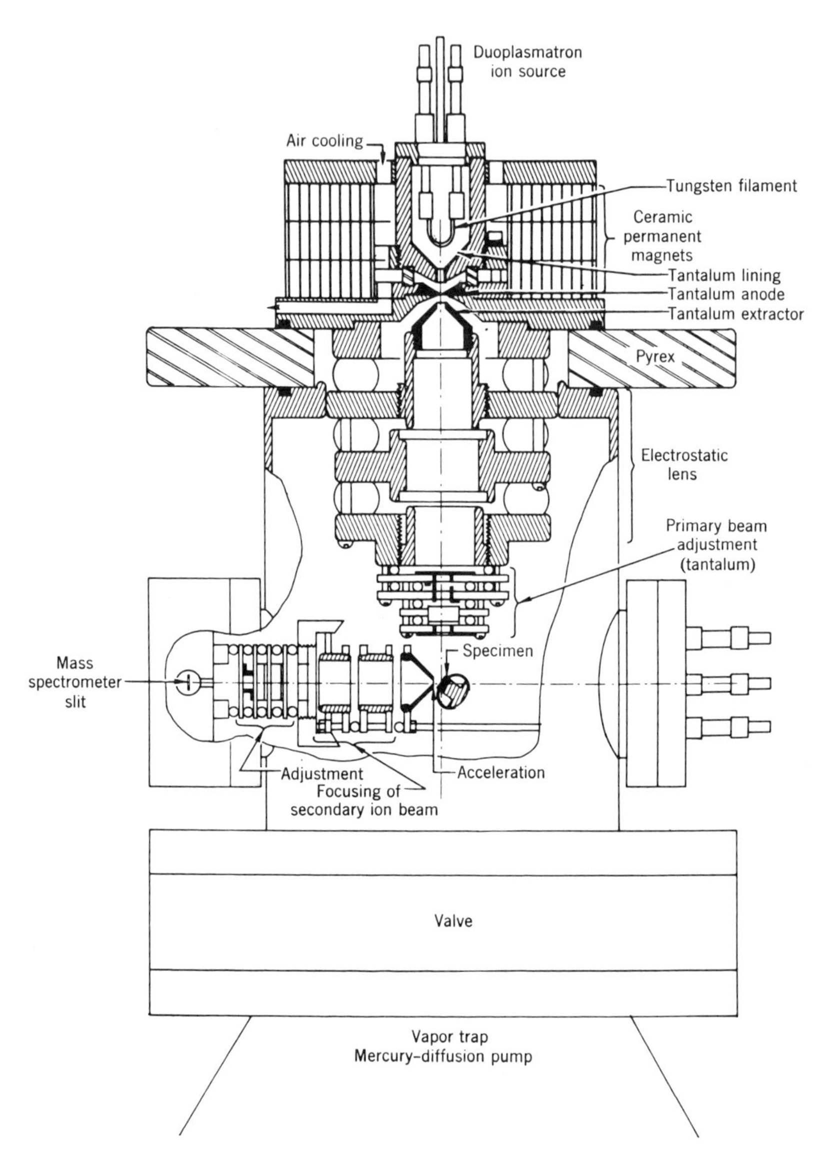

Fig. 2.24 High-sensitivity ion-beam microprobe source (courtesy G. C. A. Corp.).

Last, by pulsing the primary ion beam, the cascade spectrometer becomes an excellent tool for differentiating between the short-term and equilibrium emission of ions from a surface ionization filament, or for detecting the effect of ion "doping" on the thermionic emission properties of cathodes. The primary ion beam can also be used to study cathode response of photoemissive surfaces.

In general, pulsing techniques will also permit a distinction to be made between reflected ions (whose arrival time at the detector depends only on the trajectory transit time that can be computed) and ions that penetrate to some depth into the filament, diffuse to the surface, and are re-emitted as ions. The temperature dependence of this type of microdiffusion is usually difficult to measure by standard techniques.

The Ion-Beam Microprobe

Of somewhat more general analytical interest and applicability is the ion microprobe spectrometer reported by Barrington, Herzog, and Poschenrieder [40]. This type of instrument may well become one of the most important analytical and research tools in mass spectrometry. It is also a sputter-type instrument; with an intense primary ion beam current, a very high sensitivity is achieved. The sputter ion source assembly, utilizing a duoplasmatron, is shown in Figure 2.24. The specimen to be analyzed is placed at the focal point of a beam of xenon or argon ions that impinges on the sample at energies up to 10 keV. The secondary or sputtered ions are then focused by means of drawing out and accelerating electrodes into a double-focusing mass spectrometer. Sputtering rates of up to 200 monolayers per second have been reported, and an important characteristic of this new instrument is its ability to sputter from an area as small as 0.1 mm^2. It has a sensitivity in the ppm range, and is useful for the bulk or surface analysis of conductors and semiconductors, insulators, and virtually any solid with a sufficiently low vapor pressure.

Double-Focusing Spectrometer with Two Electrostatic Lenses

A variation of the Mattauch-Herzog arrangement that should prove generally useful has been suggested by Takeshita [41]. In the conventional double-focusing arrangement a disadvantage is that the slit or aperture of the energy analyzer cannot control the velocity spread (β), independent of the beam divergence angle (α). This limitation is removed by substituting a two-stage electrostatic field for the single one (Figure 2.25).

A study of aberrations for this system has shown theoretically that second-order aberrations of α^2 for all masses can be achieved under

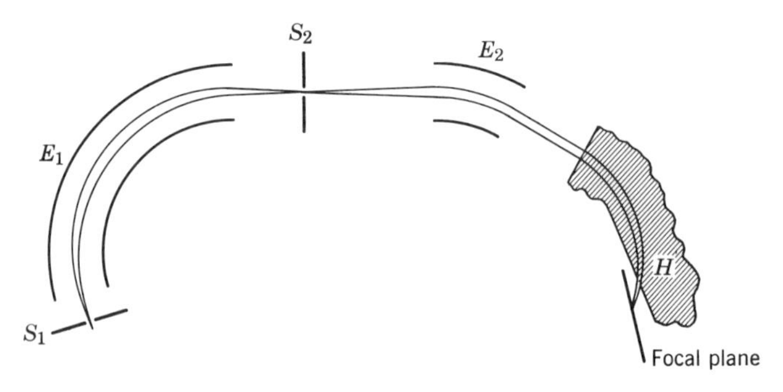

Fig. 2.25 Mattauch-Herzog spectrometer with tandem electrostatic lenses. S_1 and S_2 are source slits; E_1 and E_2 are electrostatic lenses; H is the magnetic field [41].

suitable conditions. Further, $\alpha\beta$ and β^2 aberrations can be made to vanish simultaneously at the focal plane. A numerical calculation also suggests that the total image defect is quite small for a wide range of masses.

Analyzer with Auxiliary Ion Filter

An auxiliary retardation lens has recently been added to a 30-cm radius, 90° magnetic sector, which substantially increases the abundance sensitivity of a single-stage magnet (Freeman, Daly, and Powell [42]). The lens is placed between the last resolving slit and the ion detector and it effectively rejects ions that have lost significant energy by collision with residual gas atoms in the system. The basic concept of the device is that if ions are retarded by a potential barrier of X number of volts below the nominal acceleration potential, then only those ions that have lost less than X electron volts will penetrate this electrostatic barrier.

One form of the ion filter consists of an assembly of three equidistant plates. The first plate is at ground potential and is located immediately behind the detector resolving slit. The center plate has a central aperture (covered by a 50% transmission mesh). The retarding potential is furnished to this center plate through a small bias battery that is connected to the ion source. The retardation potential thus follows any voltage fluctuations of the ion source supply. The third plate is at ground potential, and serves to reaccelerate the ions toward the detector.

At a pressure of 10^{-6} torr and with a 0.5-mm collector slit, the authors report data relating to the ion beam current of a single isotope at its adjacent mass position. Table 2.2 indicates the effectiveness of their ion filter.

This scheme for improving the abundance sensitivity appears attrac-

Table 2.2

Contributions at Adjacent Masses
Using the Ion Energy Filter ($M = 181$ region)

Mass	*180*	*181*	*182*
Lens off	2.2×10^{-4}	1	10^{-4}
Lens 6000 V	6.5×10^{-6}	1	6×10^{-6}
Improvement	34	—	17

tive, but the substantial improvement is at least partially because of the relatively high pressure. Such methods will also not solve the general problem where small-angle elastic collisions, charge exchange (to neutrals), and reflected ions all contribute to a distribution of ions along the focal plane [43].

Combination Isotope Separator—Mass Spectrometer

The combination of isotope separator–mass spectrometer system is of interest in those applications that require an intense primary ion beam that is mass resolved. Moran and Friedman [44] have used a 160-cm, 90° isotope seperator connected in tandem with a 15-cm, 60° magnetic sector mass spectrometer. The essential elements of their apparatus are shown in Figure 2.26.

The primary function of the isotope separator is to furnish intense beams (up to 30 μA) of resolved ions to a collision chamber that serves as a source for the analyzing mass spectrometer. These primary ions, which may have kinetic energies from approximately 5 to 60 keV, are focused by Einzel lenses, pass through a deceleration lens system, and enter the collision chamber. This chamber serves as the source of product ions, which are then analyzed in the mass spectrometer. This particular arrangement is reported to provide good energy resolution of the primary ions and efficient collection of the reaction products in the spectrometer. In a typical experiment the isotope separator source may be at 6000 V, and the collision chamber at 5998 V, thus giving 2 eV ions for the collision process under investigation.

This general type of apparatus is highly desirable for measuring (a) absolute total cross sections for ion-neutral interchange reactions of low velocity, (b) charge transfer processes at high kinetic energies, and (c) collision phenomena—generally for species of low abundance, which cannot be detected with the small primary-ion currents available in most mass spectrometers.

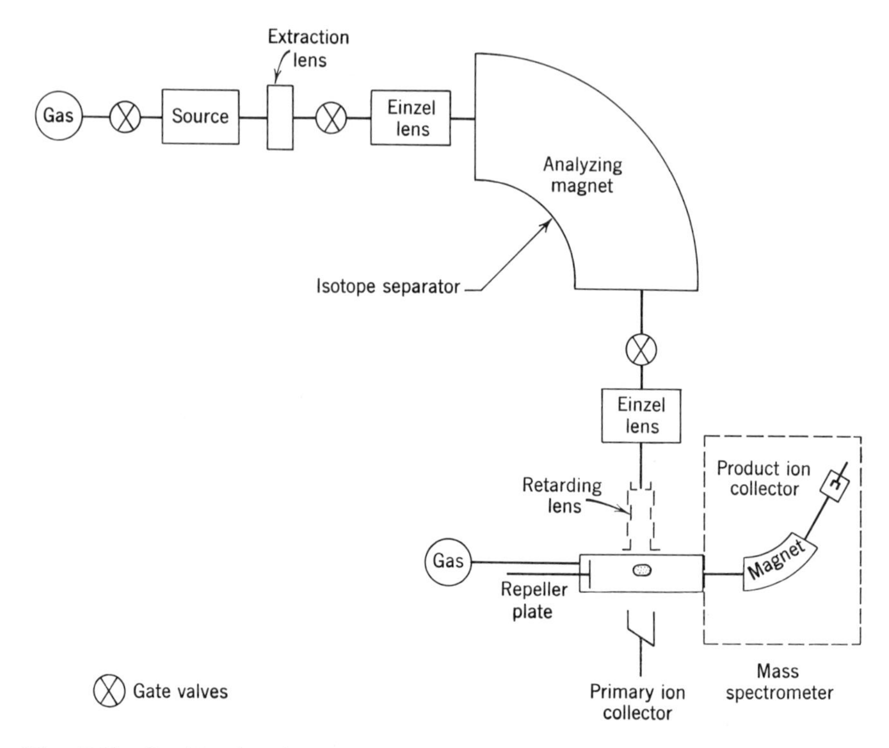

Fig. 2.26 Combination isotope separator—mass spectrometer schematic diagram [44].

Spectrometer for Positive and Negative Ions

Flesch and Svec [45] have reported the construction of a small spectrometer for the simultaneous detection of positive and negative ions, and they indicate that the instrument has served as a prototype for a similar analyzer of larger dimensions.

An ion source that consists of two Nier-type sources arranged back to back is located between two 180° magnetic sectors of 1.5-in. radii. The electron gun is maintained at ground potential and the accelerating electrodes are maintained at positive and negative potentials with respect to the gun. Thus positive ions are extracted in one direction and negative ions are accelerated oppositely. Mass spectra are obtained by magnetic scanning. In such an arrangement there are difficulties in coupling dc amplifiers to the ion-collecting electrodes that must be maintained at high positive or negative potentials. However, this type of spectrometer system appears useful in obtaining certain data on reactions and molecular bond energies that would be difficult to obtain by more conventional instrumentation.

Inhomogeneous Field Magnets

A special class of shaped magnetic lenses, used successfully for many years in beta-ray spectroscopy, is beginning to find more general applications in mass spectroscopy and isotope separations. The fundamental concept of this focusing system relates to the use of a magnetic field that decreases in intensity with increasing radius. The general appearance of shaped iron plates that could be used to produce a field of this type, often referred to as an inhomogeneous magnetic field, is illustrated in Figure 2.27.

The approximate intensity distribution across the midplane is given by the relationship $B_z = B_o(r_o/r)^n$ where r_o is the basic machine radius and n is the field index having a value between 0 and 1. Whitehead [46] has outlined the general ion-optics problem of inhomogeneous fields and its application to isotope separators. The principal advantage in employ-

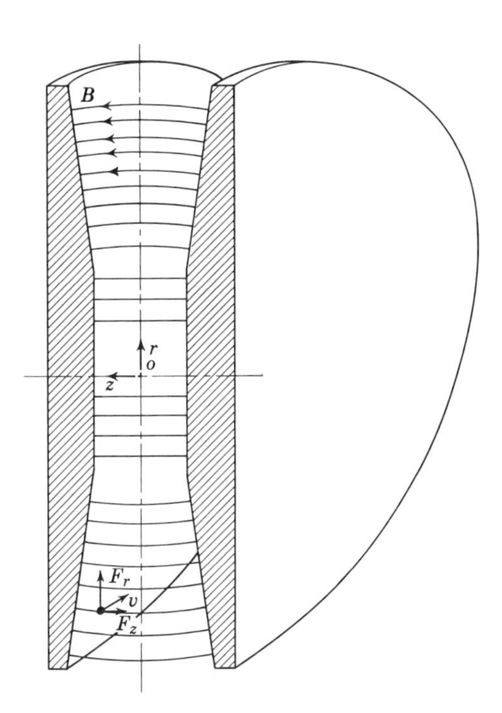

Fig. 2.27 Forces on a positively charged particle moving in an inhomogeneous magnetic field.

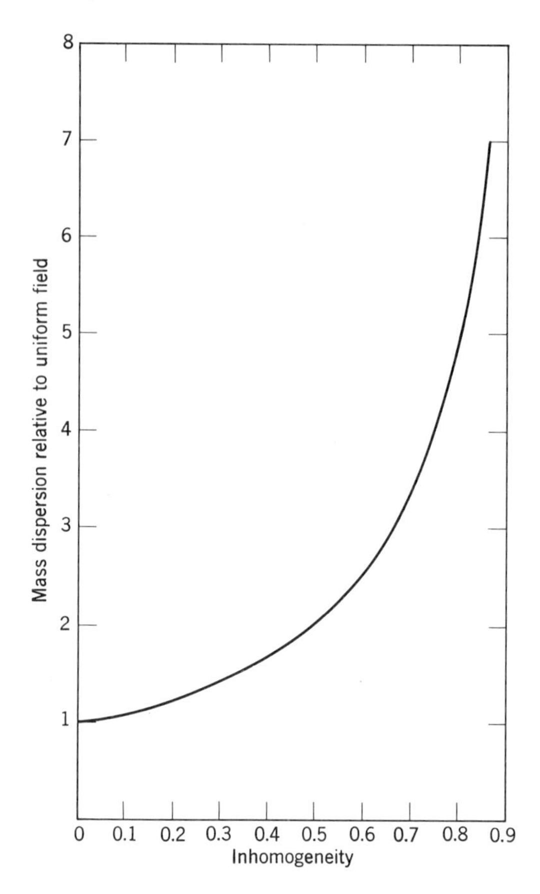

Fig. 2.28 Mass dispersion as a function of magnetic-field inhomogeneity.

ing a lens of this type is the resulting increase in mass- or energy-dispersive power. The increased dispersion is extremely valuable for it permits the use of a magnet smaller than a homogeneous field type. The relationship between the index of field inhomogeneity n and the mass dispersion relative to a uniform field of equal radius is shown in Figure 2.28.

At the Oak Ridge National Laboratory a few of the original isotope separators have been modified with contoured shim plates to produce a field with an inhomogeneity of 0.5. This system has produced good results with significant increases in the purity of separated isotopes. The only major difficulties encountered in using high-dispersion lenses is the rapid increase in image aberrations and diminishing beam transmission with increasing field index. In the case of the Oak Ridge separator the field was carefully shaped to insure both a good focus and a high transmission.

The analytical calculations that are required for the design of an inhomogeneous field magnet are not overly complicated, and several isotope separators using inhomogeneous field lenses are in operation throughout the world. It also appears likely that many more magnetic analyzers will be built in the future with inhomogeneities in the range 0.5 to 0.9.

REFERENCES

[1] A. J. Dempster, *Phys. Rev.*, **11,** 316 (1918).
[2] S. J. Balestrini and F. A. White, *Rev. Sci. Instr.*, **31,** 633 (1960).
[3] F. A. White, F. M. Rourke, J. C. Sheffield, R. Schuman, and J. R. Huizenga, *Phys. Rev.*, **109,** 437 (1958).
[4] A. O. C. Nier, *Rev. Sci. Instr.*, **11,** 212 (1940).
[5] R. Herzog, *Z. Physik,* **89,** 447 (1934).
[6] H. E. Duckworth, *Mass Spectroscopy,* Cambridge University Press, London, 1960, p. 26.
[7] R. Herzog, *Z. Physik,* **89,** 447 (1934).
[8] J. Mattauch and R. Herzog, *Z. Physik,* **89,** 786 (1934).
[9] *Ibid.*
[10] J. Mattauch, *Phys. Rev.,* **50, 617** (1936).
[11] C. M. Stevens, J. Terandy, G. Lobell, J. Wolfe, R. Lewis, and N. R. Beyer; *Advances in Mass Spectrometry,* ed. by R. M. Elliott, Macmillan, New York, 1963, p. 198.
[12] A. J. Dempster, *Proc. Am. Phil. Soc.,* **75,** 755 (1935).
[13] K. T. Bainbridge and E. B. Jordan, *Phys. Rev.,* **50,** 282 (1936).
[14] A. O. Nier and T. R. Roberts, *Phys. Rev.,* **81,** 507 (1951).
[15] K. S. Quisenberry, T. T. Scolman, and A. O. Nier, *Phys. Rev.,* **102,** 1071 (1956).
[16] M. G. Inghram and R. J. Hayden, "A Handbook of Mass Spectroscopy," National Academy of Science, National Research Council, Nuclear Science Series, Report No. 14, Washington, D.C. (1954).
[17] F. A. White and T. L. Collins, *Appl. Spectry,* **8,** 169 (1954).
[18] J. C. Sheffield and F. A. White, *Appl. Spectry.,* **12,** No. 1, 12 (1958).
[19] F. A. White, J. C. Sheffield, and F. M. Rourke, *Appl. Spectry.,* **12,** No. 2, 46 (1958).
[20] F. A. White and L. Forman, *Rev. Sci. Instr.,* **38,** No. 3, 355 (1967).
[21] J. A. Hipple Jr., and W. Bleakney, *Phys. Rev.,* **49,** 884 (1936); **53,** 521 (1938).
[22] H. G. Voorhies, C. F. Robinson, L. G. Hall, W. M. Brubaker, and C. Berry, *Advances in Mass Spectrometry* (vol. 1), ed. by J. D. Waldron, Macmillan, New York, 1959, p. 44.
[23] E. O. Lawrence and M. S. Livingston, *Phys. Rev.,* **40,** 19 (1932).
[24] H. Sommer, H. A. Thomas, and J. A. Hipple, *Phys. Rev.,* **82,** 697 (1951).
[25] E. W. Blauth, *Dynamic Mass Spectrometers,* Elsevier, Amsterdam, 1966, p. 64.

[26] L. G. Smith, *Proceedings of the International Conference on Nuclidic Masses,* University of Toronto Press, Toronto, 1960, p. 418.
[27] W. E. Stephens, *Phys. Rev.,* **69,** 691 (1946).
[28] A. E. Cameron and D. F. Eggers, *Rev. Sci. Instr.,* **19,** 605 (1948).
[29] W. C. Wiley and I. H. McLaren, *Rev. Sci. Instr.,* **26,** 1150 (1955).
[30] D. C. Damoth, *Advances in Analytical Chemistry and Instrumentation,* (vol. 4), ed. by C. N. Reilley, Wiley, New York, 1964, p. 371.
[31] W. Paul and H. Steinwedel, *Z. Naturforsch.,* **8A,** 448 (1953).
[32] W. H. Bennett, *"Mass Spectroscopy in Physics Research,"* National Bureau of Standards, Circular No. 522, Washington D.C., 1953, p. 111.
[33] E. W. Blauth, *Dynamic Mass Spectrometers,* Elsevier, Amsterdam, 1966, p. 140.
[34] U. von Zahn, S. Gebauer, and W. Paul, "A Quadrupole Spectrometer for Precision Mass Measurements." Presented at the 10th Annual Meeting on Mass Spectrometry, New Orleans, June 3–8, 1962.
[35] E. W. Blauth, *Dynamic Mass Spectrometers,* Elsevier, Amsterdam, 1966, p. 28.
[36] P. A. Redhead and C. R. Crowell, *J. Appl. Phys.,* **24,** 331 (1953).
[37] F. A. White, J. C. Sheffield, and F. M. Rourke, *Appl. Spectry.,* **17,** No. 2, 39 (1963).
[38] F. A. White, "A Cascade Magnet Having a Multiplicity of Ion Trajectories." Presented at the 14th Annual Conference on Mass Spectrometry, Dallas, May 1966.
[39] *Ibid.*
[40] A. E. Barrington, R. F. K. Herzog, and W. P. Poschenrieder, in *Progress in Nuclear Energy—Analytical Chemistry* (vol. I, ch. 5), ed. by H. A. Elion and D. C. Stewart, Pergamon, New York, 1966, p. 243.
[41] I. Takeshita, *Z. Naturforsch.* **21A,** 14 (1966).
[42] N. J. Freeman, N. R. Daly, and R. E. Powell, *Rev. Sci. Instr.* **38,** 945 (1967).
[43] M. Menat and C. Frieder, *Can. J. Phys.* **43,** 1525 (1965).
[44] T. F. Moran and L. Friedman, *Rev. Sci. Instr.* **38,** 668 (1967).
[45] G. D. Flesch and H. J. Svec, *Rev. Sci. Instr.* **34,** 897 (1963).
[46] T. W. Whitehead Jr., *J. Appl. Phys.* **36,** 3693 (1965).

Chapter 3

Methods of Ion Production

In a given experimental situation an ion source may represent either the simplest or the most complex component of a spectrometer. In most instances, however, the production of a suitable beam of ions is far from a trivial assignment, and as additional applications are found for mass spectrometry, the problems of source technology can also be expected to increase. The preparation of biological samples will call for different techniques from those appropriate to semiconductors, and gas analysis methods cannot be expected to apply to the investigation of surface monolayers of metals. The choice of a source will also be severely conditioned by the total sample size.

General criteria, however, can be listed if consideration is given to the general kinematics of ion trajectories and to the simple equation relating to the image "line width" at the detector focal plane. For a symmetric magnetic lens, a simple relationship exists between the object slit width, S_o, the half angle of rays emerging from the defining slit, α, the potential through which ions have been accelerated, V, the spread in ion energy corresponding to a potential difference, ΔV, the line width, W (at the image plane), and the radius of curvature, R, of the magnetic sector [1]. The relationship is

$$W = S_o + R\left(\alpha^2 + \frac{\Delta V}{V}\right). \tag{3.1}$$

If α is expressed in radians, a substitution of values for R, S_o, and $\Delta V/V$ will yield an approximate value for the image width. Equation (3.1) and other general considerations suggest the generally desirable properties of ion sources and ion beams:

1. The supply of ions should be sufficiently intense to be compatible with the analyzer geometry and detector sensitivity.
2. The beam should have an energy spread that is small compared with the total accelerating voltage.

3. The ion source slit width should be small compared with the radius of curvature of the analyzer.

4. The half-angle of divergence should be commensurate with the image line width—as expressed in (3.1) (unless special magnetic pole-shaping of the analyzer is employed which can minimize the α^2R contribution).

5. If the sample to be analyzed is very small, it is important that a high percentage of sample atoms become ionized.

6. Ideally, the source should be selective against unwanted ions that might appear in the same portion of the mass spectrum.

7. The source should not have a "memory" that would yield ions from the contamination of prior analyses.

8. Ion emission should be reasonable stable with time.

9. The source should minimize the chemical procedures required prior to mass spectral analysis.

Other considerations relate to beam spreading caused by coulomb repulsion (for intense beams), molecular dissociation, charge exchange, and mass discrimination. In most instances, of course, the prerequisite characteristics of the source depend upon the experimental objectives. A mass spectrometric study, for example, might be primarily concerned with the magnitude of an ion beam energy spread. Furthermore, there are situations in which a multiplicity of ions of the same mass but in several charge states provides a more positive elemental identification than can be obtained from a single atomic group.

The most important ion sources in modern mass spectrometry are reviewed below. They include electron bombardment, surface ionization, arcs, vacuum spark, ion impact, and field emission. Photoionization and laser sources are also mentioned, as they are becoming of interest in special applications.

ELECTRON BOMBARDMENT

The production of positive ions by electron impact is a technique universally employed for general gas analyses. With a suitably designed source, one can generate a copious beam of ions having a reasonably small energy spread. Furthermore, the ion beam intensity can be well controlled because the ionizing electron beam is generally space-charge limited. No other source provides a comparable stability in ion production. It is a preferred type for general analytical work as it can be utilized for nearly all gases, volatile compounds, and metallic vapors. Another important advantage of this source is that the ionization of a complex molecule can be controlled by varying the kinetic energy of the bombarding electrons. Both the number and specie of ion fragments can be altered in a manner so as to yield important data relating to the structural formula of the gas that is being subjected to mass spectral assay.

Gas inlet systems, ovens, valves, and gas mixing reservoirs are usually designed according to specific applications. Gas pressures in the ionizing region may vary from 10^{-2} to 10^{-5} torr or lower. Substances having a high vapor pressure at room temperature may be introduced either directly or indirectly into the source region, and the evaporation of inorganic solids from high temperature crucibles is a standard technique. The electron bombardment source itself, however, always possesses the common elements of (a) an electron-producing filament and electron trap, (b) high-voltage electrodes for accelerating the positive ions generated by electron impact, and (c) collimating slits to provide beam definition. A more complex source that provides additional focusing electrodes and which is typical of many gas sources is shown in Figure 3.1.

Electrons are emitted from the filament by thermionic emission and traverse the ionization chamber region. As ions are formed, they are accelerated by a drawing-out electrode, subjected to electrostatic potentials of focusing plates, and further accelerated through a potential of several kilovolts. Collimating slits effectively determine the divergence angle of the beam that enters the main section of the mass spectrometer, and beam-centering plates are also sometimes employed to increase the ion beam transmission.

It might be presumed that the energy of the ionizing electrons need

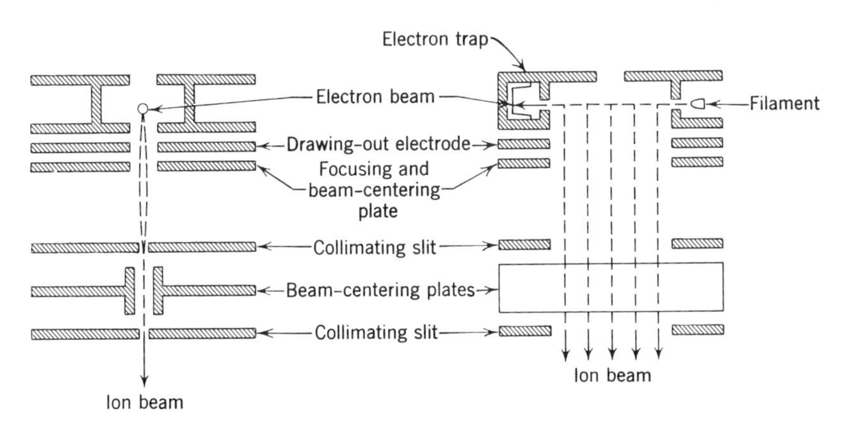

Fig. 3.1 Schematic diagram of an electron-bombardment source.

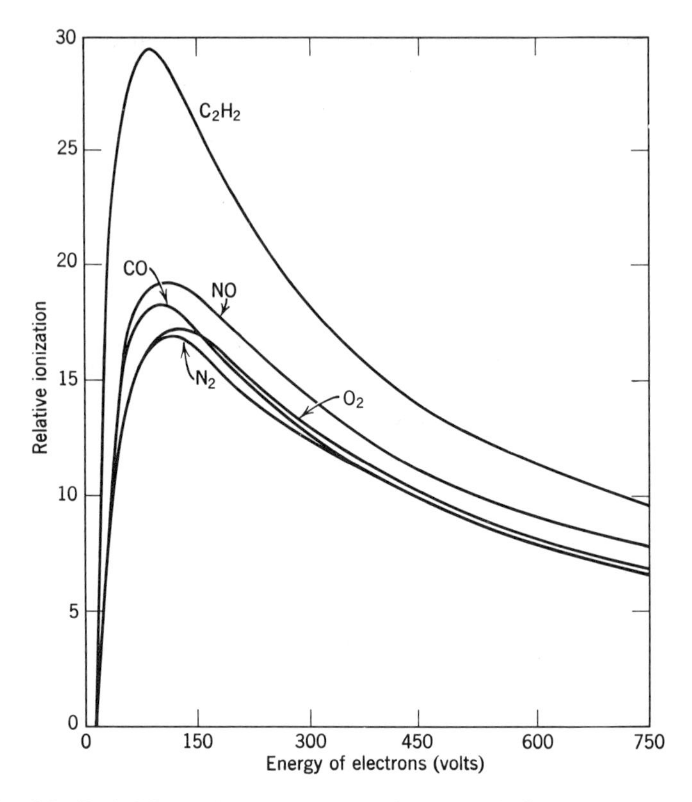

Fig. 3.2 Probability of ionization as a function of electron energy [2].

only exceed the first ionization potential of the sample gas. For achieving a maximum ionization efficiency, however, the electron beam energy must always substantially exceed this value; Figure 3.2 indicates the relative yield of ions produced as a function of electron energy [2].

It will be noted from Figure 3.2 that although there exist large differences in the positive ion yield of various gases, most ionization maxima occur in a region of 50–100 eV, and a practical operating voltage for the electron beam is about 75 V. The number of positive ions produced by the electron beam is directly proportional to the density of gas molecules at pressures $<10^{-4}$ torr. At higher pressures, the production of ions is not linearly related to gas density, owing to space-charge effects and recombination phenomena. The bombarding electrons will also produce atomic ions and molecules in many excited states. Atomic species will lose energy by radiation or by energy exchange in collisions. Molecules, however, can also undergo de-excitation by dissociation, giving rise to complex spectra that include the fragmented ions of lighter mass as well as ions of the parent molecule.

In a typical source, the electron beam is orthogonal to the trajectory of the ions, and the region from which ions are drawn into the accelerating region is small. Tungsten and rhenium filaments supply a copious electron current (of order of magnitude $\sim$1 mA) and this current is stabilized by appropriate circuitry in order to obtain a constant ion beam intensity. Electron beam stability may approach 1 part in 10^4, although this condition rarely results in a comparable stability with respect to the ion beam. The magnitude of the ion beam current will depend on several parameters—the specific gas, pressure, size of collimating slits, etc., in addition to the electron-bombarding current. Currents of 10^{-7} to 10^{-14} A are monitored in typical analyzers. The spread in energy of ions produced by typical electron-bombardment sources may be between 10 and 50 eV, but special techniques can reduce this value. For high-resolution work, however, a double-focusing system—or some type of energy discrimination—is essential.

The sensitivity of this type of source varies over many orders of magnitudes. Milligram amounts of sample may be needed, depending on the requirements of precision and other factors. The minimum gas sample is restricted primarily by the background spectrum of residual gases, the pumping system, and the detector sensitivity. Reynolds [3], using a small analyzing tube and employing a static vacuum, was able to detect xenon and argon with only $\sim 10^7$ atoms of sample. Such measurements, however, are quite beyond the capabilities of conventional analytical instruments. In fact, background gases often seriously limit the detection of trace gases because of the "memory" that an ion source retains for occluded gases on walls and surfaces. Despite such difficulties, detection limits of 1 part in 10^7 for trace gases have often been observed.

Precision values ± 0.1 to $\pm 0.01\%$ in isotopic abundance ratios can be achieved, but few investigators claim that absolute accuracies of conventional analyses approach the latter value. Utilization of standards, statistical methods, and knowledge of instrumental biases are prerequisites for achieving best values, even with an electron-bombardment type source.

Negative as well as positive ions can, in principle, be generated by electron-impact phenomena for compound species (e.g., $XY + e \rightarrow X^+ + Y^- + e$). Electron-capture processes present an alternative mode of formation. But, to date, mass spectral analysis of negative ions has been restricted to a few special problems, and detailed research studies.

Many electron-bombardment sources have been designed and been

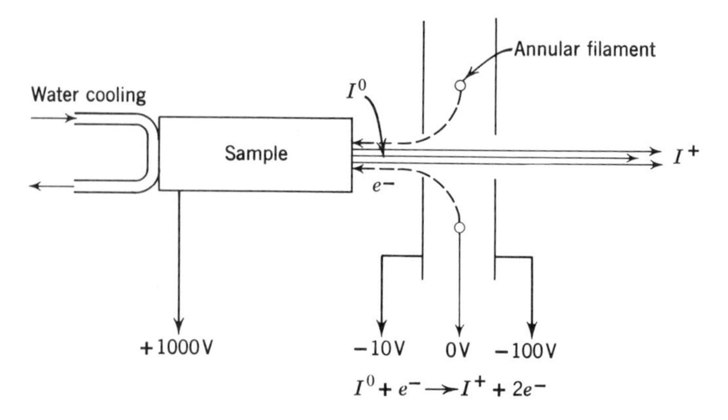

Fig. 3.3 Electron-bombardment source proposed by Honig [4].

successfully applied to the analysis of gases or metallic vapors. One specialized type of electron bombardment source has recently been proposed by Honig [4] for vaporizing refractory solids. The basic concept concerns having a single electron beam perform the dual function of vaporization and ionization—as indicated in Figure 3.3. A fraction of the neutral species evaporated by the heat of electron impact are subsequently ionized and drawn out axially into the mass spectrometer. This scheme will generate ions over a large energy range, thus requiring a double-focusing system. Nevertheless, the proposed system has merit on the basis of simplicity, and it may prove to be useful for the analysis of special samples that are difficult to analyze by more conventional methods.

SURFACE IONIZATION

The phenomenon of surface or thermal ionization has been utilized as a basis for a second and increasingly important ion source in mass spectrometry. Consider an atom or molecule that is evaporated from a heated surface. At sufficiently elevated temperatures, the emission of neutral vapor will be accompanied by positive ions (i.e., a fraction of the atoms or molecules will escape from the surface in an electron-deficient state). In such an ionization mechanism the hot metal surface is said to have a higher affinity for retaining an orbital electron of an escaping atom than the atom itself. The quantitative relationship for predicting the ratio of ionized to neutral atoms was first suggested by Langmuir and Kingdon [5] as

$$\frac{n^+}{n^0} \propto \exp\left[\frac{e(W - IP)}{kT}\right] \propto \exp\left[\frac{11606(W - IP)}{T}\right], \tag{3.2}$$

where n^+ = number of positive ions, n^0 = number of neutral atoms, e = electronic charge (1.6×10^{-19} C), IP = ionization potential (in volts), W = surface work function (in volts), k = Boltzmann constant (8.61×10^{-5} eV/°K), and T = surface temperature (in °K).

Thus the efficiency of ion production for a hot-filament or surface-ionization type source is dependent on (a) the work function of the metal filament, (b) the filament temperature, and (c) the first ionization potential of the element that is being evaporated. It will be noted that:

$$\ln \frac{n^+}{n^0} \propto T, \qquad (\text{for } IP > W), \tag{3.3}$$

$$\ln \frac{n^+}{n^0} \propto \frac{1}{T}, \qquad (\text{for } IP < W). \tag{3.4}$$

The above relationships indicate that a higher fraction of atoms evaporates as ions at high temperatures if the ionization potential is greater than the filament work function. Indeed, the surface-ionization source is ideally suited to the production of all elements of low ionization potential. Figure 3.4 shows a typical surface-ionization source assembly, complete with filament, accelerating, and focusing electrodes.

The selection of a filament material is usually limited to metals with high work functions, desirable mechanical properties, and which can be operated over a wide temperature range. Table 3.1 indicates some of the metals which have found considerable use as surface-ionization filaments. A more general listing of ionization potentials and work functions is given in Appendix 2.

Table 3.1

Metals Used for Surface-Ionization Filaments

Metal	*Work function (volts)*	*Melting point (°C)*
Nickel	4.50–5.24	1453
Niobium	3.96–4.01	2468
Platinum	5.08–6.27	1769
Rhenium	4.74–5.1	3180
Tantalum	4.03–4.19	2996
Tungsten	4.25–5.01	3410

Two developments have enhanced the usefulness of the surface ionization source in recent years. One is the greatly increased sensitivity of ion detectors. A second is the development of multiple-type filaments such as were first proposed by Inghram and Chupka [6]. The multiple-filament source (Figure 3.5) has the advantage that there can be independent control of evaporation rate of the sample material as well as the ion yield. Thus the sample filament essentially controls the emission rate of neutral vapor while the second filament is maintained at a temperature that gives a high ion-to-neutral ratio. The multiple filament is also advantageous from the standpoint of observing isotopic ratios over an exceedingly large temperature range. This allows an interpretation to be made concerning the possible contribution from impurity atoms having the same mass number.

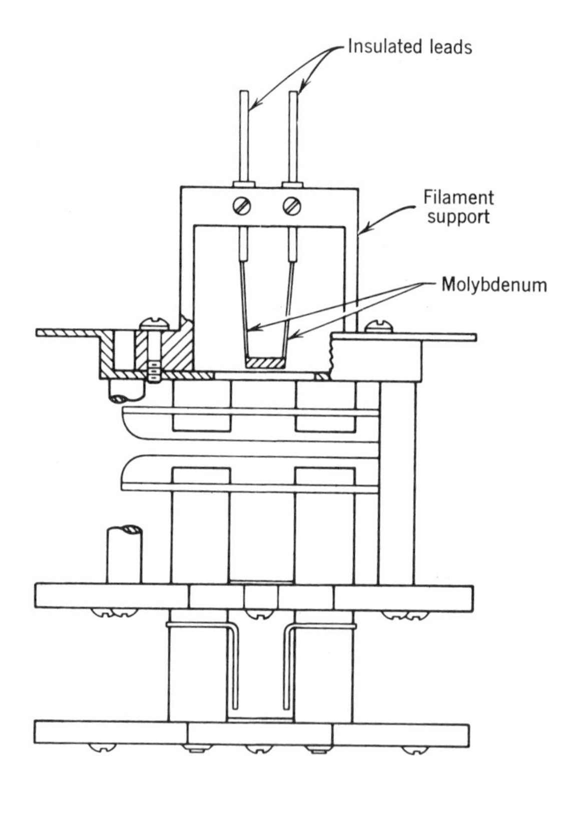

(a)

Fig. 3.4*a* Typical surface-ionization source assembly: side view.

The hot filament source is also useful for the production of negative ions. In this case the relative yield of negative ions to neutrals is

$$\frac{n^-}{n^0} \propto \exp \frac{e(EA - W)}{kT} \propto \exp \frac{1160(EA - W)}{T}. \qquad (3.5)$$

In the above expression EA is the electron affinity in volts.

The surface-ionization source is useful for an exceedingly large number of metals, and positive ion beams have been reported for elements having first ionization potentials below approximately 9 V. Negative ions are formed for elements having an electron affinity greater than about 1 V. Prime advantages of the surface ionization source include: (a) low ion energy spread, (b) simplicity in sample preparation, (c) selectivity of

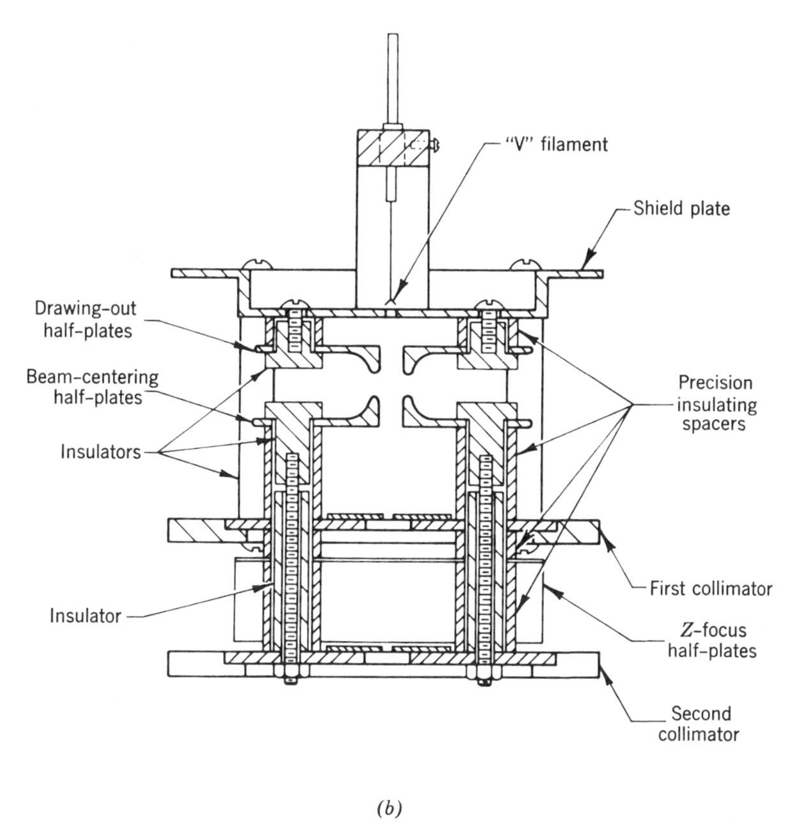

(b)

Fig. 3.4*b* Typical surface-ionization source assembly: cross-sectional view.

overlapping isobars and impurities, and (d) the fact that all ions are singly charged.

The spread in the energies of ions emitted by thermal sources is very small (<1 eV). As a result a single magnetic analyzer is all that is required for conventional analyses. In practical cases the greatest energy spread may arise from the voltage drop across the filament—caused by resistance heating (i.e., a few volts). If the surface-ionization filament is heated thermally by some indirect method, this voltage spread can be virtually eliminated. There will, however, always be some finite spread in the energies of ions leaving a thermal source, as small variations in the work function over the hot surface always exist—especially when deposition of the sample may give rise to various oxidation states.

In many instances the sample preparation will be simple. For very small samples, however, complete preheating and out-gassing of filament impurities is mandatory. As sample size decreases, the need for reagents of very high purity also becomes crucial, especially in cases of isotopic dilution problems. But for general assay and samples amounts of the order of 10^{-6} to 10^{-3} gm, the preparation of a dilute solution is relatively easy. A weak acid or distilled water solution of the element (usually a salt) is prepared, micropipetted onto the filament, and evaporated to dryness. In a majority of instances the elemental or simple metal oxide will be observed, but with some materials that have high ionization potentials it is desirable to prepare special compounds. For example, sodium borate heated on tungsten will yield molecular ions of $(Na_2BO_2)^+$, and quantitative mass spectra can be obtained for boron at mass positions 88 and 89 instead of at the atomic mass positions of 10 and 11. Detailed considerations regarding source preparation and surface ionization phenomena have been reported [7], [8].

Selectivity or the preferential ionization of sample atoms is sometimes important. It will be noted from Appendix 2 that finite and sometimes large differences exist in the first ionization potentials of many elements having overlapping mass spectra. Gaseous species are of course not ionized, as in an electron-bombardment source, so the mass spectrum is much cleaner with respect to many hydrocarbons. With a multiple-filament source an even further discrimination against impurity ions may be realized. But even small differences in ionization potentials of the metals themselves permit a degree of discrimination against impurity spectra of the same mass that cannot be achieved in other sources.

Some of the specific filament configurations that have proved to be useful are sketched in Figure 3.5. The single-flat-ribbon filament is easy to construct and it can be used for elements that are readily ionized [9]. The "canoe" type configuration was introduced in order that neutral

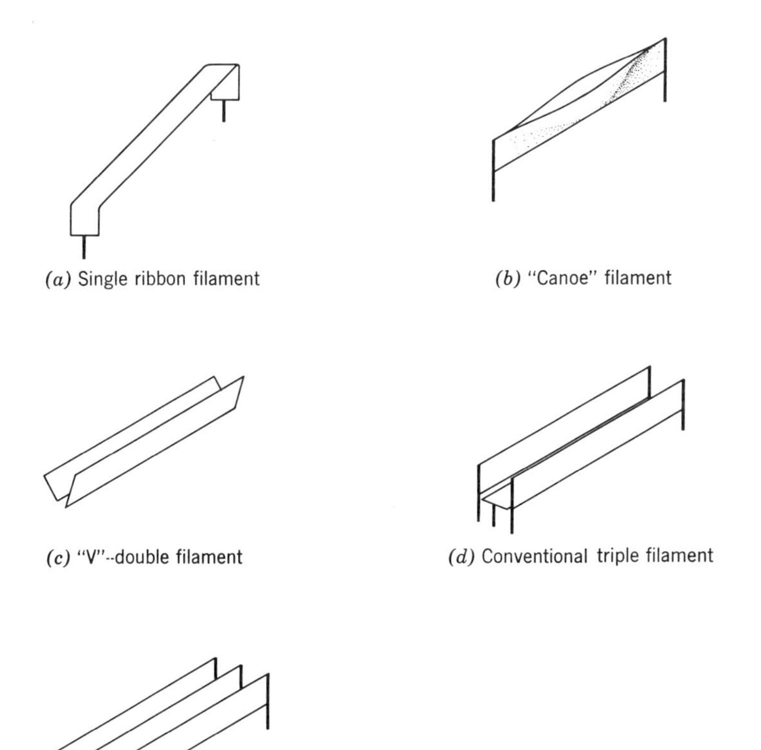

Fig. 3.5 Types of filament configuration used in surface-ionization sources. (*a*) single-ribbon filament; (*b*) "canoe" filament; (*c*) "V" double filament; (*d*) conventional triple filament; (*e*) parallel triple filament.

atoms might be afforded additional opportunity for ionization in multiple collisions.

The "V" double filament (Figure 3.5) has been used because of its simple configuration and construction [10]. It affords essentially the same independent temperature control as the conventional triple filament. A small bias voltage on the sample filament has also been shown to effectively discriminate against impurities. Consider, for example, the case in which the potential of the ionization filament (V_o) corresponds to proper focus of the mass spectrometer of an ion of mass m at the detector slit. If the sample filament is operated at a potential ($V_o \pm \Delta V$), primary ions (originating from the sample filament) will not arrive at the detector providing the mass spectrometer has adequate resolution. The ΔV can be adjusted to about 10 V, but it should not exceed the first ionization potentials of residual gases in the source region as in such a case an electron impact spectra would be superposed upon the metal spectrum. It will also be noted that if the ionization filament is operated at a very high temperature (regardless of sample emission rate), a larger number of ions will leave the ionization filament as atoms rather than as molecules that tend to be dissociated. Thus a less complex mass spectrum results.

The parallel triple filament [11] is another multiple-filament type reported to be useful in comparing an unknown sample with a standard. Some increase in ionization efficiency is also claimed because of the better solid angle and the probability of multiple collisions on the part of neutral atoms or molecules.

Investigators have also explored various possibilities for enhancing the ion emission from hot filaments by effectively changing the work function of the surface or by treating the source with a reducing agent [12]. Such techniques have proved to be highly successful in specific instances when the ion emission problem can be restricted to a particular element or metallic compound. The surface-ionization source does not possess the stability of emission of the electron bombardment type, but it has been applied routinely for sample assay in the 10^{-9}-gm range. Under favorable conditions 10^{-14} gm of sample can be detected.

DISCHARGE SOURCES

An important class of mass spectrometer sources includes specific types that utilize discharge phenomena for production of ions. Considerable literature is available reviewing this field in detail. Therefore this book's discussion will be limited to the most important characteristics of sources generally categorized as (a) gaseous discharges, (b) low-voltage arcs, (c) sparks, and (d) the vibrating arc.

Gaseous Discharges

The production of ions by gaseous discharge holds more historical than contemporary interest. Gaseous discharge techniques were employed by almost all the early mass spectroscopists—including Thomson, Aston, and Bainbridge. Ions were produced in a "glow discharge" tube in which cathode-anode voltages ranged from 10,000 to 50,000 V. Because ions are produced at intermediate pressures, differential pumping was usually employed to the ion tube and analyzer regions. Extraction of ions was through an aperture in the cathode, the cathode thus having a dual

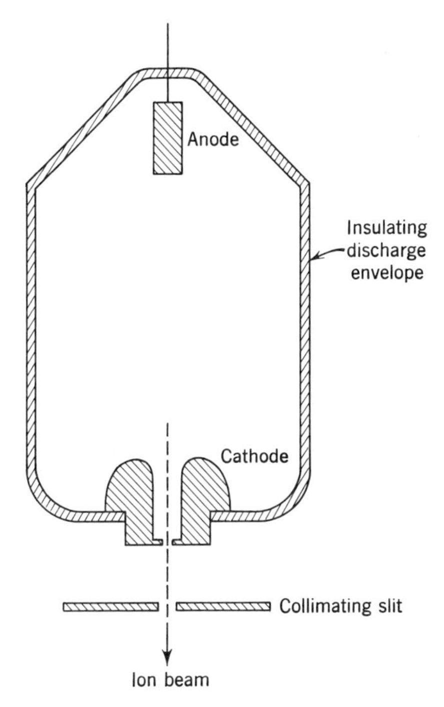

Fig. 3.6 Gaseous discharge source.

function of maintaining the discharge and accelerating positive ions into the analyzing tube (see Figure 3.6). This source is not exclusively limited to gases as several workers have introduced sample materials into one of the electrodes for subsequent volatilization and ionization. This source furnishes copious ion current of both single and multiple charge; however, its usefulness in current research is limited because of the inherent instability of the ion current and the exceedingly large ion energy spread. The latter factor essentially restricts its application to double-focusing systems.

Several variations of the basic gas discharge tube have been devised, an important one being to employ a magnetic field in a manner similar to the cold cathode ionization gauge developed by Penning [13]. This combination of radiofrequency and magnet fields serves to increase greatly the electron path lengths, with a resultant higher degree of ionization. Consider a hollow cylindrical anode located between two parallel plane cathodes (see Figure 3.7). A magnetic field is applied so that field lines are in the direction of the axis of the cylinder and an appropriate radiofrequency voltage is applied to the parallel plate electrodes.

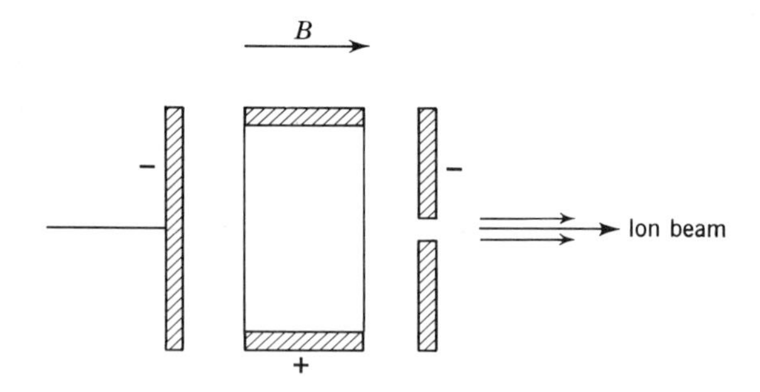

Fig. 3.7 Cold-cathode ionization source devised by Penning.

Any electrons that are present will undergo oscillating trajectories through the anode. The magnetic field lines constrain the electron paths to tight helices, enhancing the degree of ionization. At pressures of 10^{-3} to 10^{-4} torr and with interelectrode potentials of several kilovolts, a self-sustaining discharge can be maintained without a hot filament to supply electrons. This permits operation of the ion source at gas pressures about two orders of magnitude below that of the conventional type. If the anode and cathode are arranged to be perpendicular rather than collinear to the ion beam that is drawn into the mass spectrometer, the effective ion energy spread can be reduced to about 25 eV.

This same general method has been employed (but with auxiliary thermionic emission as noted below) for the production of ions in isotope separators. Ion currents up to 0.1 mA have been achieved and a high yield ($\sim$10% ions to sample atoms consumed) has been reported [14].

Low-Voltage Arcs

Low-voltage arcs in a variety of geometries have been developed for mass spectrometers, cylotrons, and isotope separators. The basic elements are shown in Figure 3.8. The essential distinction between the gaseous discharge and arc source is that the latter is not self-sustaining, but requires a hot filament to furnish a copious supply of ionizing electrons to maintain the arc. The electron beam is at right angles to the direction from which ions are extracted from the arc plasma. Improved electron-beam collimation can be achieved by means of auxiliary magnets having field lines collinear with the electron trajectories. Such a source has been termed a "duoplasmatron." While the initial discharge must usually be triggered by the introduction of a "buffer gas" or high electrode potentials, low voltages across the arc can be maintained at quite low pressures. Using a duoplasmatron, Koch [15] has reported the production of 10^{-4} A of positive ions with arc currents of 0.2 to 0.8 A in the pressure range of 10^{-3} torr. The energy spread of ions, in the most favorable cases, can be restricted to less than a volt.

Recently duoplasmatron sources have also been modified to supply low-energy negative ions. The conventional duoplasmatron cannot be used because the electron current that would be extracted from the arc would be several orders of magnitude greater than the negative ion current. This would result in a large current loading being placed on the high-voltage sources. However, if the arc axis is offset from the ion axis, electron beam loading problems are effectively eliminated and it is possible to obtain substantial currents of negative ions of limited energy spread. Aberth and Peterson [16] have reported on such a scheme; they claim it to be quite appropriate for both positive and negative ions. For the latter species an energy width at half maximum of 3 eV is claimed. Ions of H^-, O^-, OH^-, F^-, CN^-, and NO_2^- were observed at 2 keV—with beam intensities of greater than 10^{-7} A.

Low-voltage arcs are not restricted to gases. Metals can be introduced in the vapor state. In this case the entire source enclosure must be maintained at an appropriate high temperature in order to prevent large losses of sample material by condensation. Because of the large ion beam currents that can be attained, arc sources are the principle type employed for the large-scale separation of isotopes at the Oak Ridge National Laboratory. Their use in mass spectrometry for isotopic ratio measurements is somewhat restricted because of the large fluctuations in beam current.

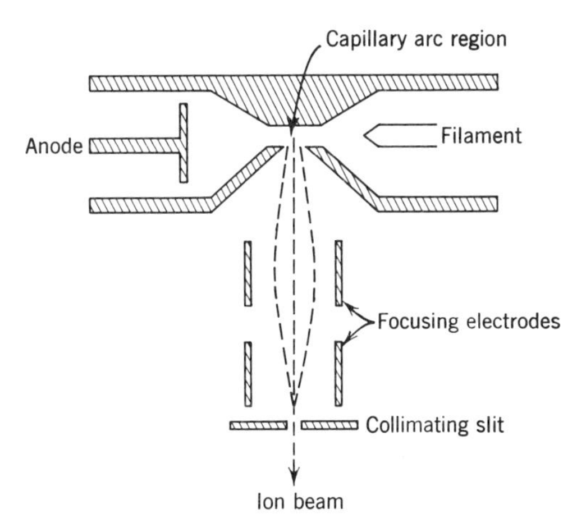

Fig. 3.8 A low-voltage arc ion source.

The Vacuum Spark

The vacuum spark source produces ions by the high-voltage breakdown across two electrodes, one of which consists of or includes the sample material. The terms *arc* and *spark* are often loosely used interchangeably, but a spark is generally characterized by a short-term, high-current flow—followed by a recovery of original electrode potentials and current quenching. The details of spark phenomena have themselves been the object of mass spectrometric study. Honig has provided a qualitative discussion of the spark breakdown mechanism, and Honig, Guthrie, Hickam, and Sweeney have presented an excellent review of spark ion sources for mass spectrometry [17].

The basic spark source is shown schematically in Figure 3.9*a*. As the potential across the two electrodes is increased, prebreakdown currents occur that are solely a function of voltage and pressure. At a sufficiently high voltage, a spark discharge will take place that results in a region of negative resistance and an interelectrode voltage reduction of several orders of magnitude. If the external circuit possesses sufficient stored energy and a low impedance, a high current arc may exist for many microseconds. At high field gradients ($\sim 10^5$ V/cm), current increases are also attributed to field emission from "whiskers" or projections ($<1\ \mu$) that exist on any cathode surface. As the voltage is further raised, some "whiskers" may vaporize and contribute cathode atoms to the interelectrode region. Electron bombardment of the anode will generate neutral atoms that will be immediately ionized by the cathode electron stream. Ions from the anode region may also cause cathode sputtering—thus aiding the process of current multiplication. During the spark, single and multiple charged ions will be produced because of the copious vaporization of the electrodes and their subsequent ionization by electrons in the discharge. In general the ion energy spread will be large, requiring a double-focusing analyzer.

The usual method of spark generation is by radiofrequency circuitry. Hickam [18] reports typical operating characteristics of such a source: 800 kc with 20-μs pulses at a maximum voltage of 30–40 kV, with a repetition rate which can be fixed at 5, 50, 500, and 5000 cps. Such sources can be used for the analysis of liquids, powders, conductors, semiconductors, and insulators. Hickam and Sandler [19] have also reported an unusual variation of the conventional spark source, which is sketched in Figure 3.9*b*. The high-velocity rotation of one of the electrodes permits a rapid "surface scan" with a spark, which is exceedingly useful in the analysis of surface monolayers (see Chapter 9).

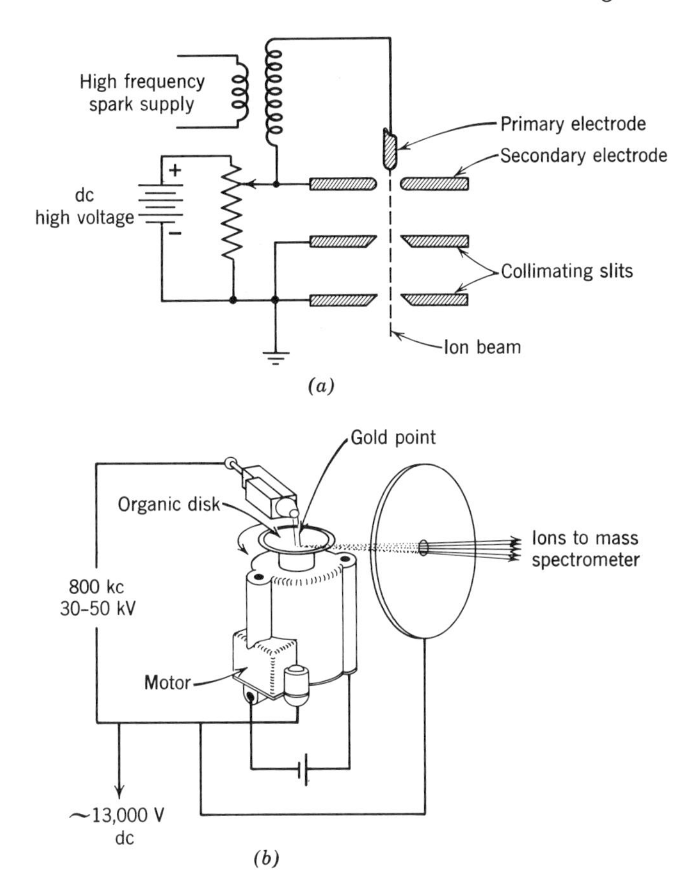

Fig. 3.9 (*a*) Vacuum-spark ion source; (*b*) vacuum-spark spinning electrode [18, adapted].

The Vibrating Arc

The vibrating arc ion source is not widely employed, although a number of investigators have been considering the applications of this low-voltage vacuum breakdown device. The essential features of this source are shown in Figure 3.10. If the sample to be analyzed is a metal of good conductivity, portions of the sample can comprise the electrodes. As indicated, the vibrating electrode system consists of a simple solenoid and dc supply (i.e., a doorbell-type circuit). When the electrodes are separated, the full voltage appears across the gap; the voltage is reduced to virtually zero and the current increases exponentially. The resultant high current passing through a small contact resistance causes a high

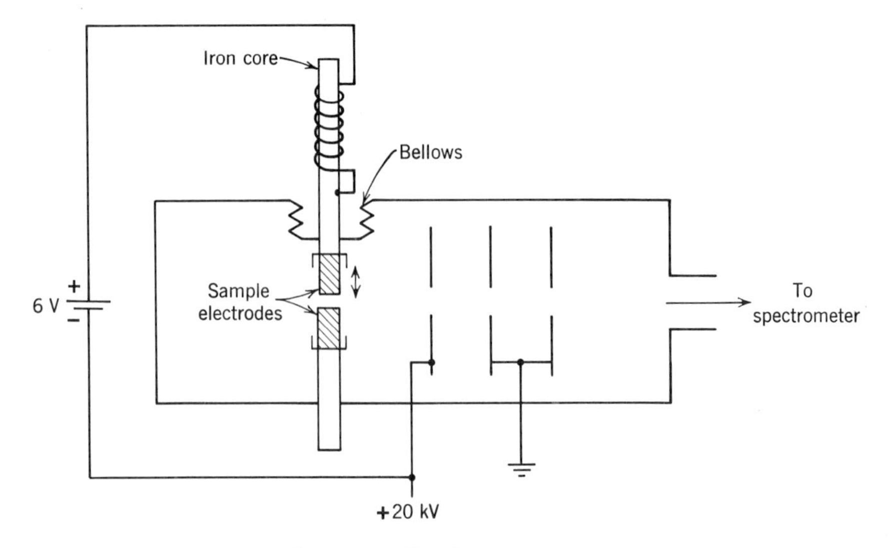

Fig. 3.10 Vibrating arc source.

power dissipation at the electrode interface. Melting of the electrode tips results in a "liquid bridge" and an instantaneous arc. Ions are observed during this arcing cycle, but the total ion energy spread is quite low. Honig has reported that doubly charged species predominate over singly and triply charged ions, and a mass spectrum displays a predominance of cathode ions when two different metals are used as electrodes.

ION BOMBARDMENT

The production of secondary ions by ion bombardment or sputtering promises to develop into one of the most important ion sources. The generation of ions by ions can be attained in gases, but the real potential for ion impact sources relates to solids. Several advantages may be stated explicitly:

1. The secondary ion energy spread is sufficiently low ($\sim$10 eV) so that a double-focusing spectrometer is not always required.
2. A sample may be analyzed at room temperature.
3. Only surface atoms are ionized ($\sim$100 Å depth, depending upon the primary ion mass and energy).
4. Positive ions are obtained for all elements having ionization potentials below about 10 eV.
5. High sample sensitivity is achieved for surface atoms.

6. The sputtered mass spectrum yields species of both atoms and compounds.
7. Such a source is amenable to programming in a pulsed mode.

Although the above characteristics make the ion impact source an exceedingly useful one, there are also several limitations for its application in conventional analytical instruments. Primary bombarding beam currents of considerable intensity may be required; if such is the case, the problem of space charge arises. In addition, if a metal is to be analyzed at room temperature, the vacuum should be high enough so that the bombarding ion beam is large compared with the time rate of arrival of residual gas atoms in the environment. The availability of pumping systems capable of achieving pressures in the 10^{-9} torr range, however, should circumvent this latter difficulty.

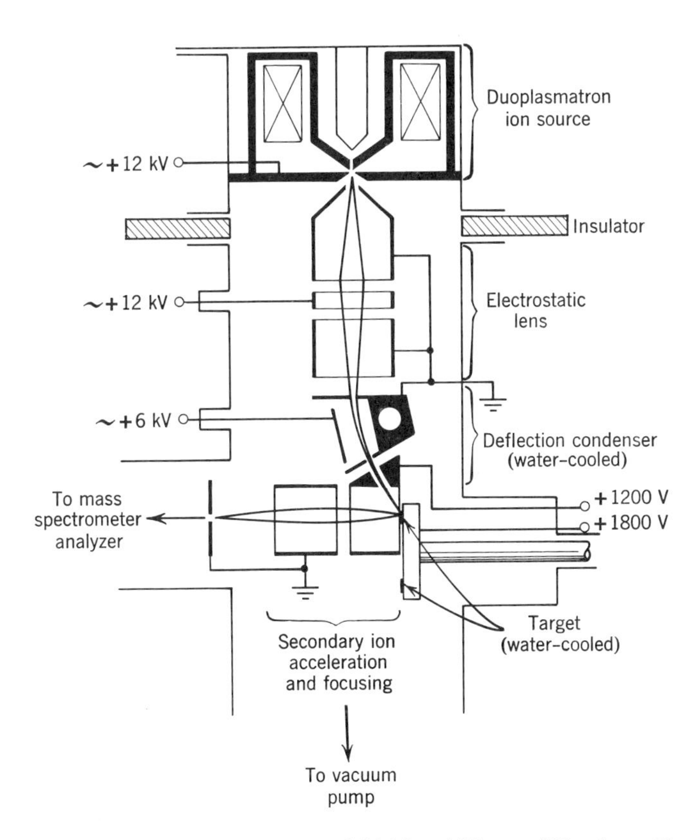

Fig. 3.11 Sputter ion source of Liebl and Herzog [21, adapted].

A review of early work on sputter ion sources has been given by Honig [20], who was the first to devise an ion sputter source for the analysis of semiconductors. A specialized ion bombardment source using normal ion impact has been mentioned in connection with the cascade ion-beam analyzer of Chapter 2; a high-intensity sputter ion source has been constructed by Liebl and Herzog [21]. This latter source is shown in Figure 3.11.

The primary intense beam of argon ions is provided by a duoplasmatron. These ions are then accelerated by a drawing-out potential, focused by an electrostatic lens, and deflected by a condenser lens. The bombarding beam intensity of about 1 mA is focused on a spot of about 3 mm in diameter. The fraction of the sputtered target that ablated in the form of ions is then focused on the mass spectrometer entrance aperture. A current of 10^{-8} A of sputtered ions was reported, with stable emission.

A highly specialized ion impact source has also been reported by Castaing and Slodzian [22], who combined a sputter ion source, a mass analyzer, and an ion microscope into a single instrument. This device extends the simple sputter source in that it provides a secondary mass spectrum with a spatial distribution. The image magnification is greater than an order of magnitude of the size of the sample bombarded area. In one sense it is complementary to an electron microprobe method. Important applications for this ion bombardment source are given in Chapter 8.

FIELD IONIZATION

If a high voltage ($\sim$20,000 V) is applied across two closely spaced concentric spheres, voltage gradients in excess of $\sim 10^6$ V/cm can be realized. In the field ionization source the inner electrode is a tungsten tip having a radius of about 500 Å. The closely spaced outer electrode is not usually a sphere, but a ring-shaped electrode through which ions can pass after their formation. This type of source was effectively developed by Mueller [23] in his field ion microscope, and it was successfully applied to mass spectrometry by Gomer and Inghram [24]. A review paper by Beckey [25] presents design details of the source, which is sketched in Figure 3.12*a*.

The small radius tip is made by special etching of a 0.1-mm tungsten wire. This tip then serves as the point source from which ions are formed and accelerated through the coaxial ring, and the auxiliary focusing lenses. An alternative geometry for a field ion source, using a razor blade tip, has been reported by Beynon, Fontaine, and Job [26]. Ionization is assumed to occur at a distance of several angstroms in front

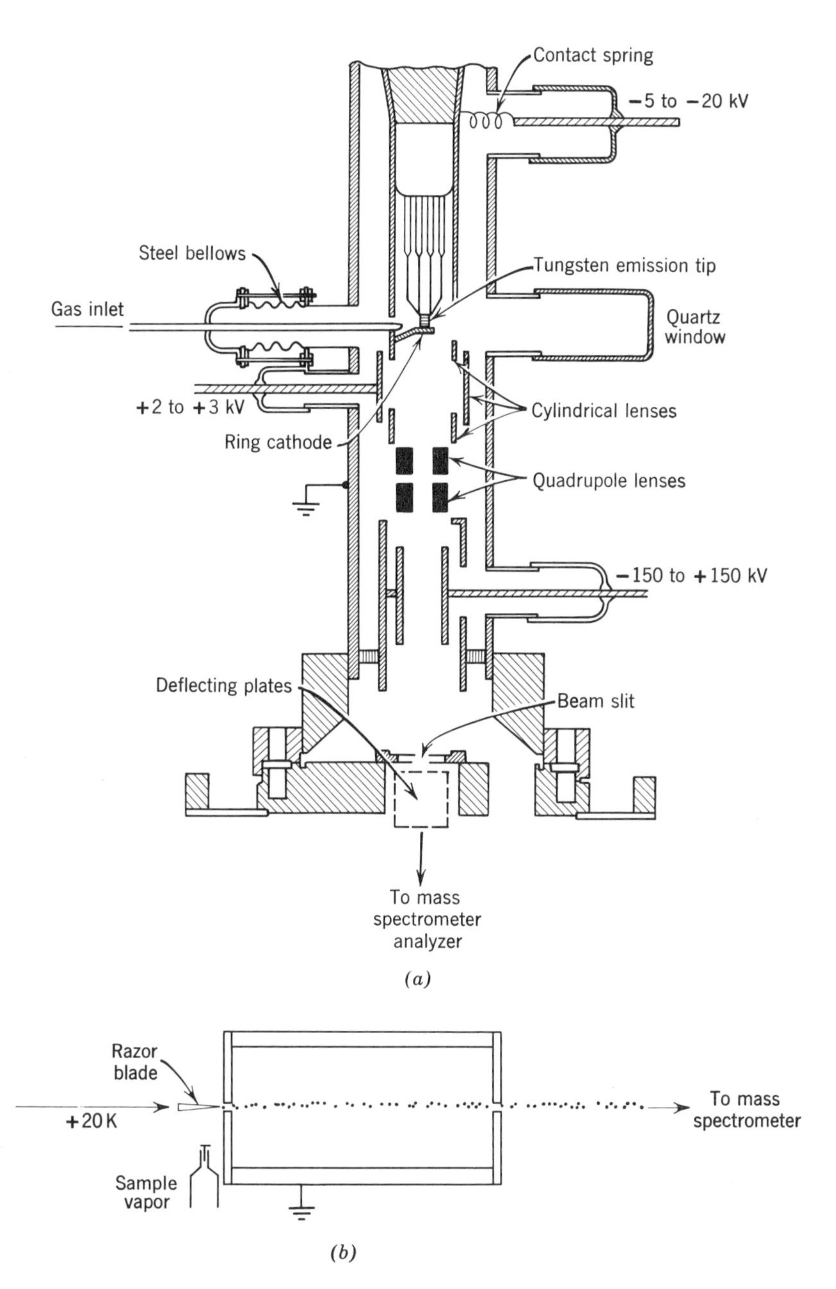

Fig. 3.12 (*a*) Field-emission ion source with wire tip [25, adapted]. (*b*) Field ion source of Beynon et al. [26, adapted].

of the tip; the minimum distance to give field ionization is given as

$$d_{\min} \cong \frac{IP - W}{\mathcal{E}}, \tag{3.6}$$

where $\mathcal{E}$ = field strength, IP = ionization potential, and W is the work function. For an organic molecule that has an ionization potential of 10 eV, a surface whose work function is 5 eV, and a field of 0.5 V/Å, $d_{\min}$ is approximately 10 Å. A theoretical explanation has been proposed for the hydrogen atoms, based on quantum mechanical considerations. At the gas-solid interface of the ionizing electrode, it is assumed that a finite probability exists for tunneling of the electron from the hydrogen atom to the metal. This electronic transfer will occur only if an electron in a ground state is raised (by the high external field) to an energy level corresponding to the Fermi level of the metal. This mode of ion formation results in an ion energy spread of only 1 or 2 V.

The gases that are to be analyzed are introduced at a pressure of approximately 10^{-4} torr or less in order to prevent electrical breakdown. The small ion energy spread of the field ion source provides better mass resolution than is possible with a spark source, and ion currents as high as 10^{-9} A have been obtained with field strengths of 10^{8} V/cm. An especially advantageous feature of this source relates to its selectivity for organic molecules. In hydrocarbon spectra parent ions are observed to be more abundant by several orders of magnitude than are dissociated ions. A less complex spectra results. This same discrimination has allowed the detection of free radicals at lower concentrations than could be observed with other techniques. Because field ionization involves small amounts of energy transfer, weakly bound molecules can be analyzed without appreciable dissociation and ion-molecule reactions can be observed. Catalytic reactions, chemisorption of molecules at metal surfaces, and the measurement of decomposition times are specific areas that are sometimes more amenable to mass analysis by this technique.

PHOTOIONIZATION

In principle, short-wavelength electromagnetic radiation appears attractive for the generation of ions in a mass spectrometer source. In practice, limitations of intensity and photon energy restrict this source from general applicability. The Planck-Einstein relationship of photon wavelength and energy is given by

$$\lambda = \frac{hc}{eV} = \frac{12{,}350}{E(eV)} \quad \text{(in angstroms)}. \tag{3.7}$$

Many ionization processes require 10 eV in energy, corresponding to a photon wavelength of 1235 Å. This is near the cutoff wavelength of optical windows that are needed to isolate the photon source and the ion slit system. Monochrometers have been used to determine ionization potentials in conjunction with photoionization sources as such sources produce simple mass spectra that are easy to interpret [27], [28].

The basic photoionization source is shown in Figure 3.13. Light from a monochrometer is incident upon gas molecules and the photon flux is measured by a photomultiplier. A window that can transmit short-wavelength radiation isolates the gas sample region from the monochrometer, or a small slit with differential pumping can be used to maintain a pressure difference between the two regions. The ion current, i, will be of the magnitude [29]

$$i = Ae\gamma I[1 - \exp(-\sigma nL)], \tag{3.8}$$

where A is a factor that includes various ion beam losses; e is electronic charge; γ is the ionization efficiency; I is the light intensity, σ is the absorption cross section of the sample gas at the appropriate wavelength; n is the gas number density; and L is the effective length of the ionization region. At low pressures where optical absorption is

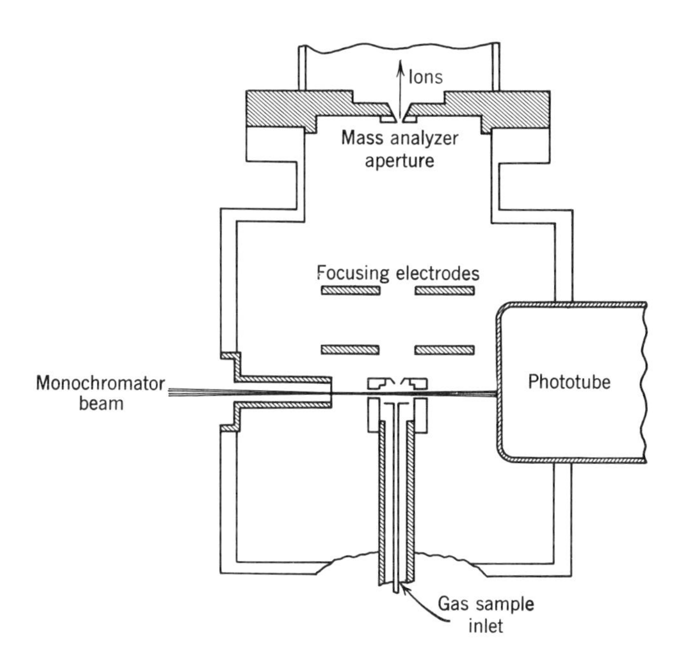

Fig. 3.13 Photoionization source [29, adapted].

negligible, the ion current is approximately $i = AeI_o \sigma nL$, and the observed ion current is directly proportional to gas pressure.

At pressures in the range of 10–100 μ, Poschenrieder and Warneck [30] have reported peak ion currents of 5×10^{-13} A and ionization thresholds for a number of mass doublets as outlined in Table 3.2.

Table 3.2

Ionization Thresholds and Corresponding Wavelengths of Several Mass Doublets

Doublet	*e/m*	*IP* (eV)	*Wavelength* (Å)
CO_2	44	13.79	889
N_2O		12.90	961
NO	30	9.25	1340
C_2H_6		11.65	1064
N_2		15.58	796
CO	28	14.01	885
C_2H_4		10.51	1179
C_2H_2	26	11.41	1086
CN		15.13	819

Ion beam intensities are several orders of magnitude below most sources, but the advent of improved light sources and high-sensitivity detectors has done much to stimulate interest in this type of source. One outstanding advantage is the same as for the field-emission type in that it minimizes the contribution of fragmentation ions to the mass spectrum. There is little question that the photoionization source will eventually find wider usage for basic studies relating to ionization thresholds, ionization efficiencies, fragmentation patterns, and the kinetics of ion formation.

LASER ION SOURCES

The first use of laser beams as a mass spectrometer ion source was reported by Honig and Woolston [31] and by Honig [32]. They demonstrated that high radiation density beams could be focused onto very small areas, and that sufficiently high surface temperatures were generated to produce vaporization and ionization from metals, semiconductors, and sintered insulators. Well-defined microvolumes (20 to 100 μ in diameter and 1000 μ deep) have been sampled in this fashion. Up to 10^{10} ions have been determined from a single laser pulse, all detected ions

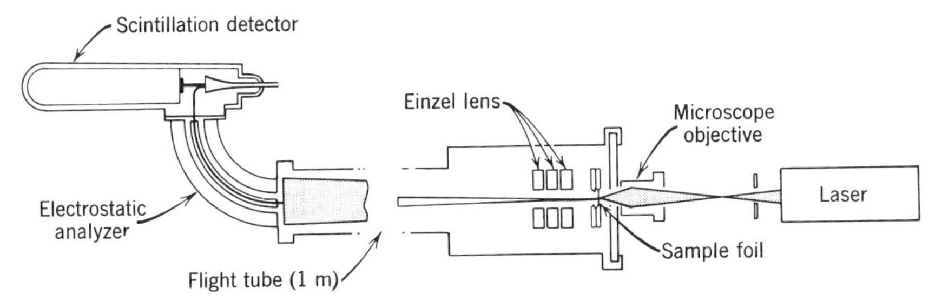

Fig. 3.14 Laser ion source in time-of-flight spectrometer [33, adapted].

being in the singly charged state. In this early work only bulk samples were used, and some difficulties were experienced with space charge.

Recently, lasers have been focused on very thin specimens and samples of 10^{-9} gm from metal foils have been successfully analyzed. A schematic diagram of this latter type of laser ion source is reported by Fenner and Daly [33] and is shown in Figure 3.14. They used a ruby laser that produced a power level of 10 mJ in 30 nsec. The beam was focused through a microscope objective onto a foil area of only 2×10^{-5} cm^2. Each laser pulse produced about 2×10^{13} atoms from a foil of 3×10^{17} atoms/cm^2. An ion energy spread of up to 500 eV was observed.

Metal foils of Li, Be, B, C, Al, Fe, Cu, Ag, Sn, Pb, and Au yielded a comparable number of ions. This foil method is claimed to give a higher signal-to-noise background ratio than does the more conventional method of evaporating bulk samples. In addition the geometry is favorable in that photons can be incident on one side of the specimen and ions can be accelerated from the opposite face. However, the very large ion energy spread requires an energy selector (in the time-of-flight system) or a double-focusing arrangement when magnetic analysis is employed.

SPECIAL TYPES

A large number of very specialized ion sources have been reported during the last decade. These include types in which ionized atoms or molecules are generated by protons, alpha particles, or fission fragments. Sources capable of producing very high ion currents, such as are employed in isotope separators, have also been scaled down to appropriate dimensions for mass spectrometric application. Many new variations of conventional sources have also been adapted for space flight instrumentation.

Recently a "cold electron source" has also been described by Testerman et al. [34], which purports to have several advantages over electron bombardment sources that use thermionic emitters. The source consists of an electron multiplier, with the initial source of electrons being generated by ultraviolet radiation or soft beta emitters. The amplified output current of the multiplier is then used as the ionizing electron beam. The source is reported to have advantages from the standpoint of reduced out-gassing, comparatively little fluctuation in ion emission, and a high frequency-modulation capability.

REFERENCES

[1] A. J. Dempster, *Phys. Rev.,* **11,** 316 (1918).

[2] S. C. Brown, *Basic Data of Plasma Physics,* Wiley, New York, 1959, p. 116.

[3] J. H. Reynolds, *Rev. Sci. Instr.,* **27,** 928 (1956).

[4] R. E. Honig, private communication.

[5] I. Langmuir and K. H. Kingdon, *Proc. Roy. Soc. (London),* **107,** 61 (1925).

[6] M. G. Inghram and W. A. Chupka, *Rev. Sci. Instr.,* **24,** 518 (1953).

[7] J. R. Werning, "Thermal Ionization at Hot Metal Surfaces," Ph.D. Thesis, University of California (1958).

[8] A. H. Turnbull, Atomic Energy Research Establishment Report No. 4295, Harwell, England (1963).

[9] F. A. White, T. L. Collins, and F. M. Rourke, *Phys. Rev.,* **101,** 1786 (1956).

[10] F. A. White, and L. Forman, *IEEE Trans. Nucl. Sci.,* **NS-14,** No. 1, 213 (1967).

[11] H. Patterson and H. W. Wilson, *J. Sci. Instr.,* **39,** 84 (1962).

[12] M. H. Studier, E. M. Sloth, and L. P. Moore, *J. Phys. Chem.,* **66,** 133 (1962).

[13] F. M. Penning, *Physica,* **4,** 71 (1937).

[14] O. Almen and I. O. Nielsen, *Electromagnetically Enriched Isotopes and Mass Spectrometry,* Butterworth, London, 1956, p. 23.

[15] J. Koch, "Mass Spectroscopy in Physics Research," National Bureau of Standards, Circular No. 522, Washington, D.C., 1953, p. 165.

[16] W. Aberth and J. R. Peterson, *Rev. Sci. Instr.,* **38, 745** (1967).

[17] R. E. Honig, J. W. Guthrie, W. H. Hickam, and G. G. Sweeney, *Mass Spectrometric Analysis of Solids,* ed. by A. J. Ahearn, Elsevier, Amsterdam, 1966, Chapters 2, 4, and 5.

[18] *Ibid.,* p. 140.

[19] W. M. Hickam and Y. L. Sandler, *Surface Effects in Detection,* Spartan, New York, 1965, p. 193.

[20] R. E. Honig, *Advances in Mass Spectrometry,* ed. by J. D. Waldron, Pergamon, New York, 1959, p. 162.

[21] H. J. Liebl and R. F. K. Herzog, *J. Appl. Phys.,* **34,** 2893 (1963).

[22] R. Castaing and G. Slodzian, *J. Microscopie,* **1,** 395 (1962).

[23] E. W. Mueller, *Phys. Rev.,* **102,** 618 (1956).

[24] R. Gomer and M. G. Inghram, *J. Chem. Phys.,* **22,** 1279 (1954).

[25] H. D. Beckey, "Field Ionization Mass Spectroscopy," in *Advances in Mass Spectrometry* (vol. 2), Macmillan, New York, 1963, p. 1.

[26] J. H. Beynon, A. E. Fontaine, and B. E. Job, *Z. Naturforsch.*, **21 A**, 776 (1966).
[27] H. Horzeler, M. G. Inghram, and J. D. Morrison, *J. Chem. Phys.*, **27**, 313 (1957).
[28] B. Steiner, C. F. Giese, and M. G. Inghram, *J. Chem. Phys.*, **34**, 189 (1961).
[29] W. Poschenrieder and P. Warneck, *J. Appl. Phys.*, **37**, 2812 (1966).
[30] *Ibid.*
[31] R. E. Honig and J. R. Woolston, *Appl. Phys. Letters,* **2**, 138 (1963).
[32] R. E. Honig, *Appl. Phys. Letters,* **3**, 8 (1963).
[33] N. C. Fenner and N. R. Daly, *Rev. Sci. Instr.*, **37**, 1068 (1966).
[34] M. K. Testerman, R. W. Raible, B. E. Guilland, J. R. Williams, and G. B. Grimes, *J. Appl. Phys.*, **36**, 2939 (1965).

Chapter 4

The Detection of Ion Beams

With the advent of improved primary sensors and stabilized circuitry, it is now possible to measure ion currents about six orders of magnitude below intensities that were generally detectable two decades ago. However, the ideal detector has not yet appeared in the laboratory.

From a purely physical point of view, the detection of ion beams seems almost easy. An ionized beam of atoms emerges from a narrow defining slit with a high degree of collimation and with an energy of about 10 keV. Thus each atom has a kinetic energy of 10,000 eV—an energy several orders of magnitude greater than the energy of a photon in the optical range or the dissociation energy of some molecules ($\sim$3 eV). It will also be recalled that 30 eV is about the average energy required to produce an electron-ion pair in a gas and that only 3.5 eV is sufficient to generate an electron-hole pair in silicon. If we wished to detect only 100 singly charged atoms in one second (1.6×10^{-17} A), and assume that each atom possesses 10^4 eV of kinetic energy, we have 1 MeV available. In the case of nuclear radiations (alphas, betas, gammas, and protons) this energy is far in excess of that needed to be detected by several transducers. For electromagnetic radiation in the optical range, the situation is even more favorable as a properly matched phototube has a 30% chance of sensing a *single* photon. In the usual mass spectrometric case the ion beam has been spatially mass-resolved so that we are only required to detect single ionizing events (without energy resolution) or suitably to integrate the charge amplification of an ion beam.

In practice, however, substantial hurdles must be overcome to (a) measure exceedingly small ion currents or (b) measure the ratio of two or more ion beam currents with great precision. Let us consider first what might be desired of the "ideal detector."

1. It should provide a one-to-one correspondence of impinging ions to amplified output pulses.

2. Output pulses should be of an amplitude that is large compared to all "noise" pulses.

3. Discrimination should be possible between ions having the same charge-to-mass ratio but having a variety of charge states.

4. Discrimination should be possible between atomic or molecular ions having a common charge-to-mass ratio.

5. Simultaneous monitoring of at least two masses should be provided.

6. The detector should provide a linear response over many orders of magnitude in intensity.

7. The primary sensor should be amenable to a convenient display or data-reduction system.

Now that we have outlined specifications that cannot currently be realized, let us review established techniques; these include photographic emulsions, integral current measurements, and a variety of ion counters.

PHOTOGRAPHIC EMULSIONS

It is difficult to conceive of twentieth-century science without the availability of sensitive emulsions for the detection of x-rays or for the measurement of the range, energy, and momentum of elementary particles. They have also been the indispensable "integrator" in astronomy, where exposure times are commonly 10^5 times those in other applications. It is this integrating function that makes these emulsions such an indispensable sensor for ion detection in mass spectrometry. In several of the commercial double-focusing instruments with ion accelerating voltages of $\sim$18 kV and high magnetic fields ($\sim$16,000 gauss), mass spectra from lithium to uranium can be simultaneously registered. The low mass spectral region can also be expanded with higher dispersion and with reduced magnetic fields.

Considerable literature exists on nuclear emulsions, both with respect to basic theory and specific application of photographic technique to mass spectrometry. The reader is referred to two excellent sources for details and extensive references in support of the brief discussion that follows. One is a comprehensive work by Barkas [1]; the second is a review article by Owens [2] that deals more specifically with photographic plates for mass spectrometry.

Sensitivities for Photons and Ions

Most emulsions consist of a suspension of silver bromide grains in a gelatin, spread on glass plates or other supporting substrates. The gelatin is somewhat permeable and allows the developing solution to find its way to the bromide crystals and to react with them. Trace amounts of sulphur in the gelatin also play an active role in the sensitizing process by making available sulfide sensitivity specks. The threshold energy necessary to render a grain developable differs for photons and ions. The band gap of silver bromide is 2.6 eV, so that the absorption of one photon having a wavelength less than about 4600 Å is sufficient to raise an electron from the valence band to the conduction band and create an electron-hole pair. But absorption of some 10 to 20 photons appears to be the threshold for producing an aggregate of silver atoms that will make a crystal grain developable.

There are less data available for ions with respect to the minimum absorption energy needed to produce one conduction electron in a silver halide grain. However, it has been suggested by several investigators that an order-of-magnitude increase in energy is required of ions over photons. The energy density to produce a detectable image is also high. Owens [3] indicates that the minimum detectable line image ($\sim 0.1\ mm^2$ in area) requires about 10^4 ions of 15 keV energy; this figure is for ions of about 100 amu on Ilford Q_2 emulsions. In a practical case it is difficult to estimate the number of ions of a given energy that are required to produce a developable grain. Grains near the emulsion surface can conceivably be made developable from the impact of a single, energetic ion. Grains at greater depths in the emulsion may require successive ion collisions, or they may be made developable by a weak luminescence of the gelatin.

Emulsion Types

There exists some choice as to the type of emulsion best suited for ion detection. A high sensitivity requires large silver halide crystals, a relatively low gelatin content, and a thin emulsion layer on the backing plate. Large grains, however, result in spectral lines having a granular appearance, so that for high-resolution spectrometry, a finer grain emulsion may be a preferred choice. The gelatin content is an important parameter, as it will determine what fraction (on the average) of the kinetic energy of the incident ion is given up to the gelatin, or to the bromide crystal in a primary collision. In the usual type of emulsion many ions will be stopped in the gelatin alone because of the very limited range of heavy ions. Ilford Q-type emulsions presumably have a high concentration of silver bromide crystals at the emulsion surface, so that there is an increased probability for producing a latent image from a direct ion impact. But even with these emulsions, a secondary blackening mechanism—attributed to a luminescence of the gelatin by ion-gelatine collisions—is believed to contribute substantially to the image density.

The desirability of thin emulsion layers becomes clear from the standpoint of ion range also. Q_2 emulsions have a grain diameter somewhat less than 1 μ (10,000 Å), which is a dimension that exceeds the range of most 10-keV ions. Of course, the grains are not neatly arranged in a plane, but a very thick emulsion serves no useful purpose in mass spectrometry. Greater background fogging can also be expected in very thick emulsion layers.

A rough indication of the sensitivity of several emulsion types is presented in Table 4.1.

Schumann plates are characterized as being gelatin-poor, and on some Schumann type plates the grains actually protrude above the gelatin. In any event relatively little gelatin is present between the incident ion and the silver halide crystal. The Q-type emulsions are the most widely used. Q_1 emulsions have fine grains and high contrast; Q_3 has larger grains and lower contrast; Q_2 emulsions are intermediate between the former two. The fine-grained x-ray plates have rather poor sensitivity, but Eastman SWR plates provide substantially the same optical densities as the Q_2 type. Most emulsion plates are available in standard sizes—10 in. in length being convenient for use in many instruments.

An interesting attempt to achieve a higher signal to noise ratio in photoplates has recently been reported by Kennicott [5]. His approach is to employ a developer which preferentially develops the latent image residing in the interior of the silver halide grains and to employ a bleach that removes some of the noise or fog that is always associated with the emulsion surface layers. Figure 4.1 shows both the usual multiple charged ion mass spectra that is obtained by a spark source in a Mattauch-Herzog instrument. It further reveals the signal-to-noise improvement obtained by removing surface fog. The three photoplates are all

Table 4.1

Emulsion Sensitivities for 10-keV Ions [4]

Emulsion type	*Relative sensitivity*
Agfa-Schumann	1.0
Ilford Q_3	0.4
Eastman SWR	0.3
Eastman x-ray	0.02

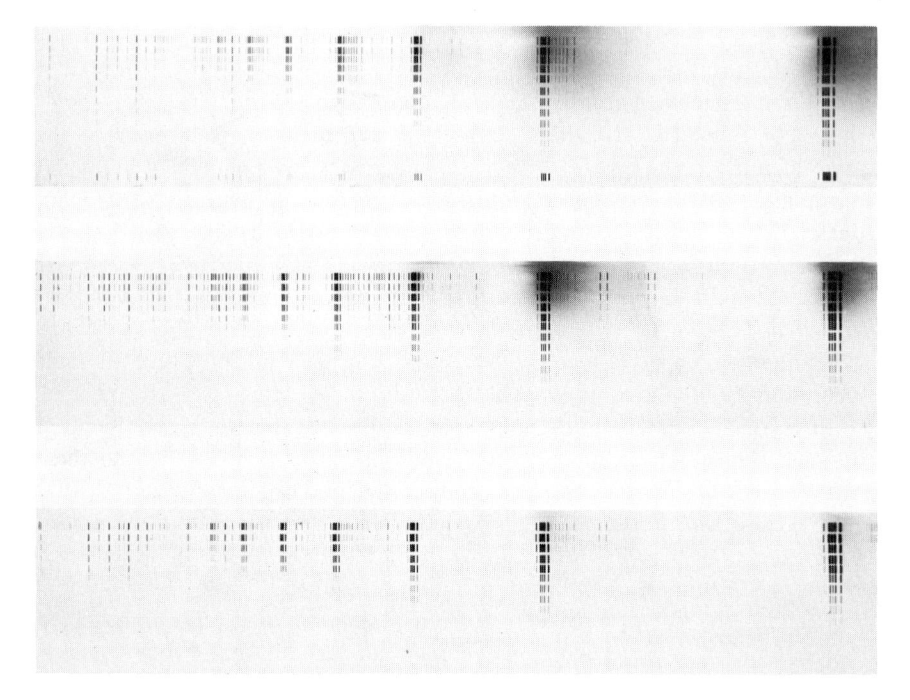

Fig. 4.1 Mass spectrum displayed on photographic plate (courtesy P. Kennicott).

spectra of tungsten; masses increase to the right. The top plate was given a maximum exposure of 100 nC. The middle and bottom plates were given a maximum exposure of 300 nC—a factor of three longer than the top plate. Successively shorter exposures decrease by a factor of three. The top plate was developed in the conventional manner using D19 developer. The middle plate was developed in a developer which produces only the image on the surface of the silver halide grains. The bottom plate was processed in a solution that developed only the image in the interior of the silver halide grains. The reduction in fog in the last two plates can be observed. And although it may not be too evident from the photograph, the signal-to-noise ratio and thus the sensitivity increases from top to bottom.

Exposure and Calibration

Other important considerations with photographic plates relate to their latitude or range of measurable ion densities, mass dependence, and calibration. In the range of ion intensities that can be recorded, photographic emulsions excel. If the minimum detectable number of ions is 10^5 and a strong spectral line is caused by 10^{14} ions, this represents a sensitivity range of 1 part per billion for a given element. There will also be a mass dependence on sensitivity, which should be taken into account; in general, emulsion sensitivity decreases for increasing mass number. Burlefinger and Ewald [6] have reported a sensitivity of Ilford Q_1 emulsions to be proportional to $(2/M)^{1/2}$. In any practical case, however, rather extensive calibrations are imperative for quantitative work. Special densitometer methods have also been devised for measuring line densities and for plotting spectral line densities as a function of exposure for various elements, and for multiply charged ions. Standards of metals or semiconductors of known composition must be available for comparison. Further, extensive experience is required to relate optical densities, as measured by any photometric system, to isotopic abundances or to chemical impurity content. A recent review paper by Owens [7] presents detailed data on the properties of photographic emulsions as ion detectors and the optimum conditions for their use.

INTEGRAL CURRENT METHODS

Simple Faraday Cage

The basic technique for detecting ion currents in the range of 10^{-6} to 10^{-13} A is to allow the beam to stop in a hollow conducting electrode. This type of beam monitor is often termed a "faraday cage"—after the eminent nineteenth-century physicist Michael Faraday, who employed a metal bucket for collecting electric charge. Figure 4.2 is a schematic diagram in which a mass-resolved ion beam, defined by a slit, passes

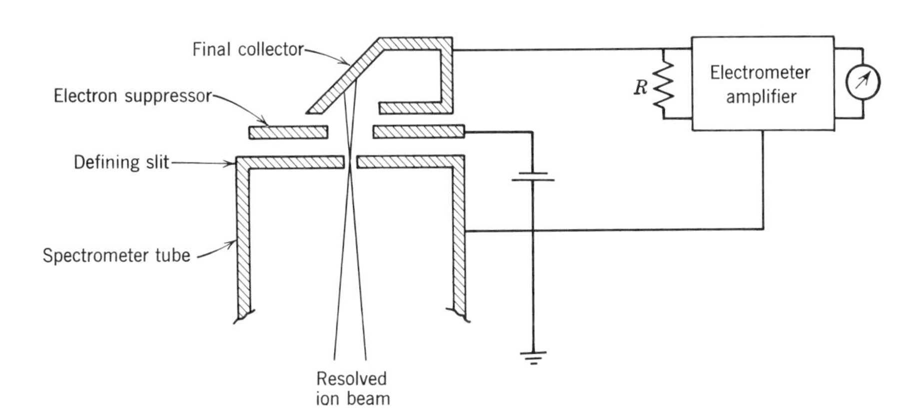

Fig. 4.2 Simple faraday cage ion-beam collector.

into the interior of a small cylinder that is appropriately insulated. The slit edges should be sharp to prevent ion reflection or scattering, and the dimensions of the cylinder are usually chosen so that secondary electrons, which are generated by stopping of the ion beam, are effectively trapped within this electrode. A suppressor electrode with a negative bias of about 20 V can also be used to prevent secondary electrons from drifting out of the cylinder. It is important that the collecting electrode be shielded from light to prevent photoelectron emission or ionization from spurious ions that may drift to the cylinder, and from varying magnetic or electric fields that can induce sizeable voltage in any electrode of small capacitance. For optimum results these electrodes should be connected directly to the grid of a suitable electrometer tube, or through a short, shielded cable of low capacitance.

The electrometer tube should have a grid current that is small compared with the minimum beam current to be measured (e.g., $\sim 10^{-16}$ A). Input currents of 10^{-12}, 10^{-10}, and 10^{-8} A would then generate a 10-mV signal if 10^{10}, 10^{8}, and $10^{6}\,\Omega$ resistors, respectively, were connected as input impedances. Customarily, the faraday cage will be connected to a commercial electrometer circuit that has a selection of input resistors, good stability, and a current gain such that the output signal can drive a recorder. If appropriate precautions are taken, and a mass scan is slow compared with the time constant of the ion collector and electrometer, a low impedance output signal will provide a reasonably accurate display of relative primary ion beam currents. When difficulties are encountered at very low beam currents, they can often be attribut' l to space charge effects, insulator leakage currents, spurious ions developed by diffusion pumps, ionizing radiation, poor shielding or grounding, and so on.

Dual Collectors

There are many instances in which greatly improved isotopic ratio measurements may be achieved by the simultaneous collection of two or more resolved ion beams. A dual collector may be used to (a) circumvent errors due to variations in ion beam production, (b) detect ratio changes arising from the preferential depletion of isobaric impurities, (c) conserve samples, (d) decrease the time required for analysis, and (e) improve precision generally by the cancellation of certain time dependent errors that are associated with the analysis per se.

There are probably as many variations of multiple beam recording as there are laboratories that employ the method. The basic scheme, however, as devised by Nier, Ney, and Inghram [8] is shown in Figure 4.3. The multiple slit arrangement permits one ion beam to proceed through a narrow defining slit to one collecting electrode; a second beam is intercepted on a plate. Both electrodes are close to parallel plate suppressors, the function of which is to turn back secondary electrons produced by the incidence of the high-energy ions. A negative bias of 20 V is sufficient to repel most secondaries, and this potential is also sufficiently low so that the primary ions are not deflected—even if slight electrostatic asymmetries occur. If the two electrodes are connected to two stable amplifiers with variable input resistors (R_1 and R_2), a voltage balance can be achieved by suitable resistor values, a null reading will indicate an ion beam ratio that is proportional to R_2/R_1, but which will also depend upon amplifier gain and feedback loop voltages. For accurate analysis, isotopic standards must be employed to achieve normalization.

It will be noted in Figure 4.3 that if the polarity of the secondary electron suppressor is reversed and made positive, some amplification can be achieved. (This method is usually used for a single collector only.) The amplification ratio is equal to $(I_p + I_s)/I_p$, where I_p is the positive ion current and I_s is the secondary electron current *leaving* the electrode, thus effectively producing an additional positive current. A multiplication of about 3 may be achieved in this fashion, but this artifice is only resorted to for marginal situations.

In very precise work account must also be taken of possible errors arising from the ejection of secondary positive ions, although this effect is usually neglected because it is small (<0.1% of the secondary electron current).

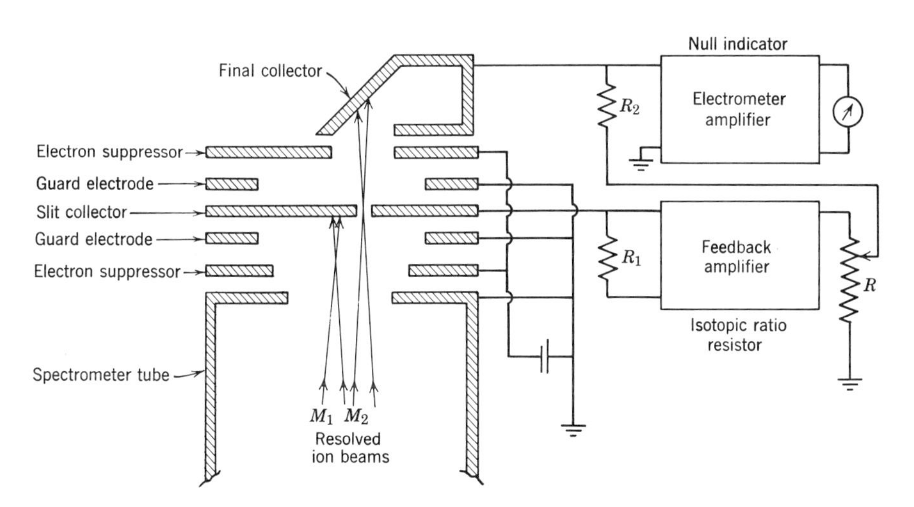

Fig. 4.3 Dual ion-beam collector [4, adapted].

While dual collectors comprise the majority of the multiple ion beam detector schemes, systems having many elements have been designed and constructed. Recently, a 180° mass spectrometer has been reported by the Lawrence Radiation Laboratory in which 94 equally spaced cups were placed along the magnetic focal plane and serve as collecting electrodes [9]. The objective was to record a complete mass spectrum from only a few pulses of a high-current ion source. The collector cups, or faraday cages, are fastened mechanically to the vacuum analyzing chamber but are electrically insulated from it. Each cup is connected to a 1000-pF silver-mica capacitor having an exceptionally low leakage current. The recording of collected charge is accomplished by a trolley pickup electrometer that is programmed to sense the potentials across the 94 capacitors. Such a multichannel scheme does not lend itself to great precision, but spectra that include ions of H^+, D^+, C^+, O^+, Ti^{++}, and Ti^+ have been monitored with this integral-current type of method.

Vibrating Reed Amplifiers

In conjunction with any of the foregoing methods high quality dc preamplifiers and amplifiers should be employed. For currents in the 10^{-13} to 10^{-15} A range, a vibrating reed amplifier [10] is required. Developed during World War II for use in nuclear research, this type represents the ultimate electrometer for integral-current measurements. The salient feature of the vibrating reed electrometer is the conversion of a dc input signal to an ac signal that can then be amplified. The "vibrating reed" is a dynamic condenser in which the input current flows onto one plate of the capacitor. The other plate is vibrated at a fixed frequency (a few hundred cycles/sec) and this periodic capacitance change generates an ac signal. The resultant voltage is then amplified by a drift-free ac amplifier. Available commercially, this device possesses long-term stability and excellent response characteristics.

Some investigators claim that a vibrating reed device can be used even below 10^{-15} A. For mass spectrometric measurements, however, 10^{-15} represents a practical limit for any type of mass scanning because of the long-time constant (e.g., $RC = 10$ sec for a 10^{-11} farad input capacitance and a $10^{12}\ \Omega$ resistor, although this can be reduced by feedback circuitry). The electrometer can also be useful as a "rate-of-drift" detector; in this case the electrometer electrode is disconnected from the input resistor, and the time rate of voltage change (as observed on an output recorder) can be correlated to the ion current. One factor affecting the sensitivity limit with this mode of operation is the charge generated by occasional alpha particles emitted near the electrometer collecting electrode (from trace quantities of uranium in the metals comprising the electrometer components). Although of infrequent occurrence, these alpha particles induce large voltage changes in a "floating" electrode because of the very large number of ion-electron pairs that are generated by such highly ionizing particles.

DC Electron Multiplication

Electron multipliers can be operated in an integral-current mode to provide the current amplification needed in the current range below 10^{-15} A. Several possible schemes are available. One approach is to allow the ion beam to strike a phosphor directly. Several phosphors can be operated in high vacua (e.g., $CaWO_4$), and good optical coupling can be made in a conventional manner to a high-gain photomultiplier. The method is seldom used, however, for low-energy ions. At high energies ($>$30 keV) an atom may produce up to 10^3 photons in a phosphor [11], but the over-all efficiency for secondary electron production has not made this method attractive. However, for survey work, (e.g., visual observation on a fluorescent screen of high energy—high current beams such as are generated in an isotope separator), phosphors are useful qualitative indicators. A phosphor can also be used with multipliers that are activated by secondary electrons accelerated to it from an ion-electron transducer (see next section). The primary reason for using either of these methods, however, is the fact that the commercial multiplier can then be used without breaking its vacuum envelope, i.e., it is thus protected from deterioration and contamination when exposed to atmospheric pressures.

The usual technique is to allow the ion beam to impact directly the first electrode (or dynode) of a multiplier and achieve current amplification via the secondary electron multiplication of many stages. The over-all gain is

$$\text{gain} = pq^n, \tag{4.1}$$

where p is the number of secondary electrons initially produced on the first dynode by a positive ion impact, q is the average secondary electron yield per dynode, and n is the number of multiplying stages. It is more correct to say that p and q represent the electrons that reach the subsequent dynodes, but with good electron optics the number of electrons produced should be only slightly greater (on the average) than those that arrive at the next stage. The parameter p is dependent on dynode material, its surface state and degree of oxidation, the kinetic energy of the ion, and specie. The mass effect is usually small for small fractional mass differences (i.e., in the high-mass range). The average secondary electron yield per stage, q, will depend upon dynode material and it will be dependent upon the interstage voltage.

Figure 4.4*a* is a schematic of one type of electrostatically focused electron multiplier that has been used for ion beam detection. Figure 4.4*b* is a so-called "venetian blind" type in which secondary electrons are cascaded from shutter-like dynodes. Either device, having 10 stages and an interstage voltage of about 150, will produce a gain of about 10^4 to 10^5. This amplification is all that is needed for input currents of the order of 10^{-16} A; the output signal can then be fed into any suitable amplifier and recorder. If constructed with a high work function material (such as Be-Cu alloy) and suitably "activated," the multiplier supplies desirable characteristics of reasonable gain, low background "noise," and reasonable stability. When it is operated in this fashion, however, great care must be taken to stabilize the voltage applied to the multiplier, and reasonable precautions must be taken to minimize

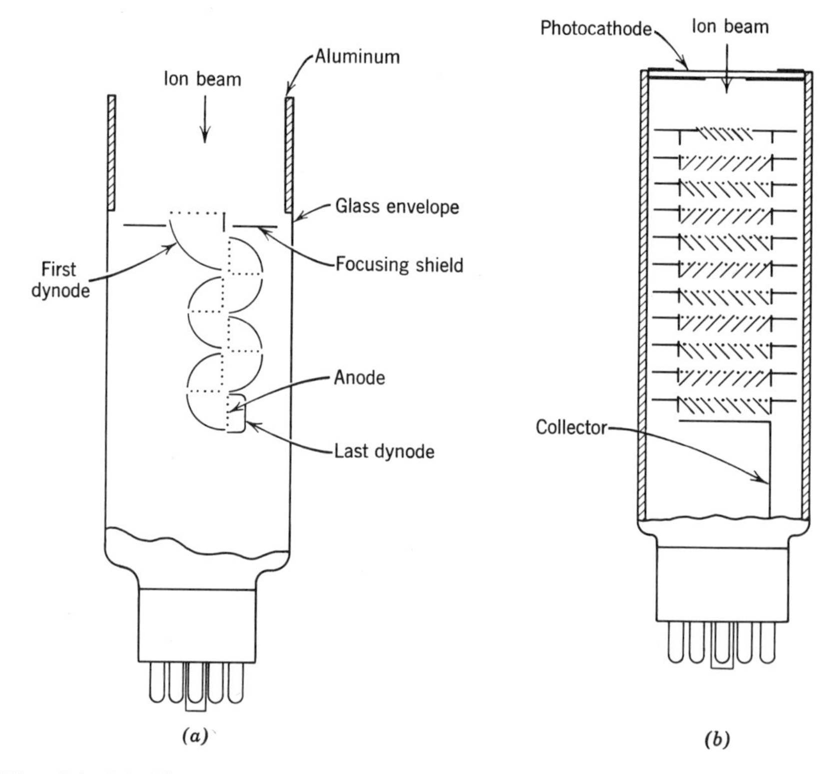

Fig. 4.4 (*a*) Electrostatically focused electron multiplier; (*b*) "venetian blind" multiplier.

dc leakage currents (arising from poor insulators or spurious sources of ionization). Almost any good commercial multiplier can be used for integral-current measurements, but the gain will usually decrease markedly after exposure to atmospheric conditions. Magnetic and light shielding is almost always required if stray fields or sources of radiation are present.

For very low current measurements many advantages accrue to the investigator using multipliers in a pulsed mode; hence, greater emphasis is given to the specific types and operating conditions for devices used in this fashion.

PULSED COUNTING METHODS

Electrostatic Focusing Multipliers

Electrostatically focusing multipliers can be successfully operated in a pulsed counting mode for quantitative mass spectral work provided that the following obtains:

1. The first dynode has a suitable ion—secondary electron yield.
2. A sufficient number of electron multiplying stages are used with reasonable gain per stage.
3. The multiplier is shielded from magnetic fields.
4. The arrival rate of ions is restricted to values commensurate with the resolving times of both multiplier and associated circuitry.

Assume that an ion generates two secondary electrons and that an average gain of 2.3 per stage is achieved for 18 subsequent stages. A single ionic charge (1.6×10^{-19} C) will then generate $2\ (2.3)^{18} \simeq 6.5 \times 10^6$ electrons. This is a sufficient number of electrons to generate a large voltage pulse across an input resistor. In the usual multiplier the total transit time for an electron will be between 10^{-7} and 10^{-8} per second and the pulse rise time will be in the 10^{-9} sec range.

Some immediate advantages arise from counting these ion-induced pulses. First, much of the "dark current" of a multiplier may be eliminated. A discriminator will usually be used to reject all but the large pulses arising from the first or second dynode. This discriminator will then bias-out all pulses arising from thermionically emitted electrons of the latter stages of the multiplier. An improvement of at least 10 in signal-to-noise ratio is usually achieved. Secondly, the multiplier is not so sensitive to voltage fluctuations of the multiplier voltage source as when operated in a dc fashion. Low-frequency noise is filtered out, and recording of pulsed currents is generally much more convenient and amenable to statistical treatment. One electron multiplier that has been exceedingly useful for pulse counting applications has been the type designed by Allen [12]. This multiplier contained 12 dynodes covered with a thin layer of Be, operated with interdynode potentials in excess of 300 V, and a reported gain of 10^5. The high photoelectric work function of Be ($\sim$4 eV) makes this metal a desirable dynode material as it is not sensitive to light in the visible region.

In order to achieve a higher over-all gain for general mass spectrometric applications, White and Collins [13] extended the number of electron-multiplying dynodes to 19 and achieved gains of greater than 10^7. A commercially available alloy of Be-Cu was employed (2–4% Be), and the many resistors needed to provide potentials to these dynodes were connected to them within the vacuum system. Each individual resistor was also encapsulated in a small Pyrex envelope to prevent the outgassing of these resistors from affecting the high vacuum of the spectrometer tube. The Allen-type dynode configuration is shown in Figure 4.5, with the dynodes being supported by an insulating slab of Mycalex. A final collecting electrode receives the large number of electrons generated by multiplication; this is connected directly to a suitable amplifier-discriminator. Figure 4.6 is a photograph of an assembled multiplier unit, fabricated of Be-Cu, and mounted on a high vacuum flange for use in a mass spectrometer.

This type of multiplier, used in the pulse counting fashion, has been widely used during the last decade. The device is somewhat large, but the generous spacing between dynodes prevents voltage breakdown, and it has a very low background current. One recent 14-stage model, reported by Marchand *et al.* [14], has even been scaled up by a factor of 1.5 (dynode dimensions) over the original Allen structure in order to obtain a higher efficiency for counting ions from a quadrupole mass filter. When used with Ag-Mg or Be-Cu materials, the number of thermionically emitted electrons is sufficiently small so that 10^{-18} A of ion beam current can be detected and measured. It is possible to reduce

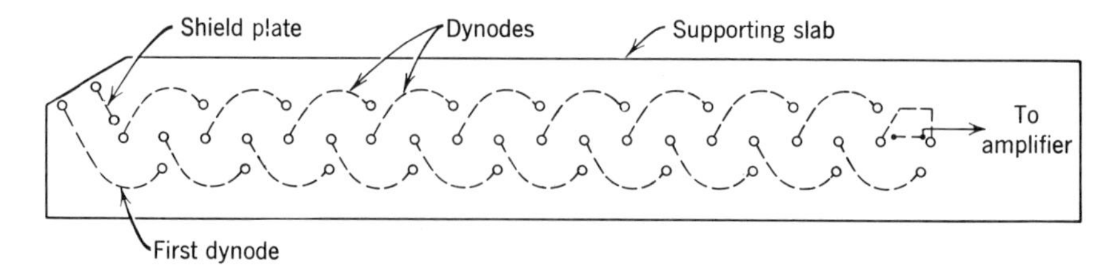

Fig. 4.5 Dynode construction of Allen-type multiplier.

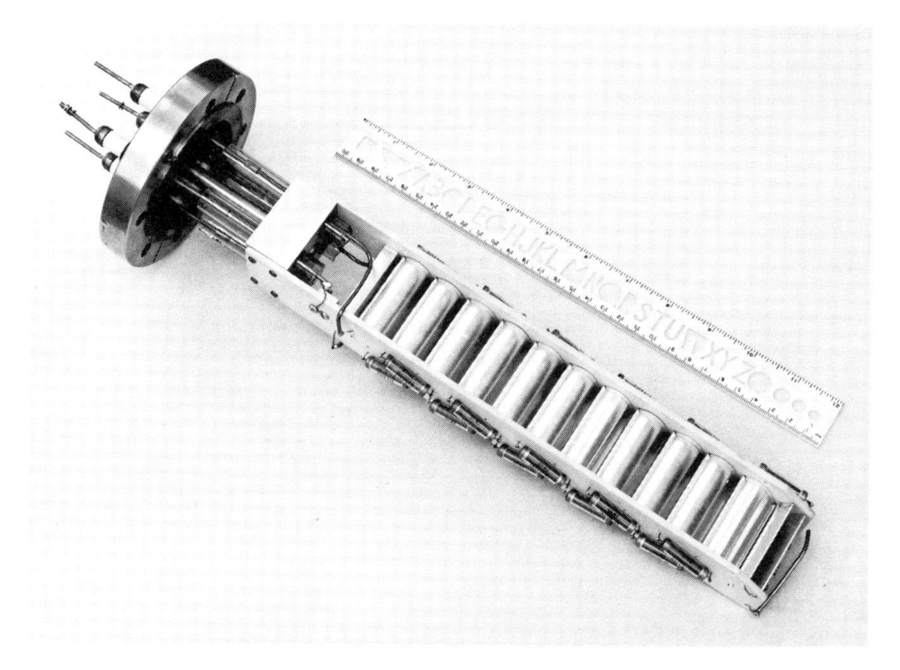

Fig. 4.6 Assembled Allen-type multiplier.

further thermionic emission from the first dynode by effectively reducing the dynode area, and several schemes have been employed. Reduction of dynode temperature may also be considered. At any temperature above absolute zero, Fermi statistics predict that some electrons in a metal will have energies large enough to exceed the surface energy of the metal. The number of electrons with energy between E and $E + dE$ is

$$N(E)dE = \frac{8\pi m \sqrt{2mE}}{h^3} \cdot \frac{dE}{e^{\frac{(E - E_o)}{kT}} + 1}, \tag{4.2}$$

where m is mass of electron, h is Planck's constant, E_o is the energy of Fermi level, k is the Boltzmann constant, and T is the absolute temperature. The form of the equation suggests that even a small decrease from room temperature should be significant. Hence, a recent experiment has been performed in the author's laboratory to cool the first dynode by means of a small thermoelectric junction [15]. A thin wafer of sapphire was used to provide high thermal conductivity between the dynode and the thermoelement, and also to achieve electrical isolation. The thermionic electron emission was indeed reduced by this technique but the background did not completely disappear as at very low counting rates (equivalent to 5×10^{-20} A) precautions must also be observed with respect to any possible source of spurious ions. These may include trace radioactive nuclides in construction materials, cosmic rays, or other effects which are usually considered negligible. Ions from ion-getter pumps can be effectively trapped, but the detector should usually not be located in the immediate proximity of either an ion or diffusion pump.

Reference is made to Figure 4.7 with respect to multiplier outputs, and the values of the interdynode resistors. It will be noted that if the multiplier is both resistance and capacitance coupled, it is possible

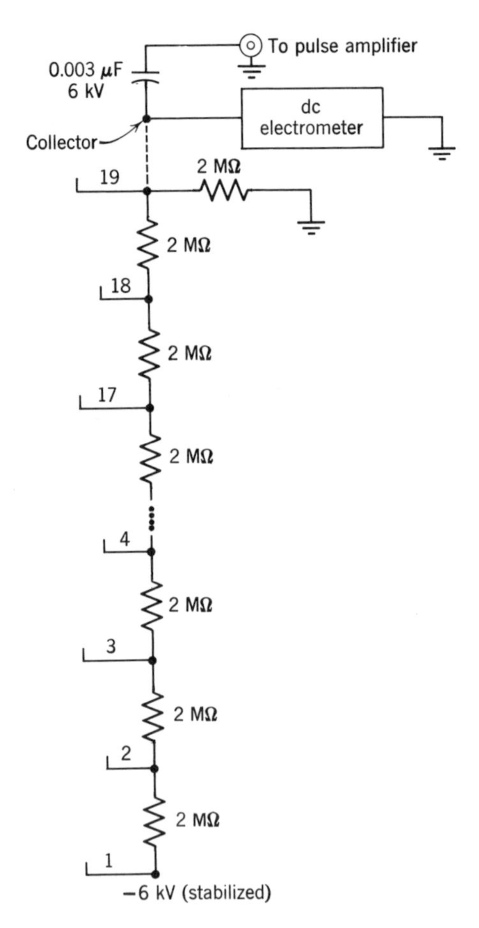

Fig. 4.7 Resistor chain and connections for an electron multiplier.

simultaneously to measure a dc-amplified current and a pulsed current. The technique is useful when a correlation is to be made of gain, or some estimate is to be made of counting rate losses. In this case the collector will be maintained near ground potential. If the multiplier is to be operated as a counter with the first dynode at ground, and the collector at a high positive potential, a high-quality coupling capacitor having a high voltage rating (e.g., 10 kV) should be used.

A precaution must be employed at high counting rates—not only to prevent counting rate losses (discussed in a subsequent section) but also to prevent an actual decrease in pulse amplitude. Consider an electron multiplier with a total voltage across 20 multiplying stages of 6000 V (i.e. 300 V/stage). Let an input ion current of 1.6×10^{-12} A (10^7 ions/sec) be amplified so as to produce an effective output current of 1×10^{-5} A (a gain of somewhat less than 10^7). This current is sufficiently large at the output so that it begins to approach the magnitude of the current in the resistors in the multiplier chain. For example, 6000 V across 20 2-megohm resistors will provide a voltage divider current of only 150 μA. This is not sufficient to prevent a voltage drop across the last multiplier resistors, induced from the amplified ion current. A separate lower impedance supply, or batteries, is thus indicated for the last resistors in the chain. Multipliers subjected to continued high currents also fatigue because of a change in the dynode surface conditions.

The amplitude of the output pulses can be predicted only if specific information is available concerning multiplier characteristics. Figure 4.8*a* is a "typical" yield curve that displays the variation of the secondary electron yield with ion kinetic energy and mass. If a singly charged atom of intermediate mass strikes a metal surface, the yield will be a monotonically increasing function of ion kinetic energy. The curve indicates the general desirability of obtaining maximum ion energy—which is increased substantially (for positive ions) if the first dynode of the electron multiplier is biased at a high negative voltage. The second advantage of negative bias for the multiplier is that the output resistor or capacitor will be near ground potential.

Figure 4.8*b* indicates why an optimum voltage generally exists for multiplier operation. Too high a voltage per stage will actually result in a decreased over-all gain. For any particular multiplier the gain per stage as a function of voltage can be observed by an oscilloscope or by a multichannel analysis of output pulses that are due only to the generation of one secondary electron on the first stage. Output pulses may also be generated from single electrons emitted from the first dynode by several means. If an optical port is available for testing, changes in counting rate may be noted by raising the temperature of the first

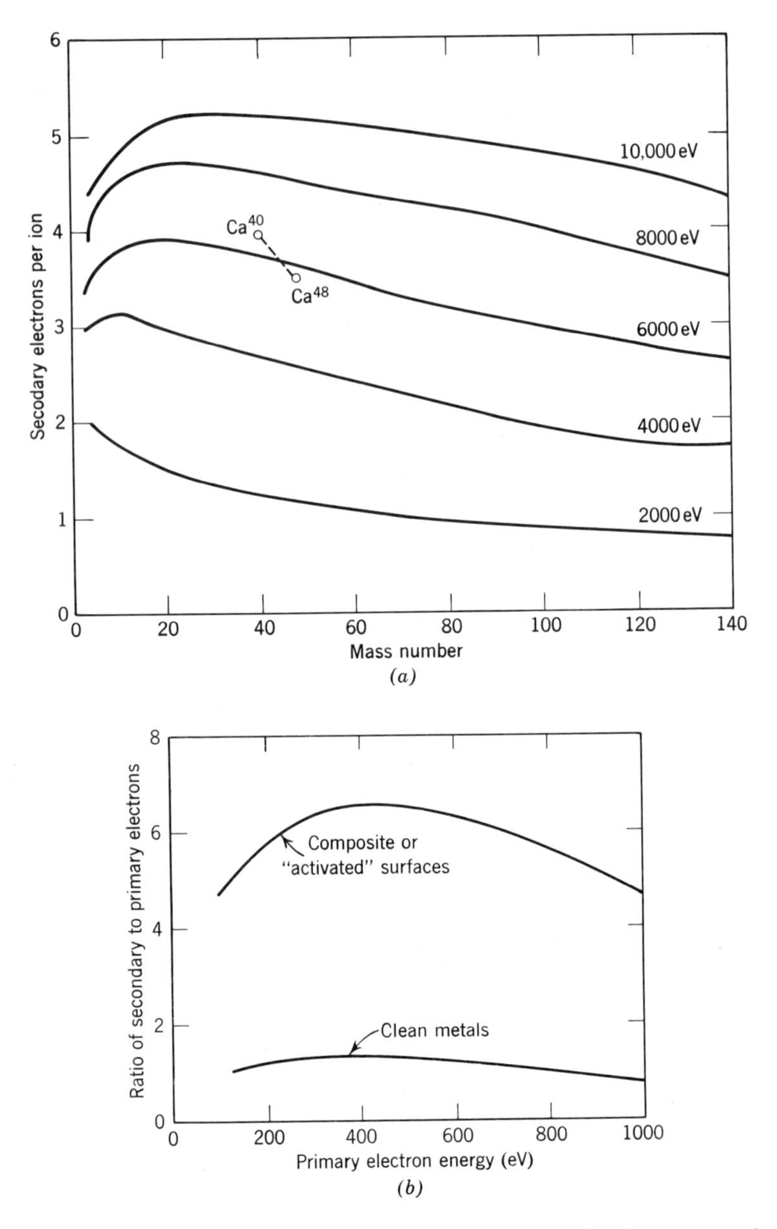

Fig. 4.8 (*a*) Relative electron yield from ions of varying kinetic energy and mass [4]; (*b*) Typical secondary electron yield from primary electrons.

dynode (by infrared heating); ultraviolet light sources may also be used to generate single electrons. The final test is to make an actual mass spectrometer evaluation, with a multiplicity of discriminator bias settings, to determine if a suitable "plateau" exists for a particular multiplier.

A primary objective in pulse counting is to achieve a very high secondary electron yield on the first dynode from the incident ion. If an initial gain of 10 could be obtained, discrimination against all types of noise could be expected, and close to a one-to-one correspondence of input ions to output pulses would be realized. In practice an ion-electron yield of 2–3 is more realistic over extended periods, but even this low value allows counting efficiencies of about 75%. Because isotopic ratios involve relative measurements only, this figure is an acceptable one. For an analysis of the statistics of secondary electron multiplication and the critical role of the first dynode gain in shaping the output pulse height distribution, the reader is referred to the excellent paper of Dietz [16].

One important phenomenon affecting pulse-height distributions has been reported by Stanton, Chupka, and Inghram [17] with respect to the number of secondary electrons generated by a complex ion. They have reported, under otherwise identical conditions, that a $(C_{14}H_{10})^+$ ion generated more than twice the number of secondary electrons as an atomic ion (Hf-178) having the same mass number. Presumably a polyatomic molecule, in dissociating at a surface, acts as a group of particles and enhances the electron yield. In certain instances it would be advantageous if this effect were very large, for then some discrimination might be made for molecular vs. atomic ions of the same mass number. In other analyses, however, a small difference might simply lead to the increased complexity of obtaining precise isotopic ratios from organic spectra.

Fortunately an increasing number of electrostatic focusing multipliers are being made available from commercial sources that are suitable as ion counters. In many instances they may be ordered with a specified number of multiplying dynodes. There will be some differences in electron transit time for various types, but in most instances the important parameters will be over-all gain, pulse-height distributions, stability, "dark current," size, and adaptability to a specific spectrometer application.

Scintillation Detector and Photomultiplier

A different approach to ion counting has been devised by Daly [18]; it has proved very useful. Daly chose to count ions from light pulses

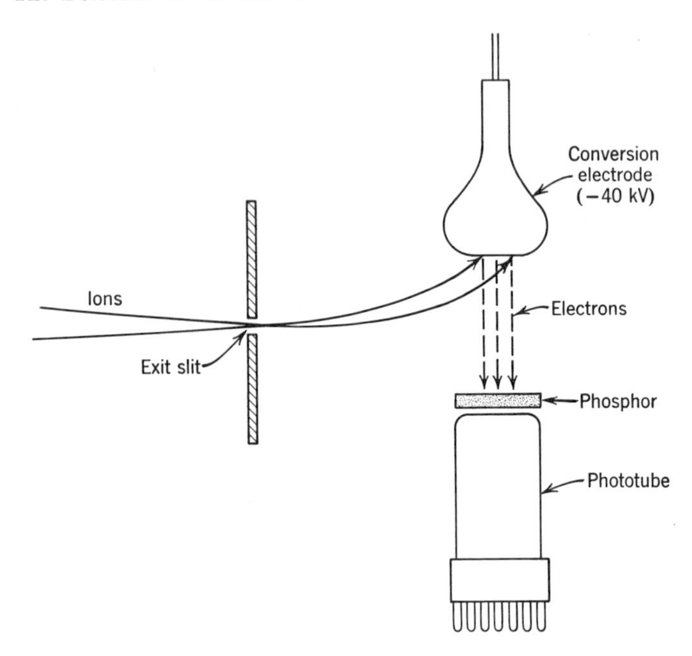

Fig. 4.9 Scintillation counting scheme by Daly [18].

generated in a phosphor in a fashion indicated in Figure 4.9. After emerging from the detector slit of a mass spectrometer, ions are accelerated and impinge upon a secondary emitting electrode that is maintained at a high negative potential ($\sim$40 kV). These secondary electrons are accelerated by the same electrostatic field and strike a plastic phosphor that is optically coupled to a photomultiplier, but the multiplier is external to the vacuum system. This scheme has several advantages over scintillation methods that were attempted by earlier investigators to detect heavy ions directly. First, the phosphor is impacted by electrons which produce less damage than does the beam of primary ions. Second, the conversion efficiency to light quanta in an organic phosphor is about 10 times larger for an electron than for a heavy positive ion [19]. Finally, a high secondary emission is obtained because of the high ion energy. As a result, it has been possible to achieve a very high signal-to-noise ratio. Daly [20] has reported that when a thallium ion releases five secondary electrons, which are subsequently accelerated through 40 kV and deliver a total kinetic energy of 200 keV to the phosphor, about 90 photoelectrons are released. It is thus possible to operate the phototube at low gain, and effective noise levels as low as 4×10^{-20} A have been achieved. A good plastic phosphor has a short decay time (3 nsec) [21] so there is no sacrifice in the ion counting rates that may be observed. Another important advantage of this method is that the activated surfaces of the photomultipliers are never exposed to the atmosphere, thus greatly extending the useful tube life.

Crossed-Field Multipliers

The crossed-field type multiplier makes use of both electrostatic and magnetic fields to determine the secondary electron trajectories. Such multipliers are especially attractive if the detection of ions must be made in a magnetic field, as purely electrostatic multipliers are rather severely defocused in a magnetic field of only a few gauss. Two important crossed-field multipliers were those developed by Smith [22] and by Wiley and McLaren [23]. The device reported by the latter investigators had a thin metallic oxide film coated on glass backing plates approximately 4 in. in length. When a high voltage is applied across the terminals of this high-resistance strip, a high voltage gradient results. The addition of a superposed magnetic field, with field lines perpendicular to the voltage gradient, caused electrons to multiply in cascade along the length of electrode. This device was subsequently developed in commercial form as the basic detector for the Bendix time-of-flight spectrometers.

Another crossed-field multiplier has been reported by White, Davis, and Sheffield [24]; it has a very high gain for its physical size. It differs from the above-mentioned type in that its dynodes are made of bulk semiconducting silicon. A schematic diagram of this multiplier is shown in Figure 4.10. In addition to serving as the mechanical dynode structure,

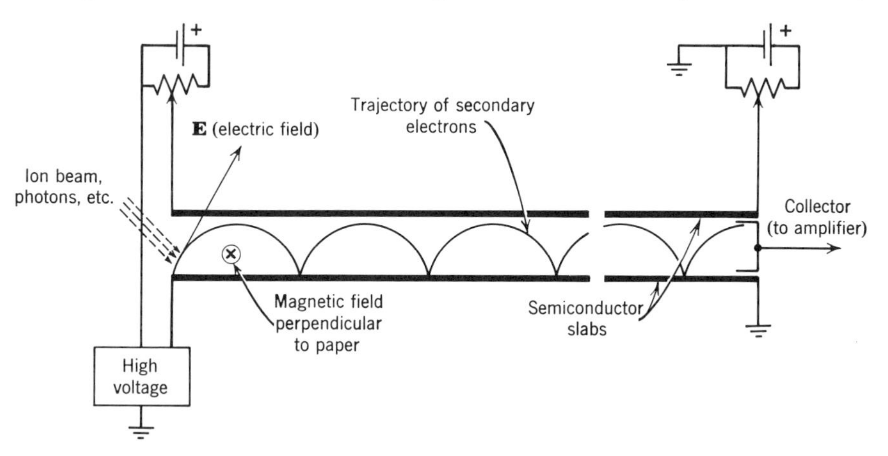

Fig. 4.10 Crossed-field multiplier with silicon dynodes.

the silicon also provides the high electrical gradient needed to accelerate the secondary electrons. High uniform potential gradients can be applied to the dynodes without "burnout," allowing the device to operate with a high gain-per-unit length. The semiconducting surfaces need no treatment or special "activation" and the multiplier can be exposed to the atmosphere indefinitely. A disadvantage is that the dynodes must be cooled to make their resistivity suitably high. This inconvenience of cooling is compensated by the distinct advantage that the device has a thermionic emission or "dark current" of virtually zero. This feature makes the device a candidate for applications that involve exceedingly low counting rates.

The two slabs that are shown schematically in Figure 4.11 and which constitute the continuous parallel plate dynodes, are of n-type gold-doped silicon. The resistivity of this material is of the order of 10^5 Ω-cm at room temperature. At an operating temperature of —50°C, however, the resistivity increases to approximately 10^7 Ω-cm—large enough to maintain high voltage gradients with only microampere currents. The dimensions of the pieces sawed from the ingot were 2×5 cm and 0.5 mm thick. In practice only 2-cm lengths are needed to obtain a gain of 10^6. The additional length merely serves as a mechanical support that connects to a copper cooling block attached to the Dewar cooling reservoir. Thin sapphire wafers between the silicon slabs and the copper block provide electrical insulation; the sapphire also possesses the high thermal conductivity needed for cooling. A 1-mm-thick sapphire piece functions as a spacer and insulator between the two parallel silicon slabs. The high-voltage wires, after receiving an ultrasonic "tinning," are soldered across the leading edges of the multiplier.

The gain or amplification of the semiconductor multiplier can be varied over a wide range. Both gain per stage and total number of stages depend on the voltage gradient and the externally applied magnetic field. For

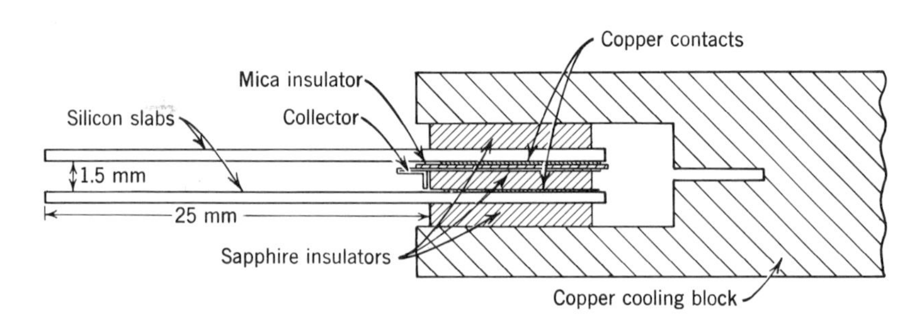

Fig. 4.11 Construction of semiconductor multiplier.

a fixed voltage gradient, an increase in the magnetic field increases the number of stages per unit length but decreases secondary-electron yield per stage. Thus the device exhibits a broad maximum with respect to variations in magnetic field. The radius of the cycloidal generating circle is equal to mE/eH^2. Figure 4.10 also indicates that the intercept of the cycloid on the lower dynode (and hence the interstage distance) can be varied by the bias voltage. For a voltage gradient of 1500 V/cm, a bias voltage of 500 V, and a magnetic field of 1000 gauss, the radius of the generating circle is approximately 0.3 mm. This corresponds roughly to an interstage distance of 1 mm.

The device has been used to scan the extended mass spectrum of a large 180° spectrometer [25] in which the fringing magnetic field of the spectrometer served as the magnetic field for the detector. A modified version of this device has also been reported by Sheffield [26], who has used bulk semiconducting glass. This material is somewhat more convenient for general applications as it can be operated at room temperature. It has, however, a higher background current than the gold-doped silicon device.

Reverse-Bias *p-n* Junctions

Devices known as *p-n* junctions hold out the promise of becoming one of the most important new types of ion detector in mass spectrometry. They have already made possible many new measurements in nuclear physics—providing excellent energy resolution for protons, alpha particles, and fission fragments [27].

Basically the reverse bias junction acts as solid state ionization chamber, with approximately one electron-hole pair being formed per 3.5 eV of energy expended in a silicon crystal. A major difficulty in adapting junctions for ion detection is the very restricted range of charged atoms in matter. Only very light ions of high energy (>30 keV) have a sufficient range to penetrate the silicon junction "dead layer," expend a large fraction of their energy in the "depletion region," and thereby generate a large amount of ionization. Post acceleration of mass-resolved ions is possible, but very high voltages would be required for heavy ions. An alternative is to cause the ions to strike a converter plate that will generate secondary electrons and accelerate these secondaries into the junction. As in the case of the scintillation counter, there is the further advantage of charge amplification in the secondary process itself.

A silicon *p-n* junction, operating under conditions of reverse bias, is an exceedingly shallow ionization chamber. The depth of this silicon ionization chamber, or "depletion zone," can be derived from Poisson's equation, the current continuity equation, and various parameters of the material. The results, however, can be expressed simply as [28]

$$x = 3.2 \times 10^{-5} (\rho V)^{1/2} \text{ cm} \qquad (p\text{-type}), \tag{4.3}$$

$$x = 5.3 \times 10^{-5} (\rho V)^{1/2} \text{ cm} \qquad (n\text{-type}), \tag{4.4}$$

where x is the depletion depth, V is the reverse bias (in volts) applied to the p-n junction, and ρ is the resistivity of the silicon expressed in Ω-cm. For high resistivity silicon (3000 Ω-cm), and a reverse bias of only 50 V, the effective depth of the depletion zone would be approximately 200 μ. The range of electrons in silicon has been found to follow the approximate relationship shown in Figure 4.12. It is thus concluded that a junction may easily be formed under conditions of reverse bias—having a depth comparable to the range of secondary electrons that are normally incident upon it—for kinetic energies up to 50 keV.

Use of such a p-n junction for detecting mass spectrometric ion beam currents was first reported by White, Sheffield, and Mayer [29], with the basic arrangement indicated in Figure 4.13. In this instance a diffuse type junction was employed. Recently, a surface-barrier junction was employed with substantially improved electronics but with the same basic scheme. Positive ions were focused onto a high-voltage stainless steel electrode (30 kV, negative) and secondary electrons were accelerated onto the junction. The interesting results are shown in Figure 4.14. The graph clearly shows seven quantized peaks as displayed in an output multichannel analyzer, and indicates the relative number of secondary electrons released by primary ion impact [30]. The first large peak is caused by ions that generated but a single electron, corresponding to the formation of electron-hole pairs produced by a single 30-keV

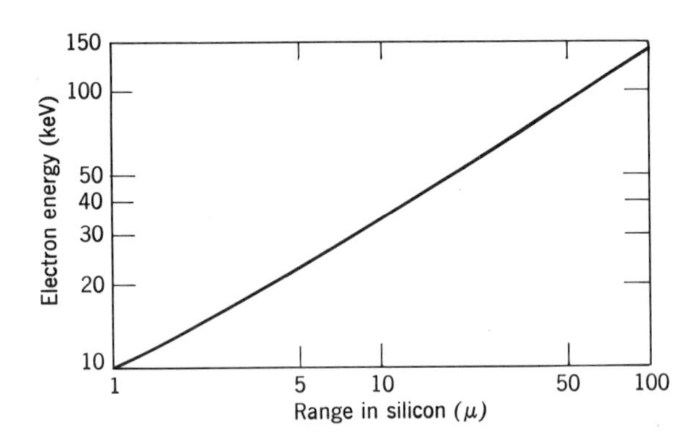

Fig. 4.12 Range of electrons in silicon.

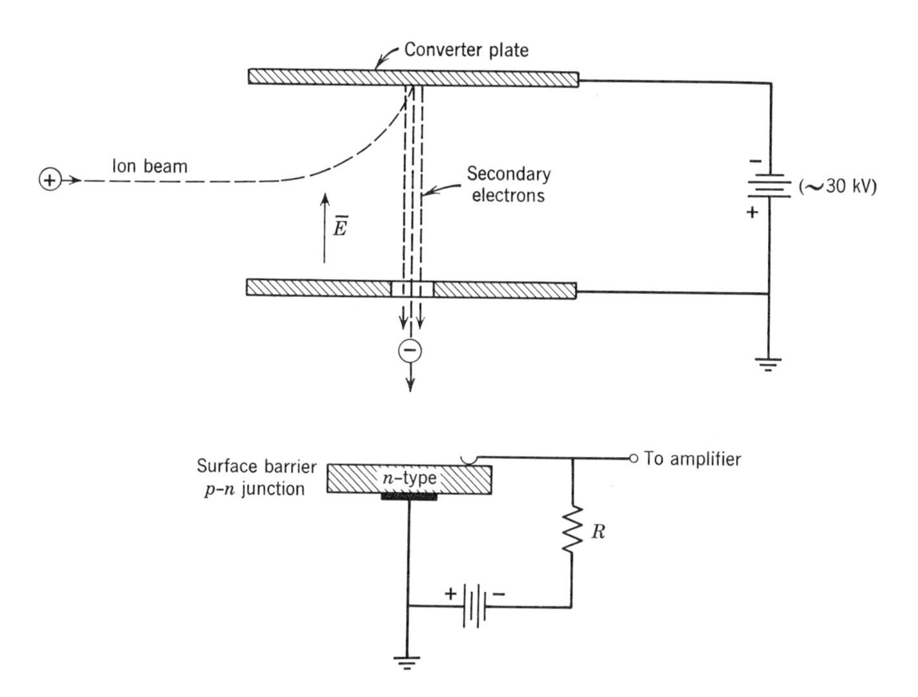

Fig. 4.13 Use of p-n junction as an ion detector.

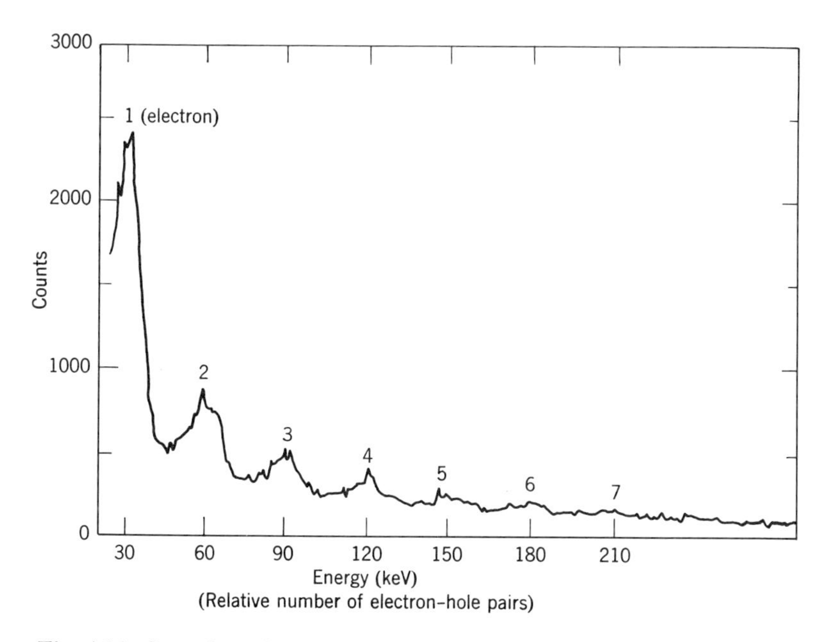

Fig. 4.14 Secondary electron yield from ions as detected in a p-n junction.

electron. Successive peaks indicate ionization energies in the junction region of 60, 90, 120, 150, 180, and 210 keV, corresponding to the impact of two, three, four, five, six, and seven (30-keV) electrons. It will be noted that this detection scheme is reasonably well biased above noise for even a single electron. For this reason I am extremely optimistic about the ultimate use of such junctions in mass spectrometry. Because of their small physical size, a large number of them can be adapted to read-out an entire mass spectrum simultaneously. If this is achieved, such a method would possess the sensitivity of an electron multiplier; it would also have the integrating capability of an extended photographic plate.

It is appropriate to point out one further potentially interesting possibility relating to this method. Combined with a suitable read-out system, it would be possible not only to obtain a complete mass spectrum from a transient phenomena (e.g., an arc) on a micro-to-millisecond time scale, but also to follow the *time-dependent* change of isotopic ratios that are so important in the interpretation of trace samples and the like. Such a two-dimensional display (time-and-mass position) from a computer would probably yield information that, at present, can only be monitored by fast scanning techniques operating at considerably reduced sensitivities.

Dual Multplier Counters

There are many instances in which it is desirable to count simultaneously the ion current from two isotopes. Although the *p-n* junction scheme appears to have great potential, electron multipliers have already been successfully applied to this problem. Two examples will be cited that represent quite different solutions.

One method has been to construct a dual multiplier of the crossed-field semiconductor type [24] with two sets of silicon slabs that may be separated by a fixed distance, corresponding to the dispersion of two isotopes. The very small dimensions of this device make it one of the best multipliers for this application. Of somewhat greater usefulness and general application is to have two such multipliers so that the spacing between them can be varied. Such an arrangement has been reported by Fenner and Ridley [31]. They measured a W^{184}/W^{186} ratio to an accuracy that approached the statistical limits of counting, even though the variation in the primary ion current was approximately a factor of two.

A second method has been reported by Dietz [32], who used conventional multipliers in a manner shown in Figure 4.15. Two isotopic ion beams emerge from a 180° magnet and enter into a pair of electrostatic

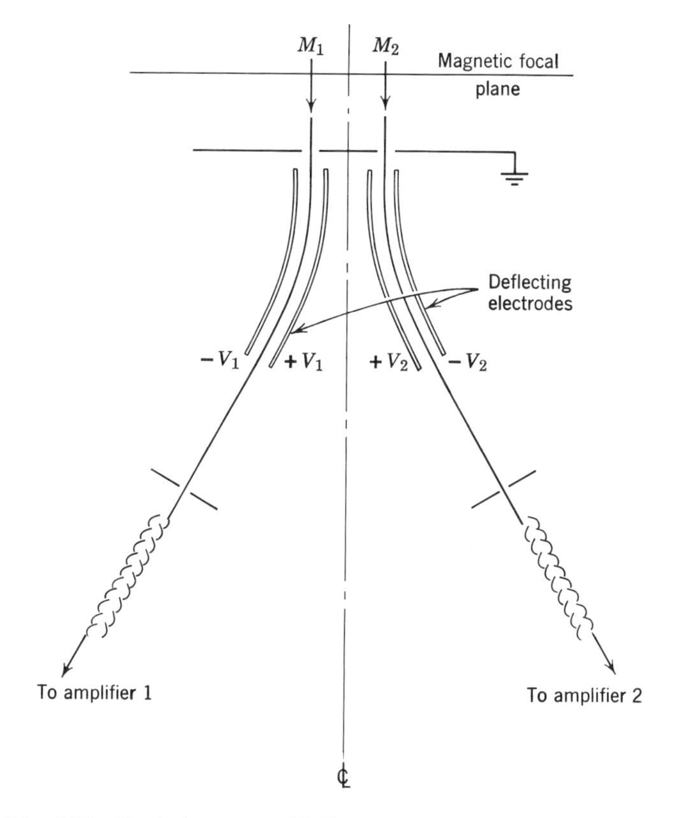

Fig. 4.15 Dual electron-multiplier counter of Dietz [32, adapted].

deflector plates. Each pair of deflector plates, including the defining slits, is independently and continuously translatable along the focal plane so that any desired pair of ion beams can be accepted into the deflector system if the separation is in the range of 0.25 to 5 cm. The two ion beams are then deflected into two electrostatic multipliers, which are also translatable (and also magnetically shielded). An added advantage of this arrangement is that the beam deflector system provides some further focusing action of the ion beams on the first dynode of the multipliers.

Coincidence Methods of Ion Detection

The preceding pulse-counting methods can be adapted in some fashion for most investigations. In the future, however, isotopic ratio measurements on samples that are exceedingly small—perhaps even in the

10^{-18} gm range can be anticipated. The question then arises: What is the ultimate signal-to-noise ratio that can be achieved with respect to the detection of single ions? The preceeding paragraphs have suggested the importance of a high ion-secondary electron yield on the first multiplier dynode, if all "noise" pulses are to be biased-out in a discriminator amplifier. Maximization of this ion-secondary ratio is, in my opinion, worthy of considerable additional technical effort and provides the best practical method for the detection of low ion currents. Consistently high-yield surfaces are, nevertheless, difficult to maintain, and the incidence of an intense high-energy ion beam will sputter off the activated surface. In practice it is also reasonably difficult to completely bias-out all noise pulses—from spurious ions, and thermionically emitted electrons.

A coincidence method provides an interesting alternative to detecting single ions, with virtually no noise—providing some sacrifice is made in the over-all counting efficiency. One coincidence method is to employ a very thin film for generating secondary electrons that can be focused into scintillators and photomultipliers, as reported by Daly [33]; the same basic scheme has also been tested without a scintillating phosphor [34]. Figure 4.16 indicates this latter scheme.

Consider an exceedingly thin metallic foil (~500 Å) to be the target of the mass-resolved ion beam, and let potentials of a few hundred volts be symetrically applied so as to focus secondary electrons into two opposing multipliers. If the range of the ion is sufficiently great so that it can actually penetrate the thin foil, it will simultaneously generate secondary electrons from both faces of the foil. Clearly the

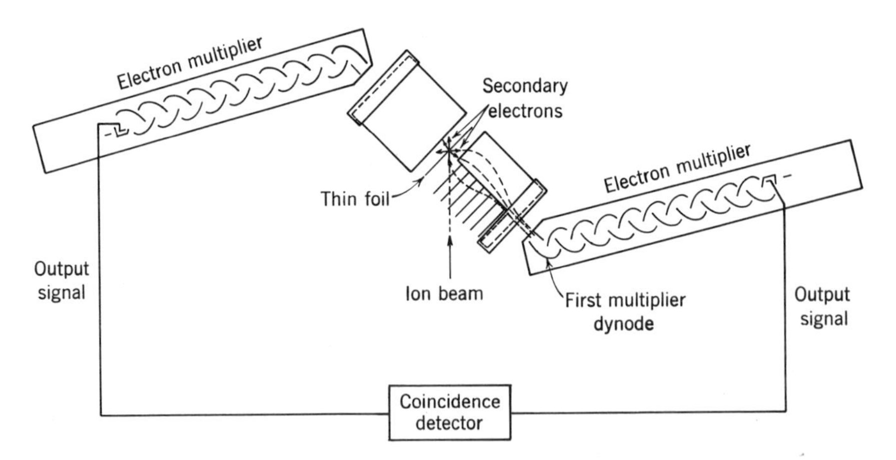

Fig. 4.16 Thin-foil method for producing coincident pulses.

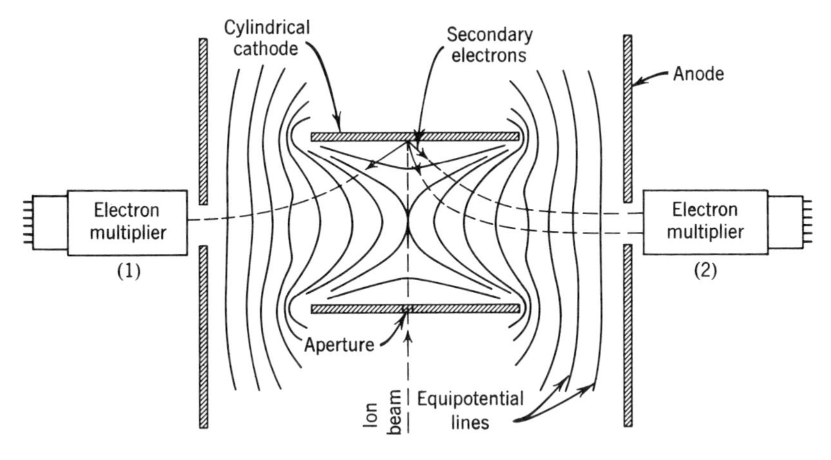

Fig. 4.17 Cylindrical ion-electron converter for coincidence pulse counting.

atom must emerge with sufficient kinetic energy (a few keV) so that secondary electron production is possible, but it is not necessary that the atom emerge as a charged particle. The atom will undergo charge exchange in the foil, but an energetic singly charged ion or neutral will give approximately the same secondary yield. Provided only that there is an adequate secondary yield (one electron on each side), a reasonable probability exists that two multipliers will furnish output pulses that are in very tight coincidence ($\sim 5 \times 10^{-8}$ sec). Such a scheme may find special use for detecting high-energy ions in a radiation environment in which the operation of a simple multiplier would be impossible. This method is attractive if the ion beam is collected over an area of several square centimeters—such as might be encountered in nuclear physics research. It can also be employed to detect single electrons in the 5-keV range. The preparation and fragility of very thin foils present distinct difficulties, and the efficiency of such a scheme is essentially low. It will be clear, however, that the tight time coincidence means that the dual multipliers can, in principle, operate without a single noise pulse in hours of operation.

An improved coincidence method has been suggested that utilizes a simple cylindrical ion-electron converter [30]. Only simple electron optics are required and the method is attractive because it can be used for detecting ions whose kinetic energies are comparable to those used in mass spectrometry. Figure 4.17 is a schematic of the coincidence multiplier arrangement. Let the ions enter normally through a small slit in the wall of the cylinder and impinge upon the opposite inner surface.

If this surface has been activated, several secondary electrons will be emitted for each incident ion. As in the usual case, these secondaries will have a mean kinetic energy of only a few electron volts. If accelerating electrodes (biased a few hundred volts positive) are placed at ends of the cylinder, symmetric fringing electrostatic fields will reach into the interior of the cylinder, effectively dividing the cylinder. Hence the secondary electrons will be focused into multiplier No. 1 or No. 2 depending, upon their initial directions of motion. Coincident output pulses will be obtained if both multipliers are triggered by electrons generated from a single ion impact. Low primary electron beam currents can also be detected by this scheme.

The inherent advantage of this method is its simplicity, with a virtual guarantee of zero noise pulses. With reasonably good secondary electron focusing, acceptable counting efficiencies can be expected. There are many situations in which the investigator is willing to trade high counting efficiency for a reduced counting efficiency with zero background. This trade-off is made in many other scientific studies. With this—or some improved coincidence scheme—it is thus realistic to predict spectrometer ion counting rates that may have significance not only in terms of one ion per second (i.e., 1.6×10^{-19} A) but also in terms of a few ions per hour ($\sim 10^{-22}$ A).

DATA ACQUISITION

Digital Display Systems and Recorders

One important advantage of using pulse counting detectors is ease of data acquisition. The isotopic ratio of two mass-resolved ion beams is simply the ratio of the counts obtained in two counting channels. In the general case "gating" of these channels is programmed with voltage modulation of the ion source, and long or short timing intervals can be selected. Either high or low counting rates can be handled by simple circuitry, and a multiplicity of ratios can be acquired to which appropriate statistics may be applied. Output pulses from an electron multiplier may be routed through a very simple or an elaborate data-recording network. Conventional components will usually include a preamplifier, amplifier, discriminator, scaler, digital recorder, count-rate integrator, and integral chart recorder. The block diagram of Figure 4.18 illustrates the interconnections of circuitry.

A digital display system operates equally well for presenting an integrated mass spectral scan, with the aid of dc recorders. In this case a motor-driven potentiometer provides a slow change in the high-voltage ion source voltage. Output pulses from the electron multiplier are routed

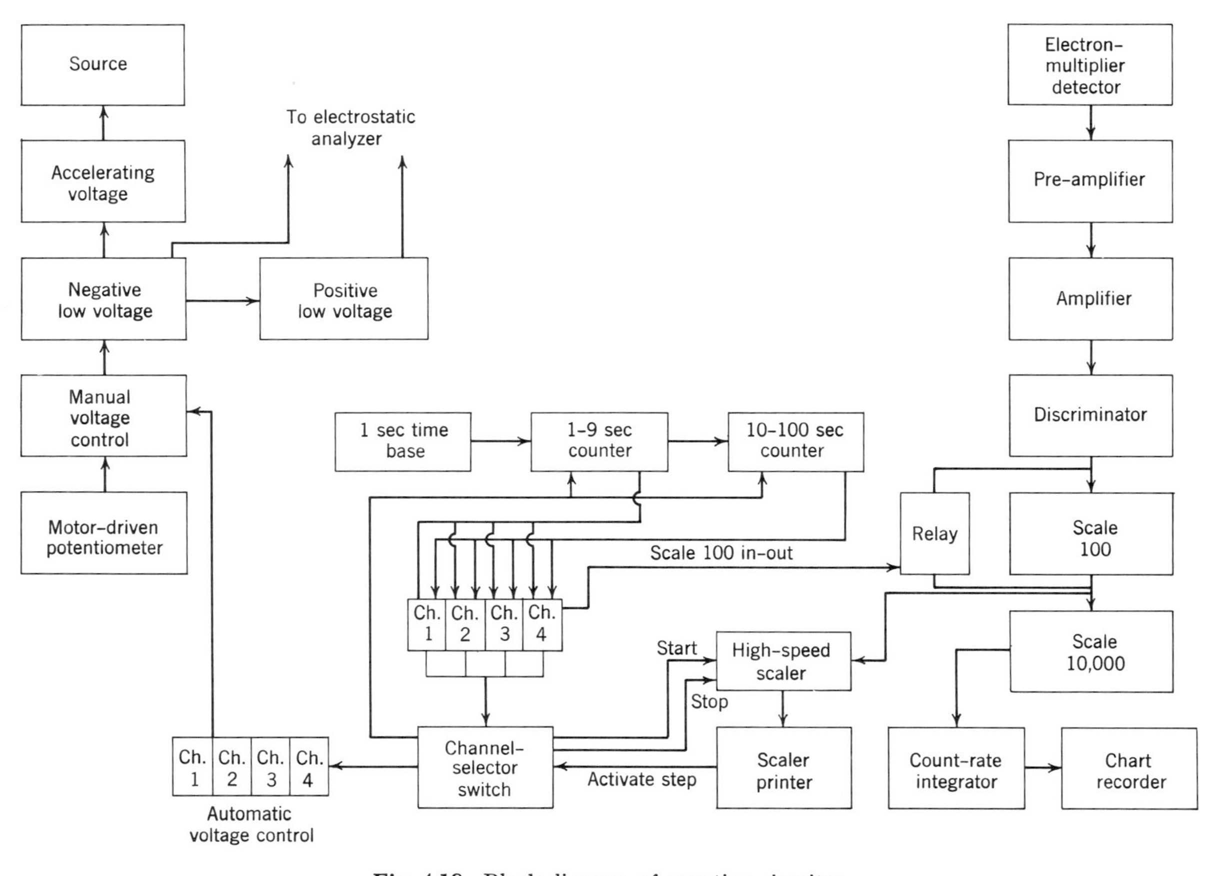

Fig. 4.18 Block diagram of counting circuitry.

through a scaling circuit into a count-rate meter, the output of which feeds into a chart recorder. Factors of 10, 100, 1000, or greater can be selected by appropriate switching from the scaling circuit so that the count rate meter accepts every pulse, or only 0.1, 0.01, or 0.001, of the actual number of output pulses. A reasonably accurate graph of mass spectral intensities can be recorded in this fashion. The commercial count-rate meters will usually be equipped with a sensitivity selection, thus increasing the range of ion currents that corresponds to a full-scale reading on the output recorder.

For precision work, when only one or two mass ratios are to be obtained, it is customary to modulate the ion-source voltage in discrete voltage increments, corresponding to the focus of the isotopes through the detector slit. As indicated in Figure 4.18, a time base is used to select counting periods during which pulses are recorded. Electronic counters and timers can provide intervals of 1 μsec to 10 sec. For most analytical work, counting periods of 1–10 sec are convenient. If a large difference exists in the intensity of two ion-beam currents, a longer counting period can be used for the ion beam of lower intensity. A "dead time" will always exist while voltage switching occurs; the minimum time will depend on how fast the ion-source circuit can stabilize. For reasonable counting intervals in a multiple-channel system ($\sim$1 sec), printing of the total counts in one channel can be performed during switching, or even while another channel is counting. Other multichannel and data-collection systems have been developed which include (a) the use of a programmed integrating digital voltmeter to sample the detector output [35], (b) a simple display device that makes use of two independent time bases provided by an oscilloscope [36], and (c) a multichannel analyzer using time-amplitude conversion for time-of-flight spectrometry [37].

Probably the most important circuit component in a high-speed counting system is the amplifier-discriminator. The electron multiplier may provide a high gain, but the output pulses from it are quite randomly distributed in amplitude and time. Electron multiplier output pulses may be from 5 to 10 nsec wide, 10 mV to 1 V in amplitude, and the average pulse rate may vary from counts of 1 to 10^6 per second. Thus the function of the discriminator amplifier is to transform the random pulse heights into output pulses of comparable amplitude and shape, so they may be counted accurately by a high-speed scaler.

A discriminator specifically designed to satisfy mass spectrometer requirements has been developed by Sawada [38]. Its general design is shown in Figure 4.19. The first amplifier section serves to amplify small signals and clip pulses of very large amplitude. This arrangement

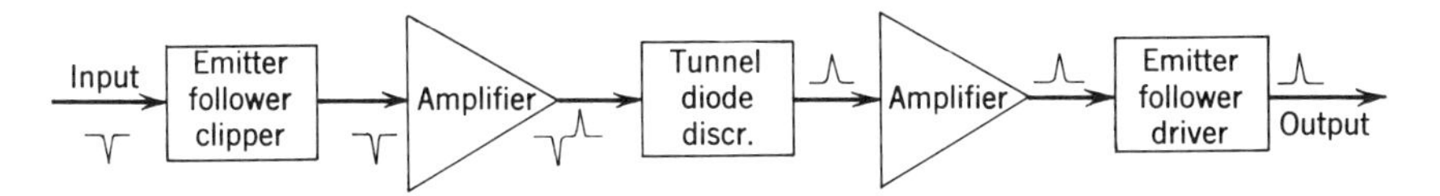

Fig. 4.19 High-speed amplifier discriminator [38, adapted].

provides a narrower range of pulse amplitudes to the tunnel diode discriminator. The diode then provides a constant output pulse for each input pulse that exceeds the discriminator threshold, and the second amplifier section furnishes the voltage and power gain to give about 2 V with a 50-Ω load.

An additional attractive feature of this unit is that the entire solid-state circuit is mounted directly on an electron multiplier base, substantially eliminating cable and connector problems. The circuit has an unusually wide dynamic range and has been used at counting rates as high as 2×10^6 Hz average rate (for random pulses).

Counting Rate Losses

Unless detector circuitry is exceptionally fast, substantial counting losses will occur at average counting rates of about 10^6/sec. Megacycle circuits and scaling units are generally available at 1, 10, and 50 mc, but these ratings are for pulses that are evenly spaced in time. The ions that arrive at a mass spectrometer detector, however, have a random distribution and many pulses are separated by times that are shorter than the resolving time of the circuitry. The true counting rate can be related to the observed counting rate by the following expression

$$n = \frac{n_o}{1 - n_0\tau}, \tag{4.5}$$

where n and n_o are the true and observed counting rates, respectively, and τ is the resolving time of the circuitry.

Now a fairly accurate measure of the resolution time, τ, can be made without recourse to sophisticated measurements. It is possible simply to select an element having two isotopes that differ in abundance by a factor of about 100. At high counting rates of the major isotope it is possible to observe a departure from the known isotopic ratio. This ratio change results from counting-rate losses of the major isotope. If R and R_o are the known and observed isotopic ratios for the major and minor isotopes, and n_o is the observed counting rate of the major isotopes, it can be shown that τ may be calculated from

$$\tau = \frac{R_o - R}{n_o(1 - R)} \tag{4.6}$$

Thus it becomes a fairly simple matter to compute counting-rate losses, provided that these losses do not exceed some reasonable value (about 5%).

Computers

The use of computers provides no guarantee of increased precision, as they cannot supply information that can be obtained otherwise if sufficient time is available. As in other applications, however, the computer can handle a large volume of data, and it provides a desirable uniformity in calculation of results. It is beyond the scope of this book to present any detailed specifics of computer applications, but a few general comments may be appropriate.

When data are to be extracted from photographic plates, an empirical function given by Judson and Hull [39] has proved useful. It takes into account the saturation transmission of the emulsion, and represents the transmission range of Ilford Q_2 emulsions as a function of exposure, thereby permitting quantitative data reduction. The equation has the form

$$E_x = Q_T A_x C_x \left(\frac{W_0}{W_x}\right) J_x + E_B. \tag{4.7}$$

Here E_x is the exposure observed in a line due to a particular isotope of an element; Q_T is the total charge of the beam monitor; A_x is the abundance of the isotope; C_x is the concentration of the element in the sample; (W_o/W_x) is the ratio of the widths of the standard and unknown lines; J_x is the over-all emulsion sensitivity for a particular element in a particular matrix; and E_B is a "background transmittance" in the region of the image line.

This basic relationship has been found useful in preparing the more specific relationships that are required for computers. Appropriate corrections must be included for background fog, if any, or for variable spectral line width and shape. The mass response of ions must also be included, which is for Ilford Q_2 emulsions approximately $M^{0.6}$ (M = mass number), according to Honig [40]. Normalization is achieved by reference to standards.

The computer will, of course, not only provide output data that correlate the intensity of a spectral line with total ion beam, but it also provides the conversion of spectral line positions into accurate masses.

At least two schemes have been considered by investigators for correlating spectral-line positions to elemental composition of ions recorded on photographic plates. In principle it is possible to calculate all conceivable (m/e) combinations of atoms and molecules, and put these on a memory tape that is recalled by the computer. Another approach is to have the computer sum the mass of atoms in all possible combinations, compare each against the obtained mass, store those that fit within the prerequisite limits, and drop those that do not. Proponents of this latter system claim this approach is fast, flexible, and avoids the need for individual tapes.

Computers have also been programmed to handle output data from electron multiplier detectors, calculate ratios, and provide statistical tests. They have also been used for component analysis. For example, a recent paper reports a FORTRAN program [41] that was developed for calculating the distribution of electrons that are emitted from the several dynodes of electron multipliers. It is based on a statistical model for electron emission from a multiple dynode structure first described by Prescott [42]. A generating function is developed that predicts the probability of an output of n electrons for dynodes of varying secondary electron yields. Computers have thus proved important not only in handling the large output of analytical data from mass spectrometers but also in furnishing an insight into the detailed performance of electron multipliers per se.

ION TRACK DETECTORS

A final comment is appropriate in this chapter with respect to detectors that might prove useful in highly specialized situations. If the total number of heavy ions to be detected is only a few hundred, it is reasonable to consider counting techniques such as are employed in nuclear physics. One approach might be to have the ion strike an electrode that produces secondary electrons, and then accelerate these electrons into a nuclear emulsion. The geometry could be essentially the same as with the scintillation or p-n junction schemes that have been mentioned. The nuclear emulsion would then be recording *individual electron tracks* (or soft beta particles). Rather than a single electron, however, one would expect to observe in a microscope several closely spaced images corresponding to a multiplicity of secondary electrons. Because of the slight differences in the direction and initial velocities of these secondaries, their arrival in the emulsion will not be absolutely coincident. Thus two isotopes might be displayed via countable dots or tracks, spatially dispersed into two rows that would represent the "line image."

A somewhat more elegant method has already been employed to detect single ions of very high energy. The detection scheme makes use of the fact that heavy and energetic ions, fission fragments, or alpha particles produce narrow trails of damage when they impact certain minerals, glasses, or polymers. Fleischer, Price, and Walker [43] have shown that these highly localized damaged sites can be rendered visible in an optical microscope by using a suitable etching that will preferentially attack the damaged region. By using different ions at various energies they have shown that, for a particular substance, there is a critical energy loss for track formation, $(dE/dx)_c$. Only if this critical value of energy absorption is exceeded, can a substance such as mica be rendered developable by ion impact. For this particular substance, and for Si ions, a $(dE/dx)_c$ of 13(MeV/mg/cm^2) has been reported [44]. This is a high value, but it should be noted that it is the *rate* of energy loss that is significant rather than the total ion energy. Perhaps heavy ions can be post-accelerated into such detectors; an optimum (dE/dx) value, and a suitable substance must then be found.

In any event these track detectors have already been useful for monitoring very low beam currents of ions having many MeV in kinetic energy. They do not have to be shielded from light and they are not fogged by the beta or gamma radiations that are usually present in particle-accelerator laboratories.

REFERENCES

[1] W. H. Barkas, *Nuclear Research Emulsions* (vol. I), Academic, New York, 1963.

[2] E. B. Owens, in *Mass Spectrometric Analysis of Solids* (Chapter III), ed. by A. J. Ahearn, Elsevier, Amsterdam, 1966.

[3] *Ibid.*

[4] M. G. Inghram and R. J. Hayden, National Academy of Science, National Research Council, Report No. 14, Washington, D.C., 1954.

[5] P. R. Kennicott, *Anal. Chem.,* **37,** 313 (1965).

[6] E. Burlefinger and H. Ewald, *Z. Naturforsch.,* **18 *a*,** 1116 (1963).

[7] E. B. Owens, *Appl. Spectry,* **21,** 1 (1967).

[8] A. O. Nier, E. P. Ney, and M. G. Inghram, *Rev. Sci. Instr.,* **18,** 294 (1947).

[9] L. Ruby, J. G. Kramasz, Jr., W. G. Pone, and T. Vuletich, *Nucl. Instr. Methods,* **37,** 293 (1965).

[10] H. Palevsky, R. K. Swank, and R. Grenchick, *Rev. Sci. Instr.,* **18,** 298 (1947).

[11] J. E. Beynon, *Mass Spectrometry,* Elsevier, Amsterdam, 1960, p. 207.

[12] J. S. Allen, *Phys. Rev.,* **55,** 966 (1939).

[13] F. A. White and T. L. Collins, *Appl. Spectry.,* **8,** 17 (1954).

[14] P. Marchand, C. Paquet, and P. Marmet, *Rev. Sci. Instr.,* **37,** 1702 (1966).

[15] J. Stein, private communication.

[16] L. A. Dietz, *Rev. Sci. Instr.,* **36,** 1763 (1965).

[17] H. E. Stanton, W. A. Chupka, and M. G. Inghram, *Rev. Sci. Instr.,* **27,** 109 (1956).
[18] N. R. Daly, *Rev. Sci. Instr.,* **31,** 264 (1960).
[19] J. B. Birks, *Proc. Phys. Soc. (London),* **A 68,** 1294 (1950).
[20] N. R. Daly, *op. cit.*
[21] W. A. P. Young, R. G. Ridley, and N. R. Daly, *Nucl. Instr. Methods,* **51,** 257 (1967).
[22] L. G. Smith, *Rev. Sci. Instr.,* **22,** 166 (1951).
[23] W. C. Wiley and I. H. McLaren, *Rev. Sci. Instr.,* **26,** 1150 (1955).
[24] F. A. White, J. C. Sheffield, and W. D. Davis, *Nucleonics,* **19,** 58 (1961).
[25] F. A. White, J. C. Sheffield, and F. M. Rourke, *Appl. Spectry.,* **17,** 39 (1963).
[26] J. C. Sheffield, private communication.
[27] G. Dearnaley and D. C. Northrop, *Semiconductor Counters for Nuclear Radiations,* Wiley, New York, 1966.
[28] *Ibid.,* p. 128.
[29] F. A. White, J. C. Sheffield, and J. W. Mayer, *Electronics,* **34,** 74 (1961).
[30] F. A. White, J. D. Walling, and A. J. Schwabenbauer, "P-N Junction Detectors and Coincidence Methods for Measuring Low Ion Beam Currents," Presented at the Fifteenth Annual Conference on Mass Spectrometry, Denver, May, 1967.
[31] N. C. Fenner and R. G. Ridley, *J. Sci. Instr.,* **41,** 157 (1964).
[32] L. A. Dietz, *Rev. Sci. Instr.,* **36,** 1763 (1965).
[33] N. R. Daly, *Rev. Sci. Instr.,* **31,** 720 (1960).
[34] D. Kraus, and F. A. White, *IEEE Trans. Nucl. Sci.,* **NS-13** (No. 1), 765 (1965).
[35] P. E. Moreland, C. M. Stevens, and D. B. Walling, *Rev. Sci. Instr.,* **38,** 760 (1967).
[36] D. E. H. Marsden, W. Forst, and K. K. Feng, *Rev. Sci. Instr.,* **36,** 1109 (1965).
[37] J. W. White, *Rev. Sci. Instr.,* **38,** 187 (1967).
[38] Sawada, F. H., *IEEE Trans. Nucl. Sci.,* **NS-12** (No. 1), 374 (1965).
[39] C. M. Judson and C. W. Hull, Paper No. 83, presented at the ASTM Conference on Mass Spectrometry, San Francisco, May, 1963.
[40] R. E. Honig, Paper presented at the International Conference on Mass Spectrometry, Paris, June, 1964.
[41] J. E. Edwards and J. I. McCarthy, Knolls Atomic Power Laboratory Report, **KAPL-M-6575** (1966).
[42] J. R. Prescott, *Nucl. Instr. Methods,* **39,** 173 (1966).
[43] R. L. Fleischer, P. B. Price, and R. M. Walker, *Science,* **149,** 383 (1965).
[44] R. L. Fleischer, P. B. Price, R. M. Walker, and E. L. Hubbard, *Phys. Rev.,* **133 A,** 1443 (1964).

Chapter 5

Nuclear and Reactor Physics

It is difficult to conceive of atomic and nuclear physics or of modern nuclear engineering without the aid of mass spectrometry. It is true that many nuclear phenomena are observable by means of alpha-, beta-, and gamma-ray spectroscopy, and a variety of other techniques. But the precise establishment of a mass scale, the positive isotopic identification of nuclides that result from the fission of the heavy elements, the measurement of nuclear cross sections, and the like are fundamental to nuclear and reactor physics—and mass spectrometry has provided many of the important data relating to these and other parameters. Thus it is doubtful if any other single analytical instrument has been more productive than the mass spectrometer in establishing the foundations of nuclear physics. This chapter, therefore, will review some of the classic applications of mass spectrometry in the nuclear field, together with a description of several recent measurements.

DETERMINATION OF ATOMIC MASSES

Before 1923 most measurements indicated a "whole-number" rule, namely, that the masses of all atoms were integral multiples of the hydrogen atom. The discovery that small differences existed from integer values led to intensive studies by mass spectroscopists and to the ultimate establishment of a modern mass scale. From the standpoint of nuclear physics, this mass scale is important because these small deviations from whole numbers relate directly to the stability of a nuclide. One can then speak of the *binding energy* of the particles comprising the nucleus. Specifically, binding energies (or "mass defects" when used with Einstein's mass-energy postulate) indicate the energy balance in a nuclear reaction. This energy balance, or Q-value of a reaction is the difference between the mass of the target and projectile atoms $(M_o + M_1)$ and the masses of the atoms $(M_2 + M_3)$, that are produced in the reaction

$$Q = c^2(M_o + M_1 - M_2 - M_3). \tag{5.1}$$

If the reaction is one of simple beta decay, the Q-value is the difference between parent and daughter masses. In the case of alpha decay, it is the mass difference between the original atom, and the daughter atom plus a helium atom (He^4). For positron decay, in order that the Q-value include all the energy release in the transition, $2m_oc^2$ must be added to the beta endpoint energy to account for annihilation radiation.

A measure of the stability of the nucleus is conventionally expressed in two forms:

$$\text{packing fraction} \qquad f = \frac{(M - A)}{A}, \tag{5.2}$$

$$\frac{\text{binding energy}}{\text{nucleon}} = \left(\frac{ZH^1 + (A - Z)_n - {}_ZM^A}{A}\right) c^2, \tag{5.3}$$

where A, Z, H^1, and n denote mass number, atomic number, and the hydrogen and neutron masses. A small packing fraction reflects high stability. Likewise, a high binding energy per nucleon suggests a stable system.

Figure 5.1 illustrates the general stability pattern of the nuclei. It is indeed this information that makes plausible the fact that fission

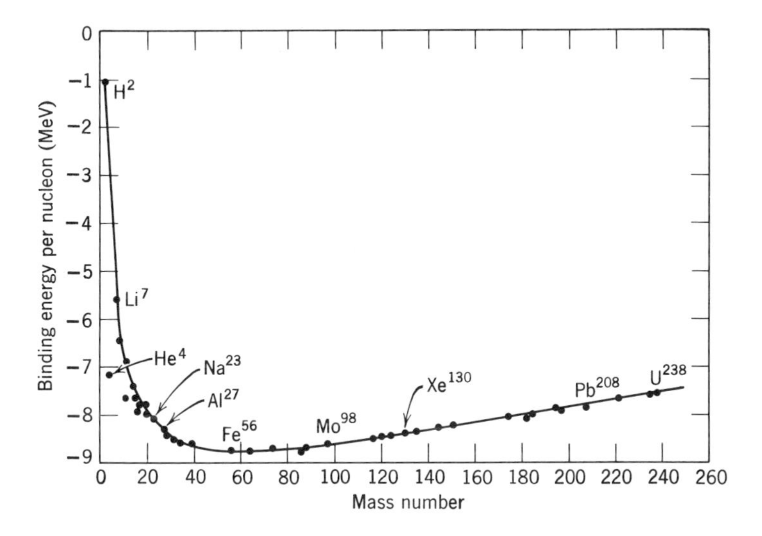

Fig. 5.1 Binding energy as a function of mass number.

energy can only be obtained from heavy nuclei, and fusion energy from a few light nuclei.

If atomic masses are measured by mass spectrometry of very high resolution, fine structure will appear in the binding energy curve. This detail is one of the major achievements of modern mass spectrometry, as it allows comparison with nuclei models that relate to average nuclear binding energy, neutron separation energy, and proton separation energy. Figure 5.2 shows that if the general curve of Figure 5.1 is examined very carefully, there exist substantial departures from a smooth plot as a function of mass number. In principle it should be possible to obtain these data by carefully measuring the mass spectrometer parameters that focus one specie to a specific image point or line.

In practice, however, the most accurate means by which this detail is obtained is by the so-called "doublet" or "multiplet" method. For example, the three species O^+, NH_2^+, and CH_4^+ all have mass number 16. But in a high-resolution spectrometer, these species would appear as distinct lines on a photographic emulsion; they would also be resolved as separate ion beams in an electrical detection system. Thus O^+ and NH_2^+ are referred to as a doublet, the nomenclature being adapted from line-splitting in optical spectroscopy. This mass or line splitting is generally small, often being of the order of 0.01 to 0.001 amu. The distance between doublets is thus quite small compared with the dispersion of integer mass numbers. Therefore a much more precise mass scale can be established,

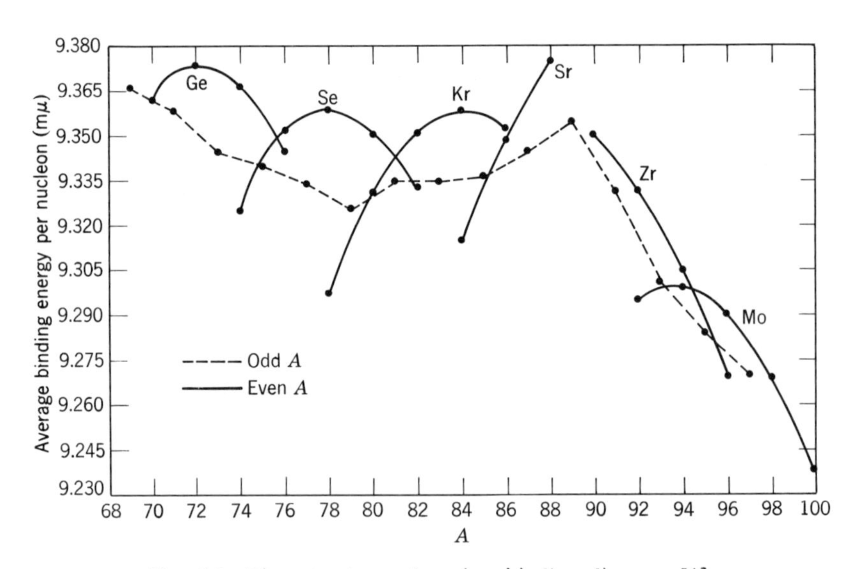

Fig. 5.2 Fine structure of nuclear-binding diagram [1].

Table 5.1

Mass Differences of Selected Doublets

Doublet	*Mass difference* ($m\mu$)*
C_5H_9–Ga^{69}	144.852
C_5H_{11}–Ga^{71}	161.370
C_5H_{10}–Ge^{70}	154.001
C_4H_8O–$Ge^{70}H_2$	117.616
$C_3H_7O_2$–As^{75}	123.009
C_6H_{10}–Se^{82}	161.545

* Mass differences are given in terms of milliunits of mass ($m\mu$); μ is now used for atomic mass unit, based on C^{12}.

and extrapolated values from known ion species can be made with an accuracy of 1 part in 10^7 to 1 part in 10^8.

Table 5.1 represents only a few representative mass doublets as measured by Ries, Damerow, and Johnson [1].

These measurements and the doublet [2] displayed in Figure 5.3 were obtained with the University of Minnesota 16-in. double focusing spectrometer developed by Professor Nier and his co-workers (see Chapter 2). This instrument has consistently provided many of the data that have established the modern mass scale. An electronic system, rather than photographic detection, is employed. Ions are produced by electron impact in an ion gun, and a 90°-cylindrical analyzer is used to select

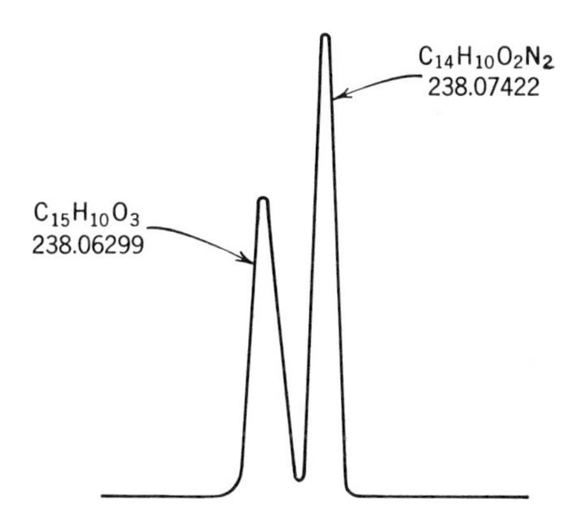

Fig. 5.3 Mass doublet as measured in the double-focusing instrument of Nier [2, adapted].

a small energy group that passes through an asymmetrical 60°-magnetic analyzer system. An electron multiplier is located behind a small detector slit. The ion spectrum is scanned over a fraction of a mass unit by varying both the ion accelerating voltage and the potential applied across the electrostatic analyzer. (An auxiliary mass spectrometer uses a portion of the 60°-magnetic sector, and appropriate feedback loops minimize errors of magnetic-field changes and drifts in accelerating potentials.) The two ion groups that are selected for a mass comparison can alternately be focused through the detector slit by a small voltage change, ΔV, which is proportional to a resistance change, ΔR, in a potentiometer circuit. The mass comparison is substantially then

$$\frac{\Delta M}{M} = k \frac{\Delta R}{R}. \tag{5.4}$$

Peak matching is effected by an oscilloscope display wherein the two doublet peaks are made to appear on alternate sweeps. Using exceedingly small slits (0.0001 in.), and a temperature-compensated resistance network, mass differences of approximately 1 part per million can be detected. It will be noted that a comparison may also be made between singly and doubly charged ions. For example, singly ionized carbon atoms (C^{12} = 12.000000), and doubly charged magnesium (Mg^{24} = 23.985045) are only slightly separated in the mass 12 region of the spectrum.

Before September 1960, atomic masses were expressed in terms of the major isotope of oxygen (O^{16} = 16.000000). Since that date, C^{12} has been the adopted standard, in part because of the convenience of relating doublets to the many species that contain carbon atoms. Table 5.2 indicates the precision that has been obtained in recent years. These values

Table 5.2

Precision Mass Measurements

H^1	1.007 825 19 ± 8*
C^{12}	12.000 000 000
C^{13}	13.003 354 4 ± 9
N^{14}	14.003 074 39 ± 17
O^{16}	15.994 915 02 ± 28
S^{32}	31.972 073 7 ± 9
Fe^{56}	55.934 936 3 ± 43
Ag^{109}	108.904 756 ± 5
U^{235}	235.043 915 ± 22

* Refers to last significant figure.

are "best" values of several investigators, and their determination includes a comparison with nuclear reaction data [3].

The importance of such measurements in nuclear physics is reflected in the active mass spectrometer programs throughout the world—the University of Minnesota, McMaster University, and the Max Planck Institute for Chemistry. With instruments of larger radius of curvature and other refinements, it may even be possible ultimately to measure mass differences that are comparable to the binding energy of the orbital electrons. If such a resolution is ever achieved, mass spectrometry may then contribute to completely new theories and observable phenomena.

NEUTRON CROSS SECTIONS

The first mass spectrometric observation of a change in isotopic composition caused by neutron irradiation was reported by Dempster [4] in 1947. He found that the abundance of Cd^{113}, normally 12.26%, had decreased to 1.6% whereas the abundance of Cd^{114}, normally 28.86%, had increased to 39.5%. Dempster's experiment furnished conclusive proof that the isotope responsible for the high absorption of thermal neutrons in cadmium was Cd^{113}, and subsequent work by his group revealed that other isotopes having exceedingly large absorption cross sections included Sm^{149}, Gd^{155}, and Gd^{157}.

There are actually several experimental methods for neutron cross-section measurements, but only the mass spectrometric technique will be discussed here. It is especially useful in cases when neutron capture leads to the formation of a new nuclide that is stable or which has a long half-life.

Unless isotopically enriched samples are available, all of the isotopes of a material will be undergoing nuclear transmutation. However, a three-nuclide transmutation expression will usually suffice in any specific experiment. Thus consider an element with three isotopes designated as $X^{(A-1)}$, X^A, and $X^{(A+1)}$, where $(A-1)$, A, and $(A+1)$ are the mass numbers. Let σ_a, σ_b, and σ_c represent the respective effective neutron-absorption cross sections of these nuclides. Under neutron irradiation, a nuclear transmutation will take place,

$$X^{(A-1)} \xrightarrow{\sigma_a} X^A \xrightarrow{\sigma_b} X^{(A+1)} \xrightarrow{\sigma_c} X^{(A+2)}, \qquad (5.5)$$

where $X^{(A+2)}$ is the daughter product of $X^{(A+1)}$.

If we let N_a, N_b, and N_c represent the respective number of atoms of isotopes $(A-1)$, A, and $(A+1)$ that exist at any time during the neutron irradiation, then the nuclear transmutations of these isotopes can be represented by the set of differential equations

$$\frac{dN_a}{dt} = -N_a\sigma_a\phi, \tag{5.6}$$

$$\frac{dN_b}{dt} = N_a\sigma_a\phi - N_b\sigma_b\phi, \tag{5.7}$$

$$\frac{dN_c}{dt} = N_b\sigma_b\phi - N_c\sigma_c\phi, \tag{5.8}$$

where ϕ is the neutron flux.

The initial conditions, at $t = 0$, are

$$N_a = N_{ao}; \qquad N_b = N_{bo}, \quad \text{and} \quad N_c = N_{co}. \tag{5.9}$$

The solutions to (5.6), (5.7), and (5.8) are then

$$N_a = N_{ao}e^{-\sigma_a\phi t}, \tag{5.10}$$

$$N_b = \frac{\sigma_a}{\sigma_b - \sigma_a} N_{ao}(e^{-\sigma_a\phi t} - e^{-\sigma_b\phi t}) + N_{bo}e^{-\sigma_b\phi t}, \tag{5.11}$$

$$N_c = N_{ao}\sigma_a\sigma_b\left[\frac{e^{-\sigma_a\phi t}}{(\sigma_b - \sigma_a)(\sigma_c - \sigma_a)} + \frac{e^{-\sigma_b\phi t}}{(\sigma_a - \sigma_b)(\sigma_c - \sigma_b)} + \frac{e^{-\sigma_c\phi t}}{(\sigma_b - \sigma_c)(\sigma_a - \sigma_c)}\right] + N_{bo}\frac{\sigma_b}{(\sigma_c - \sigma_b)}[e^{-\sigma_b\phi t} - e^{-\sigma_c\phi t}] + N_{co}e^{-\sigma_c\phi t}. \tag{5.12}$$

The above are the general nuclear transmutation equations for three isotopes. In mass spectrometry, however, the usual measurable quantity is an isotopic ratio, rather than any absolute determination of the number of atoms of a single nuclide. It is thus convenient to define

$$R \equiv \frac{N_c}{N_b}; \qquad R_o \equiv \frac{N_{co}}{N_{bo}}; \qquad R_o' \equiv \frac{N_{ao}}{N_{bo}}. \tag{5.13}$$

It will be noted that R_o and R_o' are the known or experimentally measured isotopic ratios prior to neutron irradiation. If (5.11) is divided by (5.12), and the notation of (5.13) is employed, we obtain the general relation for R, in terms of pre-irradiation isotopic ratios, effective neutron cross section, and neutron "fluence" or time-integrated neutron flux, ϕt, where t is the total irradiation time. If the time-integrated neutron flux and any two effective neutron cross sections are known, the third cross section can be determined.

There is one case that permits the use of a simple explicit expression for a neutron cross section. It is for the condition that

$$\sigma_a \ll \sigma_b, \qquad \sigma_c \ll \sigma_b, \quad \text{and} \quad R_o' \simeq 0, \tag{5.14}$$

which, in turn, leads to the simple relationship that is sometimes referred to as the "burn-up equation":

$$\sigma_b = \frac{1}{\phi t} \ln \left(\frac{1 + R}{1 + R_o} \right). \tag{5.15}$$

This expression has been used in the mass spectrometric measurement of neutron capture cross sections of Sm^{149} and Sm^{150} by Aitken and Cornish, [5].

For the case in which only a single isotope is present in the irradiated sample, and $\sigma_b \gg \sigma_c$, it will be clear that (5.15) reduces to

$$\sigma_b = \frac{1}{\phi t} \ln (1 + R). \tag{5.16}$$

Two recent measurements have been reported in which use of this very simple expression was justified because the isotope for cross-section measurement was enriched so that other isotopes were present to only 1 part per million. Forman and White [6] achieved this type of isotopic enrichment in a single separation using a four-stage mass spectrometer to provide an ultrapure sample of Sm^{147}. A similar experiment was reported by Su and White [7] relative to the thermal neutron cross section of Er^{167}. In this latter experiment Er^{167} atoms were focused on a small rhenium ribbon, similar to that used as a thermal ionization source. The ribbon was placed in the position normally occupied by the electron multiplier detector. Although only about 10^{-8} gm of sample was collected by this type of "ion implantation," the amount was sufficient to permit an accurate measure of the post-irradiation Er^{168}/Er^{167} abundance ratio. Figure 5.4 shows the initial Oak Ridge isotopic composition, the effect of mass spectrometric enrichment, and the post-irradiation abundance ratio. The thermal neutron cross section of Er^{167} was then calculated to be 699 ± 20 barns.

In connection with this particular mass spectrometric technique, it might be noted that (a) high isotopic purity affords greater sensitivity and precision, (b) "self shielding" corrections are negligible when small samples are used, and (c) no sample chemistry is required that might introduce impurities.

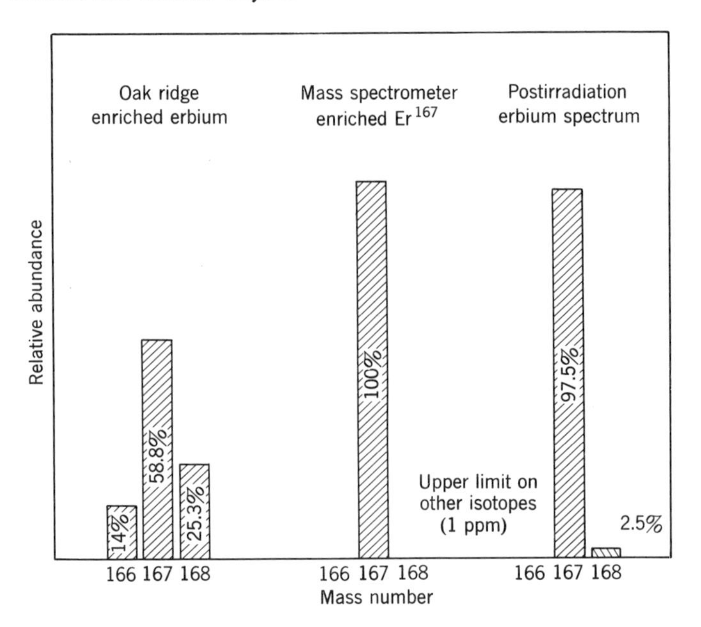

Fig. 5.4 Isotopic changes in erbium produced by neutron irradiation [7].

NUCLEAR REACTIONS

Very pure isotopes that are separated by a high-intensity spectrometer or by an electromagnetic separator are also important in general studies of nuclear reactions. It is, of course, possible to use materials of natural isotopic abundances, bombarded by neutrons, protons, or deuterons, and if a sufficient number of new nuclei are formed, reaction products can be determined by chemical methods. It is sometimes difficult, however, to make an unambiguous determination of the mass number. If a known isotope is used as a target, or if the reaction products are examined mass spectrometrically, the nuclear transmutation and the daughter products can be clearly identified.

A few of the large number of nuclear reactions which have been investigated using mass spectrometric or electromagnetic separators are listed in Table 5.3 [8].

Electromagnetic separators also permit the preparation of "target" samples in convenient form. Either metallic or gaseous species can be implanted by ion bombardment into thin metal or plastic foils. These foils then contain a highly concentrated layer of the separated nuclides

Table 5.3

Isotopes Produced by Nuclear Transmutations

Target nucleus	*Reaction*	*End product*
B^{11}	p, α	Be^8
C^{13}	d, p	C^{14}
N^{15}	p, α	C^{12}
O^{17}	d, p	O^{18}
O^{17}	d, α	N^{15}
Ne^{22}	p, γ	Na^{23}
Mg^{24}	p, γ	Al^{25}
Si^{30}	p, γ	P^{31}
Cu^{65}	n, p	Ni^{65}
Se^{74}	α, n	Kr^{77}
Sn^{120}	$d, 2n$	Sb^{120}

near the foil surface. In this fashion, isotopically pure specimens (10 to 20 μgm/cm^2) can be prepared for studies in alpha, beta, and gamma spectroscopy. Such thin targets reduce self absorption or scattering effects, in addition to making possible an unequivocal assignment of mass number. A chemical purity for the samples can also be achieved that far exceeds the purity of the highest grade of chemical reagents.

FISSION YIELDS

When neutron induced fission takes place in Th^{232}, U^{233}, U^{235}, U^{238}, Pu^{239} or other transuranic nuclides, a very large number of new nuclides are formed. Fission can also occur spontaneously, and the end products that result from a number of beta decays are distributed over many elements and isotopes. Figure 5.5 shows the general shape of the fission yield curve for U^{235} in which maxima occur at $M = 100$ and $M = 134$. These maxima correspond to the special nuclear properties of the 50 and 82 neutron configurations of fission products. Table 5.4 lists the abundance or yield of some of the more important nuclides [9] that result from the thermal neutron induced fission of U^{235}.

The pioneering studies of Thode and others revealed the importance of mass spectrometry, as well as radiochemical assay, in providing many of these data. For example, Thode and Graham [10] discovered long-lived Kr^{85} as a fission product and periodically measured its abundance relative to the stable isotopes of krypton to ascertain its half-life ($\sim$10

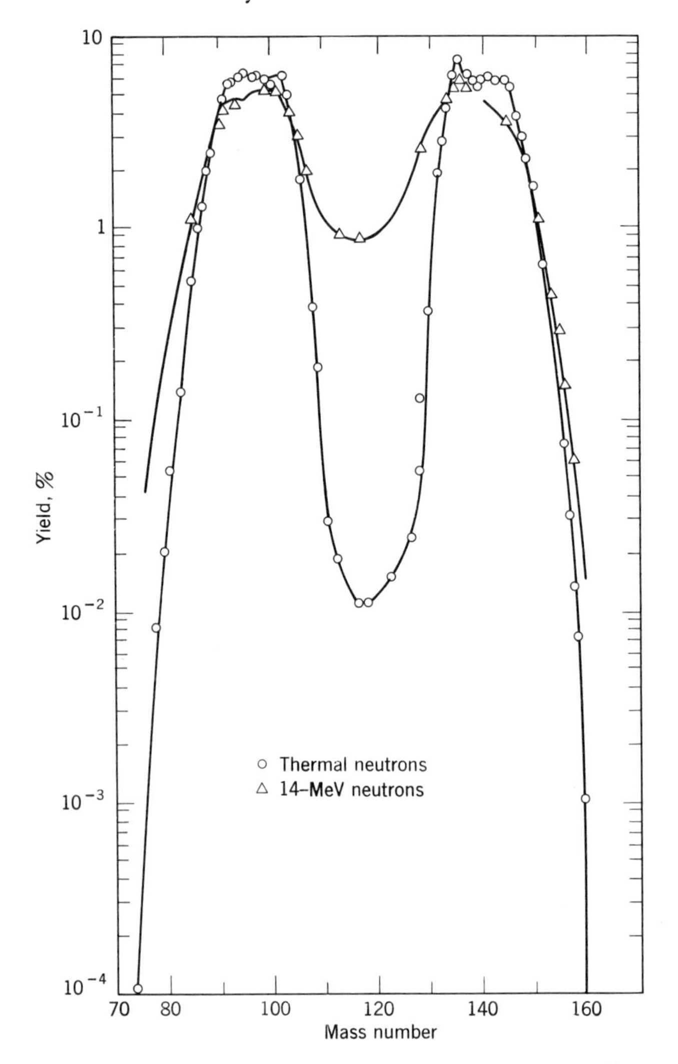

Fig. 5.5 Fission yield versus mass number [9, adapted].

yr). Figures 5.6*a* and *b*, respectively, show the isotopic composition of normal krypton and fission-product krypton.

Isotopic abundance resulting from neutron-induced fission has been studied mass spectrometrically for Kr, Rb, Sr, Zr, and Mo in the light element group. In recent years many experiments have also been designed to ascertain yields from fission products produced by high energy ions. McHugh [11] employed high-sensitivity mass spectrometric techniques

Table 5.4

Total Chain Yield from Thermal Neutron Fissions in U^{235}

Fission product	*Yield (%)*
Zr^{93}	6.45
Zr^{94}	6.40
Mo^{95}	6.27
Zr^{96}	6.33
Mo^{100}	6.30
Cs^{133}	6.59
Xe^{134}	8.06
Cs^{135}	6.41
Xe^{136}	6.46
Cs^{137}	6.15
Ba^{139}	6.55
Ce^{140}	6.44

to study fission-product distributions from the compound nucleus U^{236} ($Th^{232} + He^4$), with helium ions at 44 MeV. He found significant differences from the cumulative Xe yields of U^{235}.

At least two reasons can be cited for the continuing interest in fission yields. One relates to obtaining detailed information that can contribute

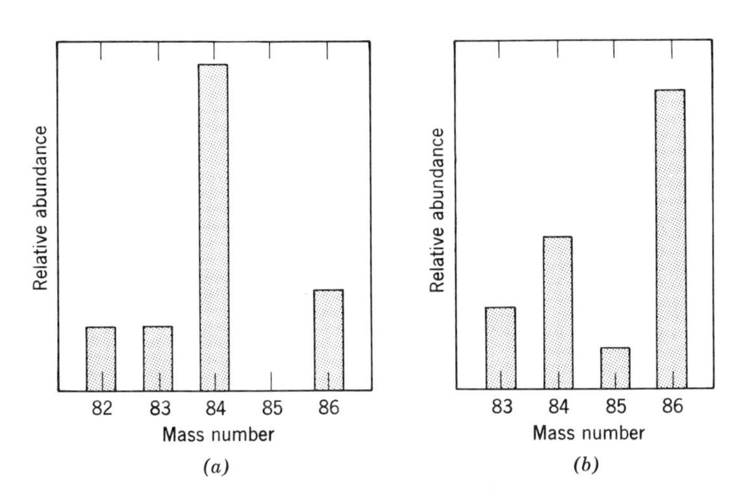

Fig. 5.6 Mass spectra of (*a*) naturally occurring krypton and (*b*) fission-product krypton as reported by Thode and Graham [10].

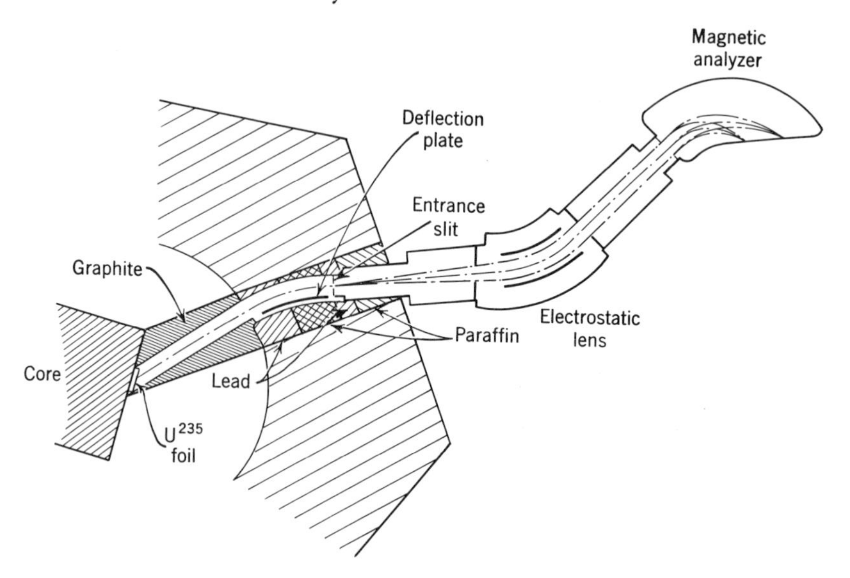

Fig. 5.7 Schematic diagram of Ewald's apparatus for analyzing fission fragments [12, adapted].

to basic nuclear theory. A second is the need for acquiring accurate normalization data for a large number of reactor engineering and physics experiments. Actually, two types of fission yields are measured: (a) independent and (b) cumulative. An independent yield relates the primary or prompt-fission products; these are nuclides that are not formed through beta decay. Yields of this type furnish important data relative to the division of nuclear charge between the primary fragments. Cumulative yields, in contrast, result from the beta decay of lower atomic number members of the fission product chain. The mass distributions can be obtained from these cumulative yield studies. Clearly, both charge and mass data are needed.

In an effort to obtain direct information regarding nuclear charge, attempts have been made to analyze the fission fragments per se. Such particles, having kinetic energies from 50 to 110 MeV and charge states from 18 to 25, require analyzing components of very large dimensions. In addition, the difficulties of shielding the apparatus from the radiations associated with high neutron fluxes are formidable. Ewald et al. [12] however, have designed a double-focusing system for analyzing primary fission fragments. A schematic diagram is shown in Figure 5.7. Because the energetic fission fragments can be analyzed within a few microseconds of their formation, a better interpretation can be made of independent yields. The collection efficiency in such an apparatus is very low, but considerable ambiguity is removed with respect to the mass and charge state of the particles prior to their decay to longer-lived species.

BRANCHING RATIOS AND DECAY SCHEMES

If a radioactive nucleus decays by simple β^- emission, a new nucleus with the same mass number will result, but the atomic number will be changed (i.e., $Z \rightarrow Z + 1$). In a number of instances, however, there will be more than a single decay mode; for example, there may be competing modes such as β^+ emission or K-capture. A radioactive sample might then proceed either to $(Z + 1)$ or $(Z - 1)$; the probability of decay for the original nucleus to these two species is termed the branching ratio ϵ, which is given as

$$\epsilon = \frac{N(\beta^-)}{N(\beta^+ + K\text{-capture})} = \frac{N(Z + 1)}{N(Z - 1)} \tag{5.17}$$

Reynolds [13] was one of the first investigators to utilize the mass spectrometer and isotopic dilution methods to determine the branching ratio, ϵ, for Cu^{64} (see Figure 5.8).

Gram quantities of copper that had undergone neutron activation were allowed to cool for a time appreciably greater than the 12.8-hr Cu^{64} half-life. The copper was then dissolved and to this solution were added the separated isotopes of Ni^{58} and Zn^{68}. The nickel and zinc were subsequently extracted from the solution and mass analyzed to yield Ni^{64}/Ni^{58}

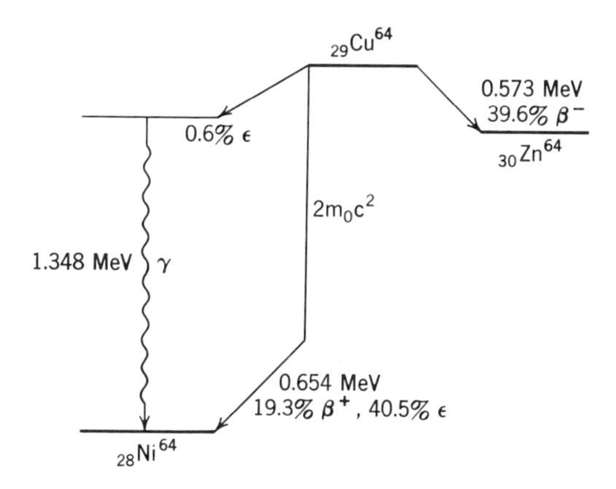

Fig. 5.8 Multiple decay scheme for Cu^{64}.

and Zn^{64}/Zn^{68} isotopic ratios. These measurements led to a Ni^{64}/Zn^{64} branching ratio of 1.62.

Counting techniques are also needed to provide detailed data if competing reactions lead to the same isotope. But the mass spectrometer is an important adjunct tool that provides highly precise information of branching ratios when decay products permit use of the isotopic dilution technique.

HALF-LIFE MEASUREMENTS

Half-lives are generally measured by observing the exponential decay of a sample, using radiotracer methods. The mass spectrometer can also be used for half-life determinations by either (a) observing the decay of the parent nucleus or (b) noting the growth of a daughter product. The latter is usually a more sensitive method for relatively long-lived species.

The work of Rider, Peterson, and Ruiz [14] will be cited as an example. In their study they measured the half-life of Cs^{137} by observing the rate of growth of stable Ba^{137} from a known quantity of Cs^{137} (using an isotopic dilution method), and obtained a value of 29.2 ± 0.3 yr. Their work involved sample preparation, the preparation of standard solutions, chemical separations, and a large number of isotopic ratio determinations by mass spectrometry. To a batch of fission product Cs^{137} an accurately known amount of Cs^{133} was added in order that the original number of Cs^{137} atoms in a master solution could be determined. An enriched Ba^{138} solution was also prepared so that the increase in the decay product, Ba^{137}, could be compared with a stable isotope. This enriched Ba^{137} solution was then standardized by the mass analysis of four separate solutions obtained by blending weighed aliquots of enriched and natural barium solutions. By mass spectrometry, the Ba^{138} concentration was determined to a standard deviation of less than 0.5%.

Barium samples (about 5×10^{-8} gm of barium) of this master solution were then taken as a function of time. Fifty-four samples were taken over a period of 100 days and the growth of the Ba^{137} daughter was measured by mass spectrometry. The data, obtained with two separate spectrometers, are shown in Figure 5.9. The slope of the line and a knowledge of atom concentrations in both the spike and master solution then permitted calculation of the decay constant and half-life.

A recent interesting development has been reported by Klapisch and Bernas [15], in which isotopes having exceedingly short half-lives (<1 sec) can be detected. The technique involves utilization of an "on-line" mass spectrometer in which (a) new nuclides are produced by energetic

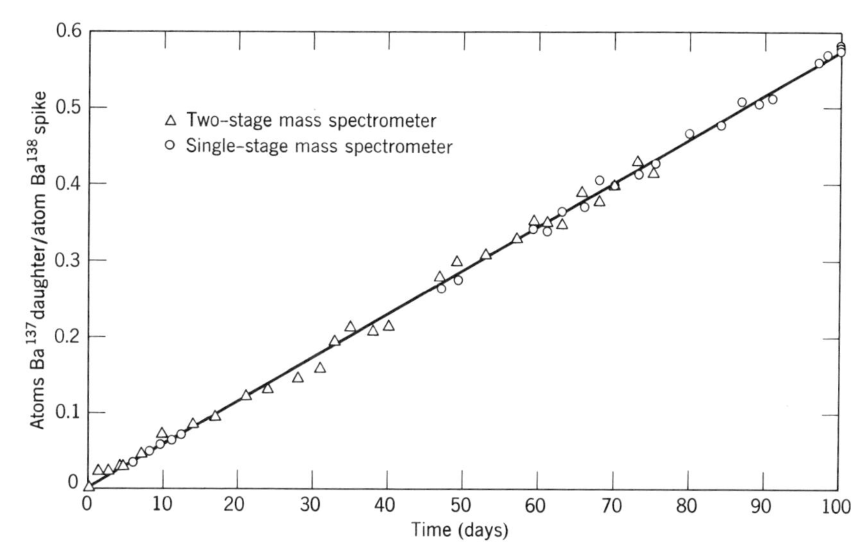

Fig. 5.9 Growth of stable Ba^{137} from Cs^{137} decay [14, adapted].

charged particles, (b) the reaction products are almost immediately ionized, and (c) the short-lived daughter products are accelerated, mass analyzed, and their decay noted after they impinge on a mass spectrometer multiplier.

The experimental approach of these investigators is indicated in Figure 5.10. A high-energy beam of charged particles (150 MeV protons from the Orsay synchrocyclotron) is directed on a specialized "target" that consists of a series of thin carbon foils. Lithium isotopes are produced by the reactions:

$$p + C^{12} \longrightarrow Li^6, Li^7, Li^8, Li^9. \tag{5.18}$$

The target is maintained at approximately 1600°C, so that the lithium atoms diffuse to the surface and a reasonable fraction leave this target as ions. Providing the diffusion time is short, the lithium atoms can be collected on the mass spectrometer multiplier.

Such a measurement was actually performed and the decay of Li^8 to Be^8 on the first dynode of the multiplier was noted by the 2-α particle decay of Be^8. Half-lives of Li^8 and Li^9 of 0.8 and 0.17 sec were reported. The general technique appears to be one of the most promising for half-life measurements which are in the 100-msec range and in which the daughter products can be ionized almost instantaneously after their formation.

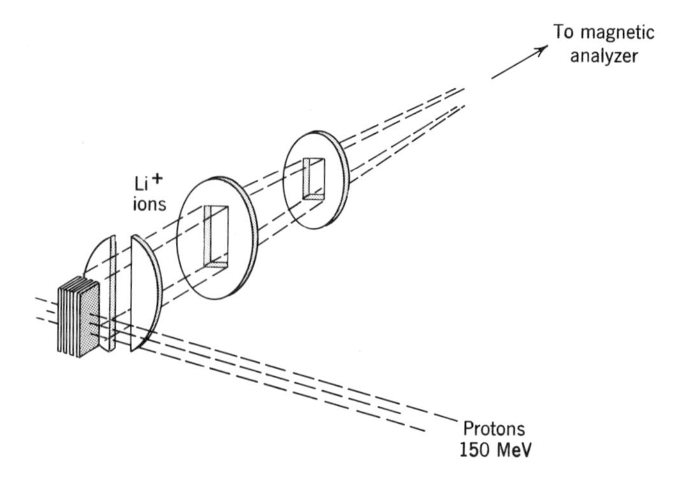

Fig. 5.10 Method of short-half-life measurements of Klapisch and Bernas [15, adapted].

Other recent "on-line" spectrometric measurements reported by the Orsay group include half-life determinations of short-lived isotopes of Rb, Sr, Cs, Ba, and La [16]. The short-lived isotopes of these elements are produced by proton induced fission.

ABSOLUTE MEASUREMENT OF ALPHA-PARTICLE ENERGIES

Magnetic spectrographs are commonly used to determine relative energies of alpha particles. By careful mapping of magnetic fields, they have also been employed to yield absolute energies. There are many difficulties associated with these latter measurements. However, it is possible to use a mass spectrometric method that eliminates the necessity of knowing either the absolute or relative value of the magnetic field along the alpha-particle trajectory. The method essentially compares the alpha-particle energy to the kinetic energy of an ion of high atomic mass traversing an identical trajectory, utilizing a common magnetic field for both the ion and the alpha particle. A schematic diagram of the experiment of White et al. [17] is shown in Figure 5.11.

Consider the case in which a singly charged ion can be accelerated through a precisely determined potential, V, which will cause the ion to assume a trajectory that is coincident with alpha particles that originate from the same source slit. If B is the magnetic field intensity, r is the radius of curvature in the analyzer, e is electronic charge, v is velocity, m is mass, and subscripts 1 and 2 refer to the singly charged ion and alpha particle, respectively, it can be shown that

$$Br = \frac{m_1 v_1}{e_1} = \frac{m_2 v_2}{e_2}. \tag{5.19}$$

Applying the usual relativistic correction to the alpha particle—rest mass m_o, energy E—it can also be shown that

$$\frac{m_1 v_1 e_2}{m_2 e_1} = c\left[1 - \frac{m_o^2 c^4}{(E + m_o c^2)} 2\right]^{1/2}. \tag{5.20}$$

Now, c is the velocity of light, $e_2 = 2e$, $m = m_o + E/c^2$, and $v_1 = (2e_1 V/m_1)^{1/2}$. An expression can then be obtained for the kinetic energy of the alpha particle of the form

$$E = \frac{4m_1 e_1 V}{m_o}\left(1 - \frac{E}{2m_o c^2} + \frac{E^2}{4m_o^2 c^4} - \cdots\right). \tag{5.21}$$

Thus, without absolute measurements of magnetic field intensity or radius of curvature, the kinetic energy of an alpha particle can be determined in terms of the mass and kinetic energy of an ion.

In the case of the measurement of a Po^{210} alpha particle, a Lu^{175} ion (mass 174.995) was used for comparison. The accelerating potential was measured by means of a precision voltage divider, with a standard cell being the ultimate reference voltage. A surface ionization source and a thin Po^{210} sample were alternately placed at the usual source slit, and a twin-channel alpha-particle detector operating in a coincidence

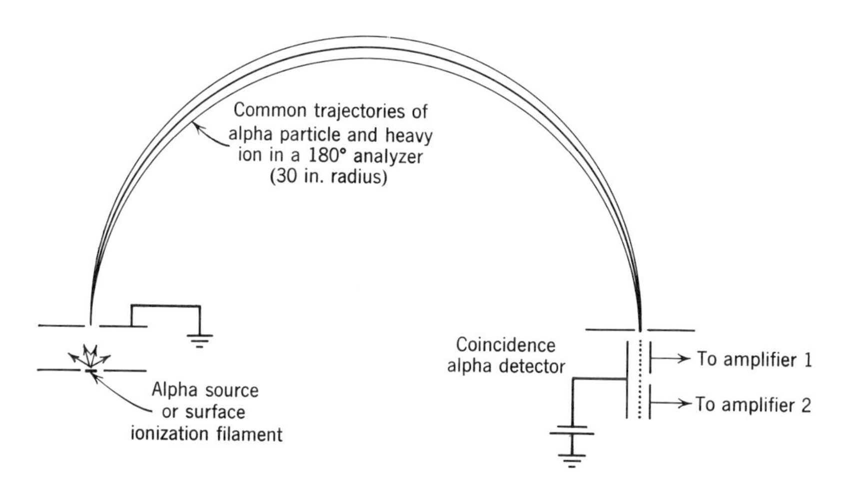

Fig. 5.11 Schematic diagram for absolute alpha-particle measurements.

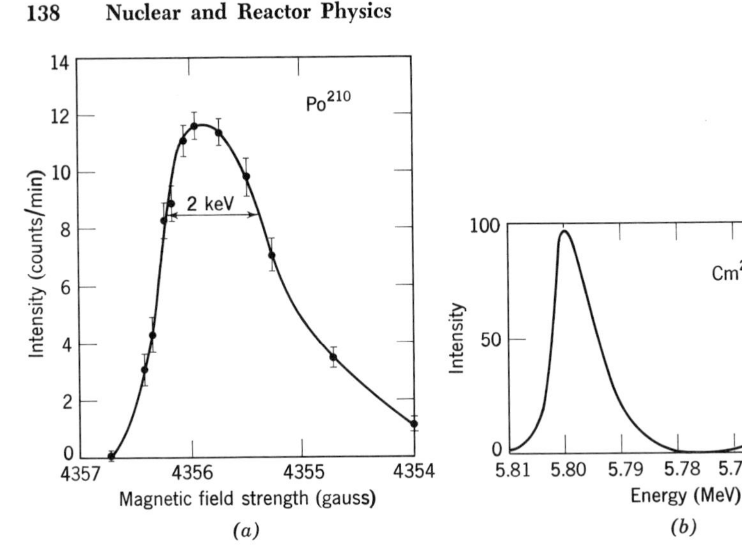

Fig. 5.12 (*a*) Resolution of Po^{210} alpha spectrum in magnetic analyzer; (*b*) alpha spectrum of Cm^{244} [17].

circuit (to minimize background counts) registered alpha particles at the detector slit. A value of 5.304 ± 0.001 MeV absolute volts was computed.

Figure 5.12*a* indicates the very high energy resolution of the mass spectrograph in this type of investigation. Figure 5.12*b* shows an alpha spectrum of Cm^{244}, with the same analyzer—but in this case the two alpha-particle energy groups were registered on nuclear emulsions.

REACTOR ENGINEERING

Neutron Flux and Temperature

The problem of measuring a neutron flux is a basic one in nuclear engineering, and the reader is referred to a recent review paper by Dessauer [18] for a clear statement relating to neutron flux measurements generally. A common analytical expression for neutron flux is simply the product of neutron density and neutron velocity

$$\phi = nv. \tag{5.22}$$

A more meaningful definition, however, replaces the simple product nv by the integral

$$\phi = \int n(v)v\, dv = n\bar{v}. \tag{5.23}$$

In (5.23) $n(v)\,dv$ represents the number of neutrons per unit volume having speeds within an interval dv that have the value v, n is the total neutron density, and $\bar{v}$ is the average neutron velocity.

The measurement of neutron fluxes is often easier if the neutrons comprise an external beam or if they are produced by an accelerator. In this case it is possible to employ neutron diffracting crystals or some other technique that allows a reasonably discrete velocity selection. If a neutron flux is to be monitored within a reactor, however, the neutrons will have a large velocity distribution. Even in a thermal reactor neutron velocities will range from near 10^9 cm/sec (corresponding to prompt fission neutrons) to thermal velocities (of the order of 10^5 cm/sec, for 0.025 eV). In either case the neutron flux must be determined by a reaction rate of a neutron detector, or a target, that responds in some fashion to the neutrons. The reaction rate, r, will then be

$$r = \sigma\phi, \tag{5.24}$$

where it must be understood that the effective cross section of the target nuclei, σ, is averaged over the neutron flux spectrum. In practice this means that a reaction rate must be monitored in terms of a sensor or target material whose neutron cross section is some known function of neutron velocity. Indeed, all of the usual neutron "activation" techniques for flux measurement (exposure of thin foils, wires, etc., and the subsequent measurement of induced activity by beta and gamma counting) are based on the availability of cross-section data. If the cross section follows some simple relationship, as in the case of B^{10} that follows a $1/v$ dependence, the flux determination is simplified.

Flux measurements can also be made by observing isotopic changes in elements provided that the isotopic monitors have high cross sections and the neutron-produced isotopes are stable or sufficiently long-lived to be measured in a mass spectrometer. Westcott's calculations [19] showed that Sm^{149} and Gd^{157} satisfied these prerequisites. Forman [20] further showed that if a comparison were made of the relative burn-up of Sm^{149} and Gd^{157} in a simultaneous irradiation, a sensitive neutron velocity or temperature scale could be established. Figure 5.13 shows the departure of these two isotopes from a $1/v$ relationship that permits effective neutron temperature to be calculated. Figure 5.14 is an approximation to the sensitive temperature scale that results if these two high-cross-section nuclides are observed in a simultaneous irradiation. Figure 5.15 shows the very large isotopic changes induced in a Gd^{157} neutron flux monitor that had been exposed to an integrated neutron flux of $>3 \times 10^{18}$ neutrons/cm^2.

In monitoring neutron fluxes or temperatures, any advantage of using

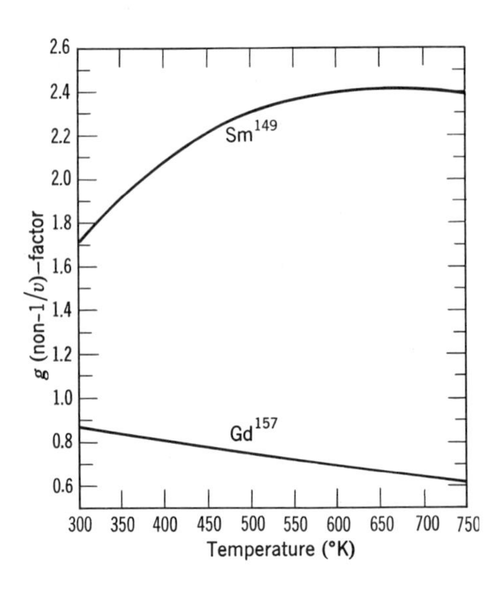

Fig. 5.13 Cross-section dependence of Sm^{149} and Gd^{157} as a function of temperature [20].

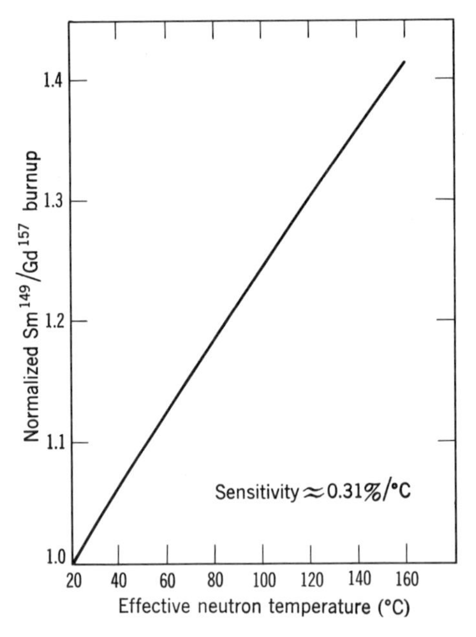

Fig. 5.14 Relative burnup of Sm^{149} and Gd^{157} versus neutron temperature in a thermal flux [6, adapted].

isotopic ratio changes over alternative techniques will depend upon the specific details of an experiment. Very small samples can be employed, however, that neither perturb the reactor flux nor require complex "incore" instrumentation; further, an integrated measurement can be obtained over an extended irradiation time. In so-called "zero-power" reactors (thermal fluxes of 10^6 to 10^8 n/cm²/sec) isotopic measurements will probably not be feasible unless the sensitivity of the mass spectrometer is greatly increased. In reactors of power generating plants, in which fluxes are in the 10^{13} n/cm²/sec range, isotopic measurements can be made easily. And in very high flux reactors (10^{15} to 10^{16} n/cm²/sec) isotopic ratio monitors may be expected to become increasingly attractive for measuring higher order reactions.

Uranium Burnup

The measurement of uranium burnup in a thermal reactor core is closely related to neutron flux determinations. In this case, however, the isotopic changes to be observed will be those of the uranium fuel. As a reactor core undergoes a continued high flux operation, the available fissionable atoms will be depleted. If the burnup is large ($>10\%$), it is convenient to assume that the capture of thermal neutrons by U^{238} is negligible compared with the number of U^{235} atoms which undergo fission, or which absorb a neutron to form U^{236}. This assumption is

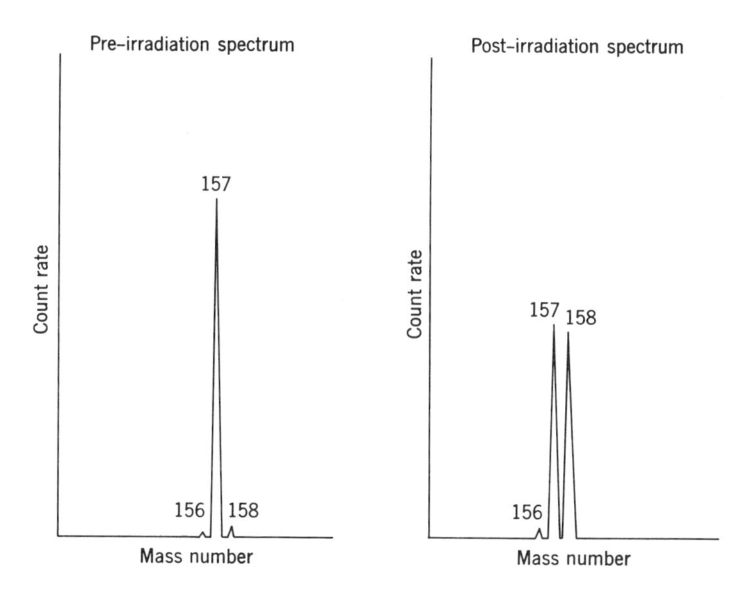

Fig. 5.15 Large isotopic change induced in a Gd^{157} neutron-flux monitor.

Table 5.5

Typical Isotopic Abundances in U^{235} Burnup [21]

Preirradiated sample		*Burnup sample*	
Isotope	*Atom (%)*	*Isotope*	*Atom (%)*
U^{234}	1.16	U^{234}	1.96
U^{235}	92.84	U^{235}	53.04
U^{236}	0.53	U^{236}	31.70
U^{238}	5.47	U^{238}	13.30
	100.00		100.00

valid as the thermal capture cross section of U^{238} is about 3 barns, but the U^{235} fission and capture cross sections are about 580 and 100 barns, respectively.

An illustration of a typical burnup calculation, using the mass spectrometric measurement of four uranium isotopes has been given by Dietz [21]. Consider the measured pre-irradiated and burnup sample described in Table 5.5.

Because the U^{238} is assumed to remain constant, each of the isotopes in the burnup sample may be multiplied by $(5.47/13.30) = 0.4113$. The normalized percent abundances will then become as shown in Table 5.6.

Thus 41.13 is the total number of uranium atoms left per 100 uranium atoms prior to the irradiation period. The number of U^{235} atoms that fissioned (% fission burnup) is $(100 - 41.13) = 58.87$; the number of

Table 5.6

Normalized Burnup Sample [21]

Isotope	*Atom (%)*
U^{234}	0.81
U^{235}	21.81
U^{236}	13.04
U^{238}	5.47
	41.13

U^{235} atoms that capture neutrons to form U^{236} (% capture burnup) is (13.04 — 0.53) = 12.51 atoms, per 100 original uranium atoms.

This type of mass spectrometric assay provides the most precise measurement of high burnup in fuel; it will also be noted that the method takes into account the significant amount of neutron capture by U^{234} to form new U^{235} atoms.

Capture-to-Fission Ratio Measurements

A specific parameter in reactor physics merits special mention. It is defined as the capture-to-fission cross-section ratio for a fissile material, such as U^{235}, Pu^{239}, or Th^{232}, and it is given the symbol alpha (α). Thus

$$\alpha = \frac{\sigma_c}{\sigma_f}, \tag{5.25}$$

where σ_c is the capture cross section and σ_f is the fission cross section. A high α value denotes that a smaller fraction of fuel atoms undergoes fission to produce useful power. Another useful form of the above equation is

$$\frac{\sigma_c}{\sigma_c + \sigma_f} = \frac{\alpha}{1 + \alpha}. \tag{5.26}$$

In the case of U^{235} the left side of (5.26) indicates the ratio of the number of U^{236} atoms formed compared with the total depletion of U^{235} from both capture and fission. This is really a burnup kind of measurement, but it is often desirable to make this determination where a relatively small number of atoms have participated in the capture or fission process. In this case the U^{236}/U^{235} ratio is made with a mass spectrometer, but the number of fissioning nuclei will be detected by a radiochemical assay of the fission products. If the fuel originally had a negligible percentage of U^{236}, and the burnup is small, the relative number of captured atoms can be observed directly. A uranium spectrum, having a small atomic burnup is shown in Figure 5.16.

The spectrum was obtained in the three-stage spectrometer, described in Chapter 2. The U^{236} peak is about 0.002 that of the U^{235} ion beam, but it is possible to measure ratios in the parts-per-million range with multimagnet systems. In fact, the distinct advantage of multimagnet systems becomes evident for this type of nuclear transmutation measurement, in which the ratio between adjacent isotopes is very large.

In the event that large uranium samples (gram amounts) can be irradiated for very long irradiation times, it is possible to analyze by chemical means and by mass spectrometry the total mass and isotopic composition of the uranium. The complete inventory of uranium can then be

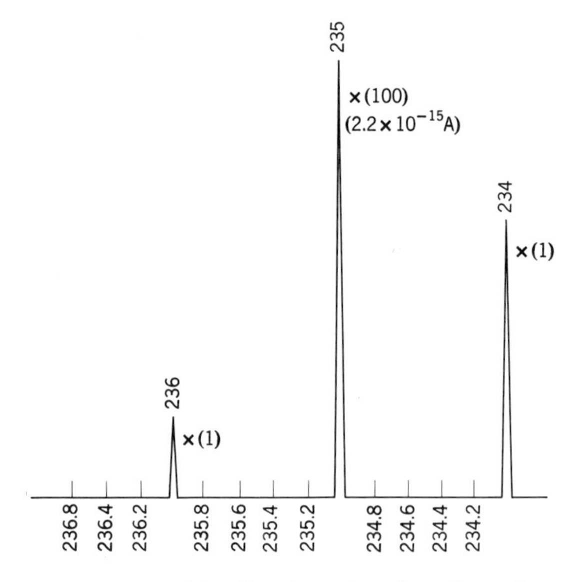

Fig. 5.16 Mass spectrum of irradiated uranium in a three-stage spectrometer.

controlled, although the shielding and processing procedures are far from routine. Okazaki et al. [22] have reported such a measurement in which a thermal value of 0.1718 ± 0.0006 was obtained for U^{235}. In this instance the number of Np^{237} atoms produced from U^{236} was also calculated, and an appropriate correction was applied.

Mention should also be made of recent measurements of alpha in reactor spectra above thermal energies (i.e., in the epithermal range). Using similar techniques, capture-to-fission ratio measurements have been reported for U^{233} and U^{235} [23], [24]. Recommended epithermal values for U^{233} and U^{235} are 0.175 ± 0.008 and 0.50 ± 0.02, respectively [25].

A final comment might be made with respect to this important reactor parameter. Quite clearly, a low value of alpha is a desirable situation for reactor cores that are designed as power generating plants. Only a small number of neutrons undergo a parasitic capture; most of the neutrons that are absorbed contribute to the fission process and yield additional neutrons. In the case in which the technical objective, however, is the large-scale production of transuranic elements (e.g., Cf^{252} and higher), a high alpha value is desirable. The buildup of successively higher atomic numbered elements will increase with increasing alpha, and a smaller number of atoms will undergo fission.

These species of higher atomic number are needed as auxiliary nuclear-power sources, for their high specific decay rates make them ideal as compact power sources of limited life. The measurement of alpha thus becomes important for all of the transuranic nuclides.

Temperature Dependent Cross-Section Measurements

A knowledge of the temperature dependence of resonance neutron capture and fission is important in determining the over-all temperature behavior of a reactor that utilizes highly enriched uranium fuel. Doppler broadening of the resonance in the neutron capture and fission cross sections of U^{235} will cause these cross sections to increase, as the operating fuel element temperature is increased.

An experiment [26] that specifically focused on the temperature dependence of epithermal fission and capture in $U^{235}O_2$ fuel rods used samples of uranium oxide irradiated at ambient temperature, and temperatures to approximately 1500°C. The sample was placed in a small cadmium-covered, neon-filled, graphite furnace that was water cooled so that the temperature of the furnace exterior remained at ambient reactor temperature. Dilute gold-aluminum alloy wires were placed symmetrically about the furnace exterior for neutron flux normalization [27]. The furnace unit was then irradiated in a neutron flux sufficient to cause a significant production of U^{236} and fission products (an epithermal flux of 1×10^{12} n/cm^2 sec). The general analytical procedures follow:

1. The relative number of neutron-induced fissions and captures in the $U^{235}O_2$ samples are determined separately by using mass spectrometric and radiochemical techniques.
2. The number of neutron capture events is determined by measuring by mass spectrometry the ratio of U^{236} to U^{235} atoms in both irradiated and unirradiated $U^{235}O_2$. The net production of U^{236} in samples irradiated at various temperatures, normalized by the neutron flux monitors, is a measure of the temperature dependence of $U^{235}O_2$ epithermal capture.
3. The number of neutron fission events is determined by either a mass spectrometric or radiochemical measurement of the Zr^{95} or another appropriate fission product. The normalized Zr^{95} concentration in samples irradiated at various temperatures is then a measure of the temperature dependence of $U^{235}O_2$ epithermal fission.

The epithermal value of α over a known spectrum, α_{epi}, is then given by

$$\alpha_{\text{epi}} = \frac{\int_{E_c}^{\infty} \sigma_c(E)\varphi(E)\, dE}{\int_{E_c}^{\infty} \sigma_f(E)\varphi(E)\, dE}, \tag{5.27}$$

where E_c is the cadmium cutoff energy, $\sigma_c(E)$ and $\sigma_f(E)$ are, respectively, the capture and fission cross sections, and $\varphi(E)$ is the neutron flux. The numerator is proportional to the rate of occurrence of epicadmium capture and the denominator is proportional to the rate of occurrence of epicadmium fission. Thus the ratio $\alpha_{epi}(T)/\alpha_{epi}(T_o)$, where T_o equals ambient temperature, is a measure of the temperature dependence of epithermal capture to fission.

Control Materials

All reactor materials must be analyzed for their purity and nuclear characteristics, whether these materials are structural, moderators, coolants, or control elements. Reactor core loadings must be analyzed for isotopic content, especially of fuels and control rods. The addition of so-called "burnable poisons" also involves mass spectrometry, because such elements can readily alter reactor reactivity and permit extended reactor life. The net effect of employing a "burnable poison" is to compensate for loss of reactivity caused by fuel exhaustion.

The rare earth elements [28] are usually mentioned as being potentially useful reactor "poisons." They have a multiplicity of isotopes, and most elements can be analyzed quite easily by a surface-ionization type of source. Samarium has a high thermal cross section and it is relatively inexpensive and abundant. Gadolinium and europium can also provide a considerable latitude in reactor control. Europium has two naturally occurring isotopes, Eu^{151} and Eu^{153}; the light isotope has a very large resonance at about 0.5 eV. Dysprosium has a continuous chain of high-cross-section isotopes and it has the advantage of short lived radioactive daughters.

Stevens [29] has plotted the europium isotope abundance as a function of total neutron absorptions per initial europium atom (see Figure 5.17). The changes in isotopic ratios and the "burnout" of Eu^{151} illustrate the importance of isotopic composition in reactor control materials and the general dependence upon mass spectrometry in reactor design studies.

Reactor Technology

The entire field of nuclear physics and reactor technology continues to rely heavily on isotopic analysis and the availability of separated isotopes. Hence I would predict an increasing dependence on precise isotopic measurements and separated materials that are made available by the Isotope Division of the Oak Ridge National Laboratory. A need exists also for small samples of very high isotopic purity. Isotopes can be used to monitor power distributions in reactor cores, and mass spectrometry provides the ideal diagnostic method in almost any system that produces a reasonable number of nuclear transmutations. Separated isotopes find use in nuclear emulsions and as targets for neutron-photon

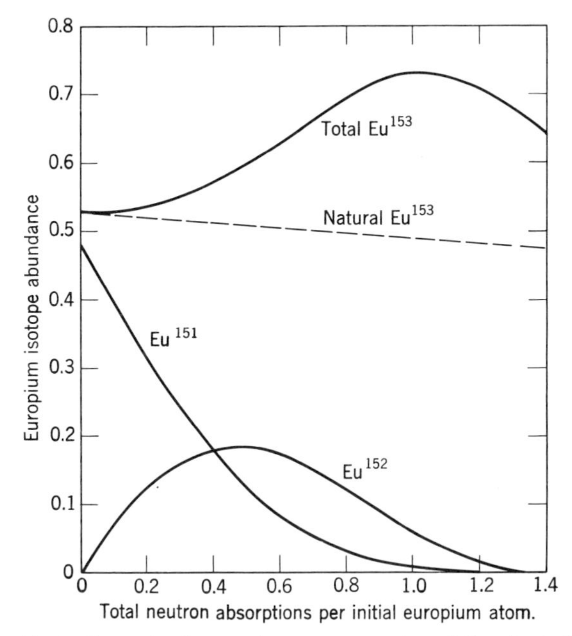

Fig. 5.17 Isotopic changes in a reactor poison [29, adapted].

and charged-particle reactions. The reactor engineer also relies on mass spectrometry to provide the input data, and to aid in problems relating to fuel loadings, fuel reprocessing, and safety. It is necessary, for example, to know accurately the total "boron inventory" of a reactor, and in particular to establish precisely the B^{10}/B^{11} ratio. This problem may seem almost trivial, but in fact it is one of the more difficult measurements associated with reactor technology. The inventory of uranium fuel is also important.

Optical emission spectrographs have often been employed for this general type of analysis. However, the isotopic standards against which such spectrographs are calibrated are established by precision mass spectrometry.

REFERENCES

[1] R. R. Ries, R. A. Dameron, and W. H. Johnson, Jr., *Phys. Rev.*, **132,** 1670 (1963).

[2] A. O. C. Nier, *Amer. Scientist,* **54,** 377, (1966).

[3] J. H. E. Mattauch, W. Thiele, and A. H. Wapstra, "1964 Atomic Mass Table." *Nucl. Phys.,* **67,** 1 (1965).

[4] A. J. Dempster, *Phys. Rev.* **73,** 829 (1947).

[5] K. L. Aitken and F. W. Cornish, *J. Inorg. Nucl. Chem.* **17,** 6, (1961).
[6] L. Forman and F. A. White, *Nucl. Sci. Eng.* **28,** 139 (1967).
[7] C. S. Su and F. A. White, *IEEE Trans. Nucl. Sci.,* 1967.
[8] H. Hintenberger, *Ann. Rev. Nucl. Sci.,* **10,** 435 (1960).
[9] "Reactor Physics Handbook ANL-5800," U.S. Government Printing Office, Washington D.C., 1963, p. 6.
[10] H. G. Thode and R. L. Graham, *Canad. J. Res.* **A 25,** 1 (1947).
[11] J. A. McHugh, Jr., "Mass Spectrometric Study of Fission from the U^{236} Compound Nucleus at Moderate Excitations," Thesis, University of California, 1963.
[12] H. Ewald, E. Konecny, H. Opower, and H. Rosler, *Z. Naturforsch.,* **19 a,** 194 (1964).
[13] J. H. Reynolds, *Phys. Rev.* **79,** 789 (1950).
[14] B. F. Rider, J. P. Peterson, and C. P. Ruiz, *Nucl. Sci. Eng.,* **15,** 284 (1963).
[15] R. Klapisch and R. Bernas, *Nucl. Instr. Methods,* **38,** 291 (1965).
[16] I. Amarel, R. Bernas, R. Foucher, J. Jastrzebski, A. Johnson, and J. Teillac, *Phys. Letters,* **24 B,** 402 (1967).
[17] F. A. White, F. M. Rourke, J. C. Sheffield, R. P. Schuman, and J. R. Huizenga, *Phys. Rev.,* **109,** 437 (1958).
[18] G. Dessauer, *Nucl. News,* **9,** 17 (1966).
[19] C. H. Westcott, "Effective Cross Section Values for Well-Moderated Reactor Spectra," AECL-1101, Atomic Energy of Canada, Ltd., Ontario, 1960.
[20] L. Forman, "The Mass Spectrometric Determination of the Thermal Neutron Cross Section of Sm^{147} Using Enriched Rare Earth Isotopes as a Neutron Temperature Monitor," Thesis, Rensselaer Polytechnic Institute, 1966.
[21] L. A. Dietz, KAPL Report M-MS-4, 1964.
[22] A. Okazaki, M. Lounsbury, R. W. Durham, and I. H. Crocker, "A Determination of the Ratio of Capture to Fission Cross Sections of U^{235}," AECL-1965, Atomic Energy of Canada, Ltd., Ontario, 1964.
[23] L. J. Esch and F. Feiner, *Trans. Am. Nucl. Soc.* **7,** 272 (1964).
[24] D. E. Conway and S. B. Gunst, *Nucl. Sci. Eng.,* **29,** 1 (1967).
[25] F. Feiner and L. J. Esch, in *Reactor Physics in the Resonance and Thermal Region* (vol. 2), MIT Press, 1966 p. 299.
[26] F. C. Schoenig, *Trans. Am. Nucl. Soc.,* **10,** 229 (1967).
[27] F. C. Schoenig, K. S. Quisenberry, D. P. Stricos, and H. Bernatowicz, *Nucl. Sci. Eng.,* **26,** 293 (1966).
[28] J. A. Ransohoff, *Nucleonics* **17,** 80 (1959).
[29] H. E. Stevens, *Nucl. Sci. Eng.,* **4,** 373 (1958).

Chapter 6

Electrophysics

On June 30, 1948, a press demonstration was conducted by scientists of the Bell Laboratories in which transistors were shown operating as speech amplifiers, oscillators, and television amplifiers. A radio receiver without a single tube was also displayed containing only semiconductor rectifiers and transistors. Within a decade, silicon transistors had been placed in the "Explorer," America's first earth satellite, and a transistor industry had mushroomed to a $100,000,000 per annum level.

Accompanying the growth of this industry was the development of a materials technology heretofore foreign to materials scientists. For the first time, impurity concentrations in the parts-per-billion range determined the operating characteristics of an electronic device. It was at this point that the mass spectrometer became a prime instrument for identifying these impurities on a quantitative basis. It has subsequently become important to the electronics and solid-state physics fields on a wider scale. There is now little doubt that mass spectrometry will be important in virtually all studies that can be circumscribed by the term *electrophysics*. The effect of impurities on thermionic emission and the role of activator atoms in phosphors; materials research in photoconductors, ferrites, thermoelectrics and superconductors; and studies of plasmas, glow discharges, and ionized gases that provide a working medium for magnetohydrodynamic power generators are all suggestive of the unusually broad range of investigations that can utilize the mass spectrometer as a diagnostic tool. The brief discussion of this section and Chapter 8 can only point up the general direction of current research and include selected specific examples.

SEMICONDUCTORS AND IMPURITY ANALYSIS

In semiconductor research, materials development, and product monitoring, mass spectrometry serves to complement x-ray diffraction, infrared spectroscopy, and electrical measurements. X-ray-diffraction techniques are generally applied to furnish information relating to crystal structures, lattice distortion, and defects. Under intensive neutron bombardment, for example, changes in diffraction patterns can be related to specific changes in lattice parameters and displaced ion centers. Infrared measurements make possible the quantitative determinations of energy levels that exist between the valence and conduction bands, whether these are caused by donor or acceptor atoms, or induced by radiation. Electrical measurements, of course, comprise the largest class of determinations—such as conductivity, Hall voltage measurements, and the measurement of carrier lifetime.

But the crucial role of mass spectrometry immediately becomes evident inasmuch as semiconductors are essentially *impurity-sensitive systems.* Unlike metals, the electrical properties of semiconducting specimens can vary over many orders of magnitude if small concentrations of foreign atoms or "defects" are introduced. As an example, the addition of one boron atom to 10^5 atoms of silicon increases the conductivity of this material by a factor of one thousand. On the other hand, the very high conductivity of copper ($\sim 10^{-6}$ Ω-cm) is substantially independent of trace impurities, as conduction results from the exceedingly large number of essentially free and mobile electrons that are available for charge transfer ($\sim 10^{23}/\text{cm}^3$). In good insulators most electrons are tightly bound and perhaps fewer than 10^{10} electrons per cubic centimeter can contribute to conduction under the influence of an electrical potential gradient. Semiconductors range between these extremes; the actual number of charge carriers varies widely, depending on crystal specimen, temperature, and impurity concentrations.

The two primary classes of semiconductors are (a) *intrinsic* and (b) *extrinsic.* The first class is composed of essentially "impurity-free" solids. Intrinsic semiconductor elements of importance include germanium, silicon, selenium and tellurium. Compounds generally classed as intrinsic include PbTe, ZnO, and Cu_2O. The degree of purification of an element will determine the conductivity or resistivity of a specimen at any given temperature. The intrinsic resistivity of silicon exceeds 10^5 Ω-cm, but even material of the highest purity rarely approaches this value. For intrinsic or impurity-free crystals, an elevated temperature or the exposure of the semiconductor to radiation provide a mechanism for conduction. Thermal excitation, for example, will provide some electrons with sufficient energy to free them from an interatomic bond or valence state. These will then be available for charge transport.

The conducting properties of extrinsic or impurity semiconductors are determined by the type and number of impurities in the crystal lattice. Impurities that provide extra electrons are termed "donors" or *n*-type.

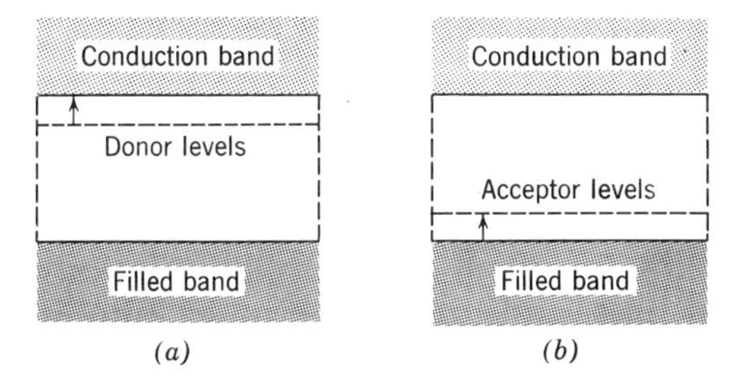

Fig. 6.1 (*a*) Donor levels introduced by impurities in *n*-type material; (*b*) acceptor levels introduced in a *p*-type semiconductor.

Electron-deficient impurities provide positive carriers and are termed "acceptors." In a germanium crystal, for example, donor impurities would include Group V elements such as arsenic, and acceptor impurities would include Group III elements such as indium. Figures 6.1*a* and 6.1*b* suggest one important feature of impurity semiconductors: the "energy levels" are often very close to either the conduction or the valence band. In many instances the energy gap is so small that room temperature is sufficient to "ionize" the impurity center and thus to (a) excite an electron into the conduction band or (b) have a Group III atom contribute a positive "hole" in the valence band.

The presence of certain impurities, of even a few parts per billion, is known to provide such levels, and these markedly affect the electrical properties of silicon, germanium, and other materials. Figure 6.2 is a typical plot indicating the large changes in conductivity that result as a function of impurity doping [1]. The graph also indicates the large increases in conductivity that accompany increases in the absolute temperature.

The sensitivity of the mass spectrometer for identifying trace elements in semiconductors has been reported by many laboratories and investigators. Table 6.1 is only a selected list [2] of detection limits for impurities of two important semiconductor materials, silicon and gallium arsenide.

Clearly such detection limits are beyond the analytical techniques of classical chemistry. But the mass spectrometer can also provide more specific information than average impurity concentrations in semiconducting materials. With the use of appropriate source techniques, it should be possible to monitor surface impurities, diffusion zones of *p-n* junctions, and detect inclusions or material inhomogeneities. I firmly believe that by use of the ion-beam microprobe and other methods, highly

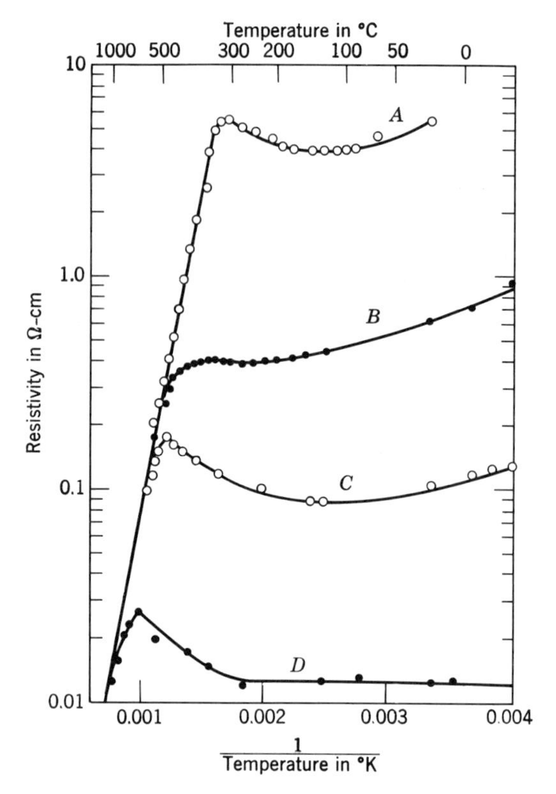

Fig. 6.2 Effect of impurity concentrations on the electrical conductivity of an extrinsic semiconductor. Curves *A, B, C,* and *D* show successively increasing impurities [2, adapted].

detailed data can be obtained with respect to junction profiles and impurity concentration gradients. Possibly the precipitation of III–V impurities, known to occur in silicon when impurity concentrations reach their solid solubility limit, may even be conveniently detected mass spectrometrically. When this type of information is compared with the usual electrical measurements that are made on a semiconductor specimen, it is then possible to remove considerable ambiguity from the analysis that electrical measurements alone would provide. Hall voltage measurements, for example, furnish information with respect to majority current carriers, but assumptions must usually be made with respect to the minority carrier concentrations.

Mass spectrometric techniques have already been employed to analyze the effect of impurities upon the luminescent properties of semiconducting

Table 6.1

Detection Limit of Impurities [2] (Parts per Million, Atomic)

Impurity element	*Silicon*	*Gallium arsenide*
Li	0.003	0.01
B	0.003	0.002
Al	0.01	0.003
P	0.01	0.003
S	0.05	0.05
Zn	0.005	0.002
Ag	0.006	0.006
In	0.03	0.005
Au	0.01	0.001
Pb	0.002	0.002
U	0.001	0.001

compounds [3]. Purity is also important in achieving high-efficiency phosphors. For example, in very pure ZnS, as little as 0.0000001% copper activator has been reported to be sufficient to give green emission [4]. In this case the activator impurity for light emission is analogous to the donor or acceptor impurities that are responsible for electron or hole charge transport in semiconductors. Yet trace quantities of iron or nickel (poisons), approximating 0.00001%, significantly decrease the phosphor efficiency.

THE PLASMA STATE

The plasma state is sometimes referred to as the "fourth state of matter" when compared to the more familiar solid, liquid, and gaseous forms. The appropriateness of such an identification, however, might well be questioned, as plasma scientists and cosmologists point out that a plasma is by far the most abundant state of matter. Greater than 95% of our universe is estimated to exist in some type of plasma. Our own sun and the countless massive stars completely dwarf the infinitesimal bits of meteoric matter and the cold planets of our solar system. The term "plasma" was adopted from the medical glossary by Nobel prize winner Irving Langmuir in the early 1920's. A true plasma is a neutral ensemble in which all individual particles are charged, with the assumption that the populations of negative and positive species are approximately equal. Today the term is used rather loosely in connection with any gases that are partially ionized.

If we accept the broader definition, it is immediately evident why plasma research is being pursued so vigorously. A more detailed understanding of the plasma state is needed to interpret naturally occurring phenomena, and to obtain insight in such diverse fields as supersonics, electrical power systems, communications, and nuclear fusion. Spectacular plasma displays in nature include atmospheric lighting, the aurora borealis, solar flares, and the countless number of brillant stars. Localized plasmas are also generated by objects moving at supersonic speeds. They result from the irreversible thermodynamic transformation of gases from shock waves, and from meterorites and man-made satellites entering the earth's atmosphere. Just how a plasma zone is born is quite complex. Presumably the gas molecules acquire energy in several modes, which results in energetic vibrational and rotational states. For sufficiently energetic shock waves, complete dissociation and some ionization will occur. The most dramatic evidence of this particular ionization phenomenon is the "blackout" of all communication signals from manned satellites upon re-entry to earth.

A much more general communication problem is associated with the highly ionized portion of our upper atmosphere. This ionosphere, a naturally occurring blanket of charged particles, extends from an altitude of approximately 60 to 300 km. Recent measurements indicating electron densities greater than 10^6/cc are now attributed to photoionization by solar radiations.

Engineering examples of plasmas are many—fluorescent lamps, arcs in high power switching circuits, and the thermionic generator. This latter device is a plasma-filled diode, designed to convert heat directly into electrical energy. The plasmas of nuclear fusion seem to be in a special category. Nuclear reactions can lead to massive amounts of energy release but the basic hurdle in thermonuclear work is plasma containment. For the nuclear reaction energy is significant only if a plasma of very high energy can be confined for a time sufficiently long so that a reasonable number of atoms can participate in the fusion process.

Mass spectrometric studies relating to plasmas may be focused on (a) the measurement of ion species and charge state, (b) estimates of ion and metastable atom densities, (c) the identification of trace elements, (d) the measurement of ion energies, and (e) electron-ion recombination and diffusion processes.

An investigation by Bohme and Goodings [5] illustrates a typical instance in which ionic species from a plasma are extracted and allowed to pass into a quadrupole mass spectrometer. Figure 6.3 is a schematic representation of their apparatus. A large metallic sampling chamber

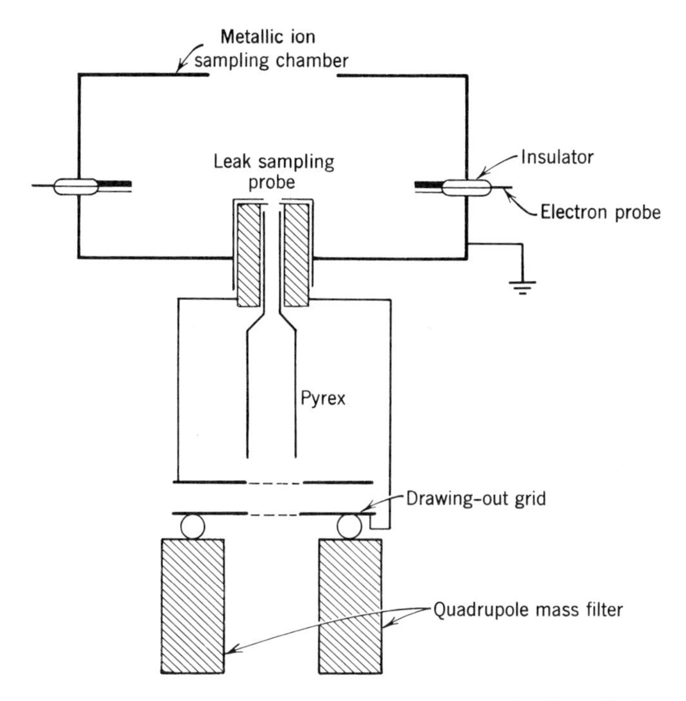

Fig. 6.3 Mass spectrometric sampling probe for investigating discharge plasmas [5, adapted].

surrounds the "leak probe," which serves as the reference potential. This potential may be varied with respect to the potential of the plasma for the preferential attraction of positive ions. The ionic species effuse through an orifice in the sampling probe and are drawn by a grid (at approximately 300 V) that extends across the quadrupole mass spectrometer entrance aperture. Additional double probes permit a simultaneous measurement of electron charge densities. With this technique these investigators were able to make estimates of the number densities of ions and electrons, and determine the percentage composition of N_1^+, N_2^+, and N_3^+ in a low-pressure plasma ($\sim$10 μ).

Another recent plasma analysis has been reported by Sauter, Gerber, and Oskam [6]. Their experimental apparatus included not only a quadrupole mass spectrometer for measuring ion density but also a light spectrophotometer and/or interference filters. They were thus able to obtain a correlation between the time dependence of light emission intensity, fractional light absorption, and positive ion density. Their detection system included integrating circuits for the photomultiplier light sensor and the ion electron multiplier, so that a continuous read-out could be made of light and ion intensities. In this fashion they were able to measure the time-dependent abundance of He^+ and He_2^+ in a helium after-glow plasma. Decay rates on a time scale of only a few milliseconds were measured.

Jefferies [7] has reported that a 2-inch radius 180° spectrometer for plasma physics experiments has been adapted to measure rapid changes in gas composition. One unique feature of his apparatus is a form of Wien filter that extracts the ion beam from the magnetic field boundary into a field-free region so that a high-gain electron multiplier can be used. The concept of this extraction filter is that the separation of the electrostatic plates (as a function of the distance from the magnetic boundary) increases with the decreasing magnetic field intensity (Figure 6.4). This crossed-electric-magnet field extraction filter thus provides a net force of zero on the ion, that is $-eE = Bev$. The spectrometer is capable of a full mass scan (electrostatic) from 2 to 100 amu in 100 msec, and a resolution of 50 is reported with 0.020 inch beam-defining slits. Thus as a diagnostic tool for plasmas, this type of instrument compares favorably with quadrupole mass filters.

The question may be asked whether the ion temperature (or energy) of an ion species can be measured by mass spectrometry. In a plasma the electrons will nearly always be the high velocity species. If, for example, the electrons and ions have equal energies, we have

$$\tfrac{1}{2}m_i \bar{u}_i^2 = \tfrac{1}{2}m_e \bar{u}_e^2 \tag{6.1}$$

or

$$\frac{\bar{u}_e}{\bar{u}_i} = \left(\frac{m_i}{m_e}\right)^{1/2} \tag{6.2}$$

Here m_i and $\bar{u}_i$ represent the mass and average energy of the ion and m_e and $\bar{u}_e$ denote the electron parameters. Even for the lightest hydrogen atoms the electrons will have relative speeds of $(1836/1)^{1/2}$ compared with an ion. We also recall that a 10,000°K ion has a kinetic energy of about 1 eV. Thus it might appear simpler to measure electron temperatures and, indeed, standard techniques are available for electron temperature measurements. But there usually exist appreciable differences in electron-ion temperatures even though a Maxwellian velocity distribution may exist for the heavy particles. These differences can be attributed to energy transfer in collisions and to heating or cooling processes that may influence only one specie of the plasma. Currently, the most reliable technique for ion temperature measurement is by an examination of the thermal Doppler broadening of optical spectral lines. This technique has limited accuracy (a) at high temperatures when resolution is poor at short wavelengths and (b) at high densities when pressure broadening

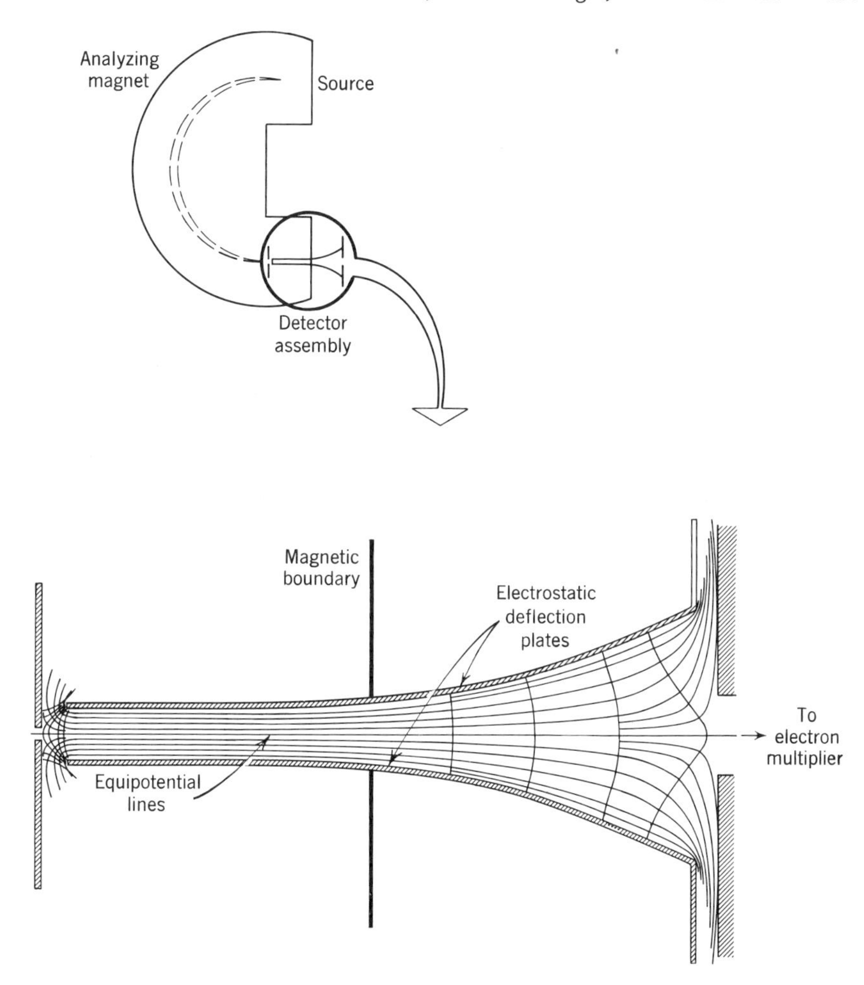

Fig. 6.4 Ion-extraction filter by Jefferies [7, adapted].

results. It is my conjecture, therefore, that high-resolution mass spectrometry, utilizing electrostatic lenses of very large radius of curvature, may be utilized for studies in ion kinetics. An extraction voltage of several hundred volts could be applied to a 1-eV ion, permit a magnetic mass analysis, and the particle could subsequently be passed through the electrostatic lens for a precise measure of ion velocity.

CORONA, GLOW DISCHARGES, AND SHOCKWAVES

Mass spectrometric diagnostic studies have been made of ionic species that are formed in corona discharges in air, at pressures up to one atmosphere. Fundamental interest in such work is focused both on understanding the mechanism of the discharge, per se, and in relating ion-molecule reactions that are presumed to take place in both upper and lower regions of the atmosphere. Shahin [8] has reported interesting results using a corona tube and the mass spectrometer arrangement shown in Figure 6.5.

A quadrupole mass spectrometer was operated at a frequency of 1.87 Mc/sec, with a variable ac voltage that allowed scanning the mass range up to 250. The ions emerging from the discharge tube were focused by various electrostatic lenses into the mass spectrometer, and were finally detected by an electron multiplier. In order to maintain the discharge tube at high pressure, a hole of 30-μ diameter separated the discharge and accelerating region, and high-speed vacuum pumps were employed at both the accelerating and analyzing regions. Studies of corona discharges were made in air, nitrogen, oxygen, and water-vapor mixtures. Results pointed up the importance of trace quantities of water vapor in nitrogen and/or oxygen when these gases are subjected to corona, and the predominance of $(H_2O)_nH^+$ ion clusters in the mass spectra revealed these groups as positive charge carriers. Values of n as high as 8 were reported at water concentrations of 0.65 mole% at atmospheric pressure.

Discharge phenomena have also been explored by Dawson and Tickner [9], using mass spectrometry. In their work the negative glow of a dc glow discharge was used to study H_2-D_2 exchange by an ionic chain reaction.

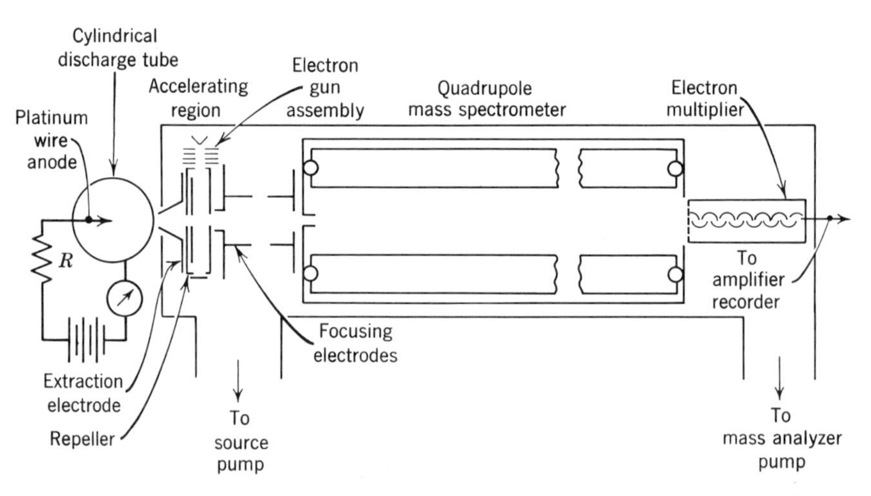

Fig. 6.5 Quadrupole mass spectrometer system for the analysis of ions from a discharge tube [8, adapted].

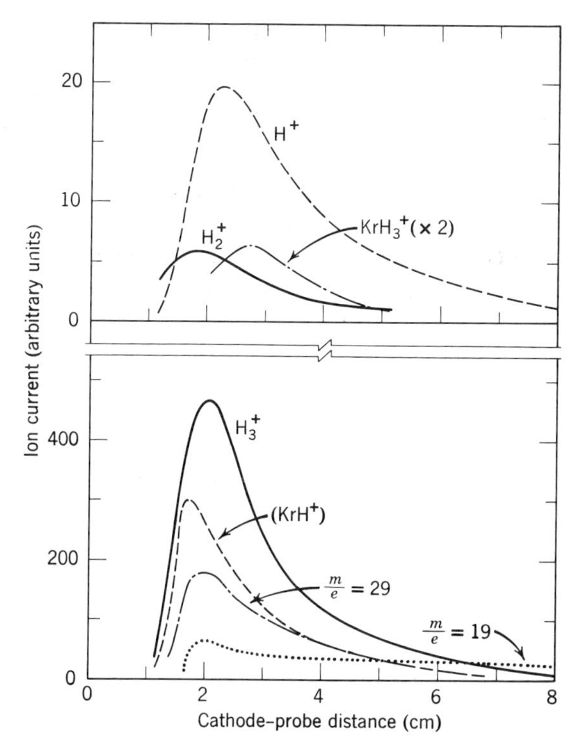

Fig. 6.6 Ions observed in a negative glow for a ($H_2 + 7.5\%$ Kr) mixture at 0.3 torr [9, adapted].

The H_2-D_2 exchange in the discharge was measured by sampling and analyzing the product gas downstream from the discharge region. In addition, small percentages of rare gases were added, in order to assess their role in a radiation-induced reaction. Trends which were noted with the addition of small quantities of krypton are shown in Figure 6.6. Results such as these provide valuable information as to reaction rates, and the detection of KrH_3^+ in the discharge is reported to suggest that this ion specie may be important in discharges and radiation chemistry at high pressures.

The importance of the mass spectrometer as a diagnostic tool for analyzing shock waves is also increasing. A conventional shock tube (typically 2–3 m in length) consists of a metal tube partitioned into sections by a thin metal diaphragm, behind which the pressure of a gas (e.g., hydrogen) can be raised to a value of 500–10,000 lb/in.^{2} On the other side of the diaphragm the gas under study may be maintained at about 1 torr. When the "driver" gas reaches a pressure that ruptures the diaphragm, the expansion of the high-pressure gas generates a shock wave in the low pressure gas, and the latter may be heated to temperatures $\sim$10,000°K; its pressure may also rise to an atmosphere in a period of 10 to 1000 μsec [10].

The kinetics of such a phenomenon must therefore be studied on a very short time scale. Typical of the use of mass spectrometry in such studies is the work of Diesen [11], in which molecular fluorine was dissociated in a shock tube. The shock tube was coupled directly to a time-of-flight spectrometer. The analyzed mass spectra were displayed on an oscilloscope, and the time-resolved ion peaks were recorded by a high-speed drum camera. This system provided a spectral display every 25 μsec with a time resolution of 2 μsec. Calculated parameters in the experiment included gas concentrations, axial velocities along the tube, terminal Mach number, mole flow rates, and specific heat ratios of the general diluent gases. Thermal dissociation phenomena of molecular fluorine were reported in the temperature range of 1650 to 2700°K, and dissociation-rate constants were compared for some rare gases. Clearly, the diagnosis of such fast transients and gaseous kinetic problems would be impossible without mass spectrometric techniques.

GASEOUS DIELECTRICS

Because of the importance of gaseous dielectrics to the electrical industry, the mass spectrometer has been used extensively in a search for gases that have high thermal stability, and which can maintain a high electric-field gradient. It is generally believed that the high electric strength of a gas or a gaseous mixture is related to its tendency to attach electrons [12]. Hence the mass spectrometer has been used to measure the relative cross sections of gases for electron capture and to study specific phenomena associated with the formation of negative ions.

However, as a high electron-capture probability may exist over a very small electron-energy range, it is important that reasonably monoenergetic electrons be used to examine capture processes in detail. A technique that has proved very useful for obtaining a reasonably monoenergetic electron beam is the retarding-potential-difference method. The basic scheme, as applied to mass spectrometry, was first reported by Fox et al. [13], and it has since been subject to refinements by many investigators. The basic concept is illustrated in Figure 6.7. Consider a heated filament that is emitting electrons and which is maintained at some negative potential V_1, and let the electrons be accelerated toward an ionization chamber within which ions may be formed. In the absence of any grid G, electrons will have an energy spread arising from a spread

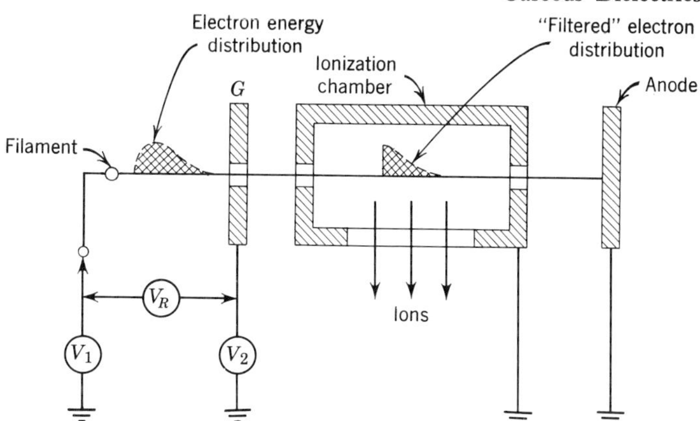

Fig. 6.7 Retarding potential method for obtaining monoenergetic electrons [13].

in thermal energies, a voltage drop along the filament, contact potential differences, and the like. If a grid is biased slightly negative, so as to furnish a retarding potential V_R, the low-energy electrons from the distribution will be removed. The change from the original distribution is shown in the ionization chamber. The energy band of electrons that is accepted into the chamber can be made arbitrarily small, and in practice, a 2-eV energy spread can be reduced to 0.2 eV or less.

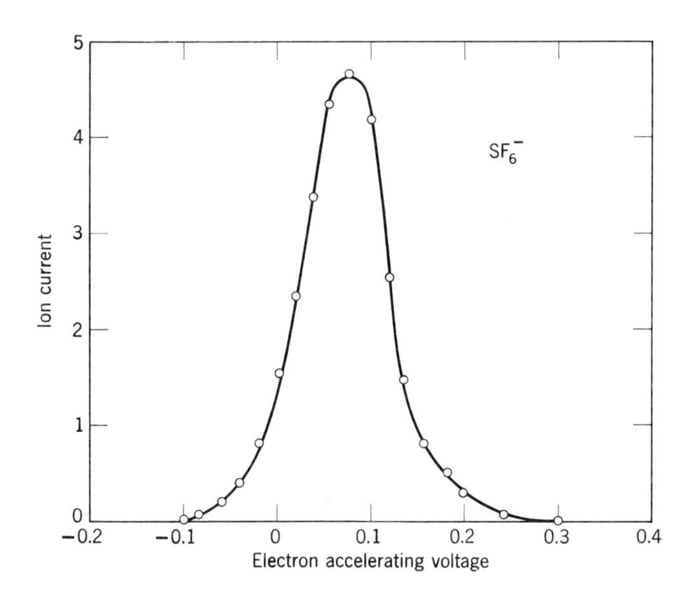

Fig. 6.8 SF_6^- ion current as a function of electron accelerating potential [14, adapted].

Using this technique, Hickam and Fox [14] reported on electron attachment leading to the formation of SF_6^- and SF_5^- in sulphur hexafluorides. They showed that in the case of resonance capture leading to the formation of SF_6^-, the capture process occurs at less than 0.1 eV and only over an energy range estimated to be approximately 0.05 eV. One of their measurements, shown in Figure 6.8, led to an estimated cross section for this resonance capture of 10^{-15} cm^2. It is typical of the highly specific information that can be obtained in a mass spectrometer, as opposed to a total ionization chamber—which does not permit mass analysis of the gas under investigation or relate the effect of impurities on negative ion formation.

The negative ion spectrum of $C_8F_{16}O_6$ at low electron energies has also been reported by Hoene and Hickam [15], and the appearance potentials measured. Positive and negative ion fragmentation of the molecule was analyzed and further evidence was obtained that electron attachment plays an important role in the high electric strength of this perfluorocarbon.

ELECTRON-IMPACT IONIZATION

Extensive use of the mass spectrometer has been made in the measurement of electron-impact ionization cross sections generally. A review of high-temperature experiments has been given by Inghram and Drowart [16]. Frequent use is made of the Knudsen cell technique in which (a) vapors from a closed cell are sampled through a small orifice, (b) the vapors are fed into an ionization chamber, and (c) the ions resulting from electron impact are analyzed and their intensities are recorded.

Recent measurements by Cooper et al. [17] have been reported of the relative cross sections for electron-impact ionization of Fe, Co, Ni, and Ag, using a crossed-beam technique with mass spectrometric analysis of the ions. Samples of the pure metals, ranging from 0.5 to 4.5 mg, were placed in a tungsten crucible with a 1-mm diameter orifice. For 60-eV electrons, *relative* cross sections of 34, 39, 33, and 35 were obtained, respectively.

A general theoretical method for estimating cross sections based on mass spectrometric data has been given by Stafford [18].

LASERS

An increasingly strong coupling exists between the field of mass spectrometry and various aspects of laser technology. In Chapter 3, mention was made of the usefulness of the laser as an ion source. The further development of laser sources is indicated on the basis of several considerations. First, lasers present us with an unusual range of power densities. The focused output of a 50-kW infrared burst from a neodymium-in-glass laser is reported to have a radiant power density of the order of 10^{12} W/cm^2. This radiation density is sufficient to study ionization and dissociation phenomena in gaseous media, as well as to observe the ion production arising from vaporized solids. The fact that a laser can produce a well-defined frequency in the ultraviolet, optical, and infrared regions of the electromagnetic spectrum also suggests that unique research potentialities exist in the laser-spectrometry field. Second, new laser probes will undoubtedly be developed for localized chemical analysis, analogous to the probes of spark-source mass spectrometry. Because of space coherence, lasers can be focused to a spot having a diameter approximating the wavelength of the laser radiation itself. This means that highly localized zones, less than a micron in extent can, in principle, be subjected to an analysis relating to atomic or molecular species. Third, there has been a wide-scale effort to shorten the duration time of laser bursts. Techniques include the use of electro-optical devices, such as polarizer-Kerr cells, whose transparency depends on the magnitude of an applied electric field, which can be actuated in less than 0.1 μsec [19]. This means that lasers can be considered as attractive candidates for time-of-flight mass spectrometry.

An additional area in which mass spectrometry can be expected to interact with laser development is in materials research. Ruby, a single-crystal aluminum oxide containing a small amount of chromium oxide ($\sim$0.05%), was the first material in which laser action was demonstrated [20]. The chromium atoms, in their trivalent ionic state, provided the prerequisite energy-level scheme for population inversion and the emission of coherent light at 6943 Å. Today, an exceedingly wide range of solids, semiconductors, and gaseous media are being explored. Table 6.2 lists only a few laser materials and their spectral outputs [21].

It will be noted that the rare earths are an important class of elements that provide laser action in various host crystals and lattices. These concentrations are usually so large that mass spectrometric techniques are not needed for quantitative assay. Nevertheless, at low concentration levels the mass spectrometer is the most convenient analytical tool. It might also be noted that a few isotopes of the rare earth elements listed in Table 6.2 have exceedingly large neutron-capture cross sections. Hence if lasers are ever used in high-neutron-radiation environments, one might expect that the separated isotopes which have the lowest neutron cross section will be selected.

Table 6.2

Ions Giving Laser Action In Solids

Ion	*Emission* $\lambda(\mu)$
Cr^{3+} (in Al_2O_3)	0.694
Dy^{2+} (in CaF_2)	2.36
Er^{3+}	1.61
Gd^{3+}	0.312
Ho^{3+}	2.05
Nd^{3+} (in $CaWO_4$)	1.06
Sm^{2+}	0.708
U^{3+} (in CaF_2)	2.61

The emission of recombination radiation from p-n junctions in some Group III–V compounds has given rise to devices such as the semiconductor injection lasers. In this area alone considerable research is being focused on the effects of impurities, optimum doping concentrations, and so forth. For example, Dobson [22] has reported results on gallium arsenide lasers with respect to characteristics of threshold and efficiency. He also noted that not all impurity atoms in gallium are electrically active and that the difference between electrically active concentrations and atomic concentrations increases at higher doping levels.

Research in gaseous lasers may not currently demand sophisticated mass spectrometric instrumentation. Nevertheless, the major gaseous constituents of these systems (helium, oxygen, neon, argon, krypton, xenon, etc.), and trace impurities can be conveniently monitored by standard mass spectrometric methods.

Finally, laser research that may require mass spectrometry as a diagnostic tool relates to ionization breakdown in gases produced by focused laser beams. Recent experimental studies suggest that laser-induced ionization of gases takes place in gases normally completely transparent to optical frequencies. The possibility of electrical breakdown being attributed to the presence of a random free electron is ruled out because of the typically small laser focal volume (10^{-6} to 10^{-9} cm^3). The presence of even one electron in such a volume would then imply a very high mean charge density. On the other hand the electrons of all gases are too tightly bound to be ejected by the energy of a single photon (e.g., the energy of a ruby laser photon is only 1.78 eV). Gardner [23] thus suggests that the photoionization that has been observed in both gases and metallic vapors might be attributed to a combination of direct and indirect (two-photon resonance) processes. At least for metallic vapors of low ionization potentials such as cesium and rubidium, he suggests that at high atomic densities ($\sim 10^{22}/cm^3$) there is a non-negligible probability that two or more atoms can interact "simultaneously" with the same photons or with one another. For a further substantiation of such a theory, many studies relating to gaseous mixtures and compounds will probably demand mass spectrometric assays.

PHOTOELECTRIC AND THERMIONIC EMISSION

Only brief mention can be made of the increasing importance of mass spectrometric techniques in basic studies relating to photoelectric phenomena and thermionic emission. As further discussed in Chapter 9, a knowledge of the surface layer is imperative if meaningful and reproducible data are to be obtained with respect to photocathodes and thermionic emitters. In the 1930's for example, some excellent work was reported on photoelectric thresholds, but there were wide discrepancies in the values that were reported by various investigators. A primary hurdle was the one of obtaining adequate and reproducible vacuum conditions, and having any detailed knowledge concerning the atoms comprising the actual photoemissive surface. Today, however, it is possible not only to achieve operating pressures in the range of 10^{-10} torr, but the mass spectrometer can also measure the partial pressure of the few remaining atoms and molecules.

Further, the mass spectrometer can be required to do much more. It is possible, for example, to use the mass spectrometer as a source for depositing surface atoms of specific chemical specie and to do this on a quasi-quantitative basis. As an example, a photocathode could actually be fabricated for research testing, using one or more ion sources (as indicated in the section on the cascade ion-beam analyzer in Chapter 2). The photo-response could then be examined simultaneously with the deposition of desired surface atoms, and transient changes in the work function of the surface would be detectable. In other words there appears to be only the usual experimental difficulties in combining mass spectrometry, spectrophotometry, and the measurement of photoelectric currents. Such experiments definitely appear to be worthy of consideration as it is now known, for example, that the maximum work function variation does not always necessarily occur either when a monolayer is reached or at a definite fractional coverage. Rather, the maximum variation is a function of the absorbate and substrate materials, and the particular type of substrate surface [24].

The same general arguments appear valid for the study of thermionic emitters. Also, in the particular case of gas-filled thermionic converters, and the direct conversion of heat to electricity, the mass spectrometer is a useful basic laboratory tool. Current problems in this area of engineering relate to the close interelectrode spacing of the thermionic elements, space charge neutralization, diffusion phenomena, and optimization of suitable gaseous mixtures. In order to relate these parameters to specifications for stable and long-lived thermionic converters, further mass spectrometric studies are anticipated. It is also interesting to note that at least one laboratory [25] is using nuclear fission products to bombard noble gases as a means of generating ions to counteract space charge. In this case the experimental emitter furnishes both electrons for direct energy conversion and ionization products that tend to cancel the effects of electron space charge.

ELECTRON EMISSION FROM ION IMPACT ON SINGLE CRYSTALS

Experiments are continuing with respect to secondary electron emission from ion impact and, in particular, with the yield of secondary electrons as a function of ion specie, ion energy, and crystal orientation. Only two recent measurements are cited below.

Ewing [26] has used protons in the energy range of 50 to 225 keV to study the yield of electrons from a (100) tungsten crystal surface. He determined a maximum value of the yield to be 1.51 ± 0.03 electrons per incident proton at an energy of 125 keV. This yield represented an 11% decrease from a polycrystalline surface under otherwise similar conditions.

The secondary electron emission has also been measured for alkali halide single crystals bombarded by helium and argon ions having energies from 20 to 600 eV. In the work by Kondrashev and Petrov [27]. single dielectric crystals of LiF, NaCl, KBr, and CsI were bombarded by an ion beam of $\sim 10^{-10}$ A. In order to avoid charge build-up, the crystals were heated to 400°C. Figure 6.9*a* indicates their results for helium, and Figure 6.9*b* show their results for argon ions. In the case of low-energy helium ions the secondary electron yield is considerably greater for the KBr crystals. It has been reported by these investigators that a comparison of the data indicates that the secondary electron yield is related to the "forbidden" band width of the bombarded materials. The larger the forbidden band width, the weaker will be the ion-electron emission.

These studies are typical of the types of secondary yield studies that could be programmed with mass spectrometric instrumentation when

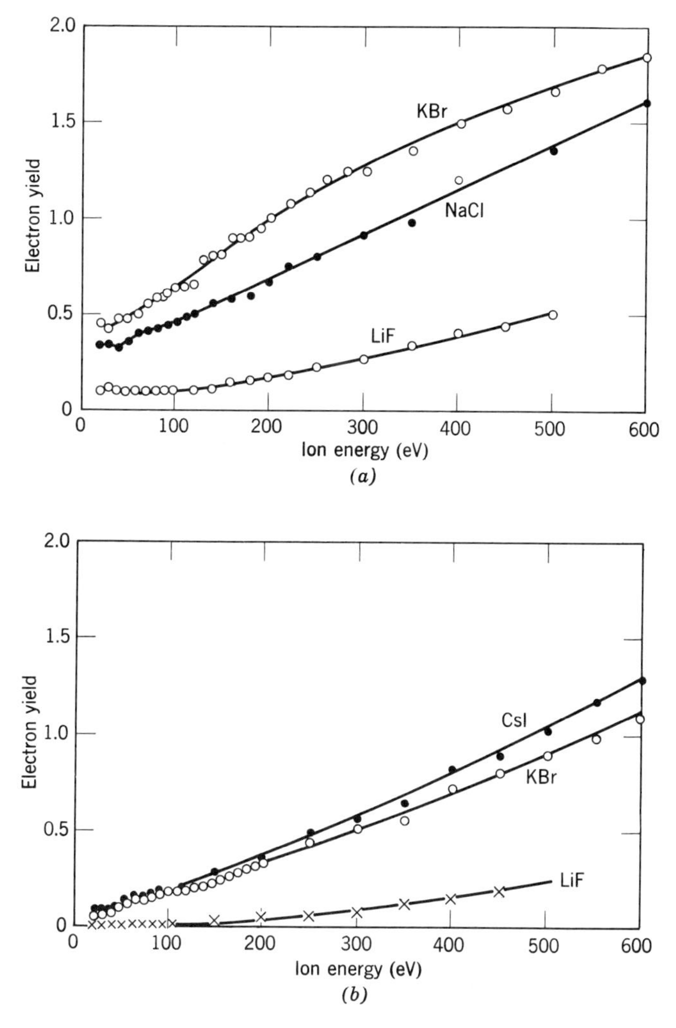

Fig. 6.9 (*a*) Dependence of secondary electron yield on the energy of He^+ ions on KBr, NaCl, and LiF [27]; (*b*) dependence of secondary electron yield on the energy of argon ions on CsI, KBr, and LiF [27].

ion energy, ion specie, and charge state can be completely identified. Bombarded crystals could be placed in the usual positions of the ion detector or electron multiplier. For reproducible results, however, it is imperative that the target be located in ultra-high vacuum. In the case of Ewing's work [26] the tungsten surface was cleaned by flash heating in a chamber of 2×10^{-10} torr, and the secondary yield was measured about 2 min after cleaning. This time was short compared to an estimated 2 hrs for a monolayer of residual gases to form on the target surface.

EFFECT OF ISOTOPIC COMPOSITION ON ELECTRICAL RESISTANCE

Quite distinct from the effect of impurities in semiconductors are studies that attempt to explain electrical conduction anomalies in metals. In particular the question has been raised as to the possible effect of isotopic composition on metallic conduction. In general no effect is expected, and few studies have been made in this area. The availability of large quantities of separated isotopes, however, now permits such measurements to be pursued.

The effect of isotopic composition on the resistance of solid metallic lithium has been reported by Leffler and Montgomery [28]. They studied the electrical resistance of separated Li^6 and Li^7 in the temperature range 4–300°K, and they also examined "isotopic alloys" comprising varying proportions of the two isotopes. Lithium was selected because of its simple crystalline and electronic structure. Further, the large relative mass difference of the lithium isotopes was desirable. The investigators' objective was to provide an experimental test of theory relating to the interaction of conduction electrons with lattice vibrations, and interatomic force fields, using isotopic mass as a variable parameter.

They clearly observed an effect of isotopic mass on resistivity and confirmed theoretical predictions that the electrical resistivity of solid isotopic alloys of any composition (including pure isotopes) can be obtained from knowledge of the temperature dependence of the absolute resistivity of a single alloy—by means of a scaling factor. This scaling factor is simply the square root of the ratio of the average isotopic masses applied to the resistivity and to the temperature. They also concluded that "isotopic impurities" do not seem to act as additional scattering sources, but serve only to slightly affect the lattice-vibrational spectra. In brief, their work is an example of the use of an isotopic perturbation to investigate the very basic mechanisms of electronic transport.

PROPERTIES OF THIN FILMS

There are other interesting resistivity measurements that relate to other effects. One is the change in resistivity from "damage" that accompanies heavy ion irradiation. This type of investigation is supplementary to "neutron-damage" irradiations. In one sense it is also the "recoil-nuclei" part of neutron damage. With ion bombardment it is also possible to gain more control over an experiment and to reduce the irradiation time by several orders of magnitudes. Because of the short range of heavy ions, however, this class of damage investigations is limited to thin films. But by suitable masking of a sample, and through ion-beam programming, it is possible to obtain a uniform ion bombardment density on a small specimen.

One recent measurement of this type has been reported by Teodosic [29], in which high-purity silver films (on glass substrates) were bombarded with silver ions. He reported that for all film thickness, ranging from 150 to 400 Å, and for all incident ions (ranging from 20 to 500 keV), a common type resistance change can be observed. His results for a 150-Å film bombarded by 20 keV Ag ions is shown in Figure 6.10. There is an initial decrease in resistivity that reaches a minimum, followed by a subsequent increase after continued ion bombardment. The apparatus used was the Aarhus Isotope Separator (University of Aarhus, Denmark).

This "resistivity-versus-dose" relationship has not been completely explained, but the phenomenon is attributed both to the production of defects and to some change in the probabilities of "specular reflection" of conducting electrons on the upper and lower boundary surface of the thin layers. Hopefully such studies will contribute to general theories of electrical conduction and radiation damage.

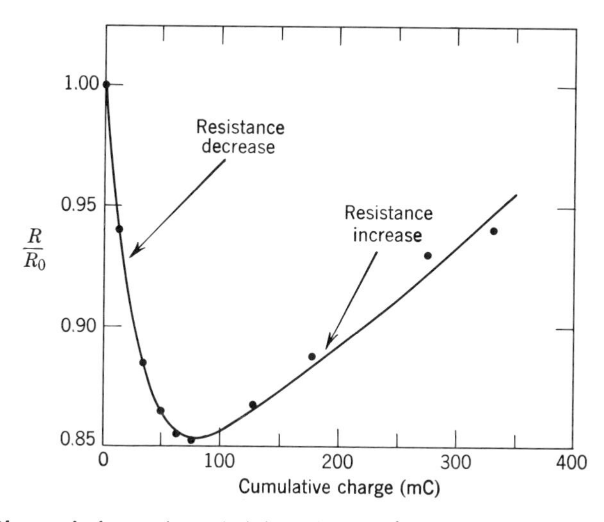

Fig. 6.10 Observed change in resistivity of a 150-Å silver layer bombarded by 20-keV Ag^{+} ions [29, adapted].

Of greater applied interest is the sputtering of metals or alloys in the presence of a reactive gas for the deposition of a variety of various thin films. Schwartz [30] has shown that by controlling the fraction of the reactive gas in the inert sputtering gas, it is possible (especially with refractory metals) to obtain phases that range from dissolved gas in the metal to mixtures and finally to semiconductors and insulators. The use of sputtered tantalum for thin-film passive components and integrated circuits has generated research in this area. Important in this phase of electronic technology is the mass spectrometer for the analysis and control of the gas composition in sputtering systems. The scope of this single aspect of electronic component technology, as identified by Schwartz, includes the following:

1. Resistivity measurements of refractory films such as tantalum compounds, and films of titanium, niobium, and zirconium.
2. Measurements of partial pressures of oxygen, nitrogen, methane, and carbon monoxides.
3. Interpretation of dissociation phenomena.
4. Investigation of superconducting transitions.
5. The detection of hydrogen, water vapor, and other contaminants in sputtering systems.
6. Detailed examination of rate controlling processes at the substrate.

The analysis of the ambient atmosphere during the sputtering of tantalum films has also been reported by Sosniak [31]. Large variations in the partial pressure of hydrogen, water vapor, and methane were noted and attributed to ion-bombardment decomposition of absorbed surface layers and desorption from the electrodes and walls of the bell jar. Deliberate additions were made of N_2, CH_4, and H_2 during sputtering in argon, and significant differences were found in the effect of the glow discharges on these gases, with the gettering of CH_4 and H_2 exhibiting a complex time dependence. Tantalum films sputtered without deliberate gas additions were found to be of the beta-tantalum form, even when the discharge-generated background was reduced to $\sim 1 \times 10^{-6}$ torr by a long pre-sputtering period. This observation is in general agreement with the findings of Read and Altman [32], who showed that Ta films sputtered without a deliberate addition of reactive gases frequently possess a different crystal structure than does body-centered tantalum. The β-phase Ta is characterized by a resistivity of 180 $\mu\Omega$-cm and a very low negative or positive temperature coefficient. For other metals such as niobium, however, Sosniak [33] obtained thin film resistivities within 50% of the bulk value by using an extended pre-sputtering clean-up.

Finally, the fabrication of insulating films as resistors or dielectrics for capacitors as well as semiconductors, will probably call for the widespread application of "on-line" spectrometry for processing and quality control.

REFERENCES

[1] M. J. Sinnott, *The Solid State for Engineers,* Wiley, New York, 1958, p. 362.

[2] Technical Data Sheets, Associated Electrical Industries, London. Values indicate measurements from sample analyzed in MS-7 spectrometers.

[3] A. J. Ahearn, private communication.

[4] F. D. Rosi, *Industr. Res.,* 61 (November 1964).

[5] D. K. Bohme and J. M. Goodings, *Rev. Sci. Instr.,* **37,** 362 (1966).

[6] G. F. Sauter, R. A. Gerber, and H. J. Oskam, *Rev. Sci. Instr.,* **37,** 572 (1966).

[7] D. K. Jefferies, *J. Sci. Instr.,* **44,** 587 (1967).

[8] M. M. Shahin, *J. Chem. Phys.,* **45,** 2260 (1966).

[9] P. H. Dawson and A. W. Tickner, *J. Chem. Phys.,* **45,** 4330 (1966).

[10] J. B. Hasted, *Physics of Atomic Collisions,* Butterworths, Washington, D.C., 1964, p. 86.

[11] R. W. Diesen, *J. Chem. Phys.,* **44,** 3662 (1966).

[12] W. M. Hickam and D. Berg, *J. Chem. Phys.,* **29,** 517 (1958).

[13] R. E. Fox, W. M. Hickam, T. Kjeldaas, Jr., and D. J. Grove, *Rev. Sci. Instr.,* **26,** 1101 (1955).

[14] W. M. Hickam and R. E. Fox, *J. Chem. Phys.,* **25,** 642 (1956).

[15] J. von Hoene and W. M. Hickam, *J. Chem. Phys.,* **32,** 876 (1960).

[16] M. G. Inghram and J. Drowart, "Mass Spectrometry Applied to High Temperature Chemistry," in *High Temperature Technology,* McGraw-Hill, New York, 1960.

[17] J. L. Cooper, G. A. Pressley, Jr., and F. E. Stafford, *J. Chem. Phys.,* **44,** 3946 (1966).

[18] F. E. Stafford, *J. Chem. Phys.,* **45,** 859 (1966).

[19] A. K. Levine, *Amer. Scientist,* **51,** 14 (1963).

[20] *Ibid.,* p. 19.

[21] *Ibid.,* p. 23.

[22] C. D. Dobson, *Brit. J. Appl. Phys.,* **17,** 187 (1966).

[23] J. W. Gardner, *Int. J. Electron,* **22,** 123 (1967).

[24] E. P. Gyftopoulos and J. D. Levine, *J. Appl. Phys.,* **33,** 67 (1962).

[25] General Motors Research Laboratories: note in *Physics Today,* **19,** 100, (1966).

[26] R. I. Ewing, *Amer. Phys. Soc. Bull.,* **11,** 816 (1966).

[27] A. I. Kondrashev and N. N. Petrov, *Soviet Phys.—Solid State,* **7,** 1255 (1965).

[28] R. G. Leffler and D. J. Montgomery, *Phys. Rev.,* **126,** 53 (1962).

[29] V. Teodosic, *Appl. Phys. Letters,* **9,** 209 (1966).

[30] N. Schwartz, in *Transactions of the 10th National Vacuum Symposium,* Macmillan, New York, 1963, p. 325.

[31] S. Sosniak, *J. Vac. Sci. Tech.,* **4,** 2, 87 (1966).

[32] M. H. Read and C. Altman, *Appl. Phys. Letters,* **1,** 51 (1965).

[33] Sosniak, *op. cit.,* p. 93.

Chapter 7

Semiconductor Devices by Ion-Beam Implantation

Few uses of mass-resolved ion beams appear as exciting as that of producing semiconductor devices. Many papers are now appearing in the literature, but the full implication of this technique for the semiconductor industry cannot accurately be predicted. However, there is little doubt that ion-implantation methods in semiconductor technology will be important. Solar cells are currently being fabricated by this method, and ion-beam-produced diodes have characteristics indistinguishable from diffused devices.

Mass spectrometry has been so closely identified with analytical impurity determinations in bulk materials that a question may arise as to its late application for chemical doping. From a commercial point of view, at least, there seems to be a logical answer: the semiconductor industry has been doing quite well without it! Advances in diffusion technology and material purity have been reflected in sharp decreases in unit costs, and improvements in quality control and automation have been truly remarkable. Nonetheless, this chapter endeavors to summarize the salient aspects of ion doping, and its possible advantages in special situations.

JUNCTION FORMATION BY ION BEAMS

The "doping" of solids by ion beams is not new. We are indebted to a large number of European investigators who during the last two decades have impregnated metals with inert gases, radioactive atoms, or separated isotopes for a multiplicity of purposes. Koch (University of Copenhagen), Nielsen (University of Aarhus, Denmark), Kistemaker (Laboratory of Mass Separation, Amsterdam), Bergström (Research Institute for Physics, Stockholm), Bernas (Laboratory of Nuclear Physics, Orsay, France) are only a few of those who have employed their isotope separators in important ion physics studies. The work of Davies at the Chalk River Laboratory (Canada) has also been of a pioneering nature, relating to the penetration of energetic ions into both amorphous and crystalline solids. A few United States scientists envisioned the latent possibilities of ion beam doping at an early date (notably Shockley [1] and Lark-Horowitz [2]), but only recently have serious experimental studies been pursued in this field.

The author's first interest in this area was sparked by discussions with W. M. Gibson of the Bell Laboratories, who kindly supplied a single crystal of silicon for ion bombardment in a 30-in-radius spectrometer [3]. The result of this single experiment was ambiguous, but the advantages of highly quantitative doping seemed clear. Junctions comparable to a diffused type were reported by Alvager and Hansen [4], (who used the isotope separator at the Argonne National Laboratory) and by Ferber [5], who used a Van de Graaff generator as an ion source. At the Aarhus University Conference on Electromagnetic Separators in Denmark (June 1965), five papers dealing specifically with ion implantation were presented. There is now little doubt that new ion-implanted crystals provide us with a new class of materials. Both deep and shallow junctions have been produced by impregnation of silicon by phosphorus and boron ions. Furthermore, some of these junctions have been large area devices (silicon solar cells $\sim 2\ cm^2$) having efficiencies exceeding conventional types [6]. A typical ion implant of phosphorus in silicon, as reported by Manchester et al. [7] is shown in Figure 7.1. An electromagnetic isotope separator at Oak Ridge was used to dope the silicon specimen, and both singly and doubly charged phosphorus ions were used to obtain ion energies up to 80 keV. The junction depth is broadly defined as the distance from the surface to the region where the implanted active impurity concentration equals the background doping of the substrate (i.e., the point of junction formation.)

The potential advantages of ion beam doping are hardly trivial:

1. There is no known method that allows the doping of chemical additives with comparable chemical purity.
2. Ion implantations can be made, in principle, on a three-dimensional basis by proper programming of the bombarding ion beam.
3. Gross heating of the crystal, which is necessary for diffusion, is eliminated and the introduction of surface impurity atoms is minimized.
4. Junction formation in devices is not limited to systems having favorable diffusion constants.
5. Under favorable conditions, the junction boundary may be made more sharply defined.

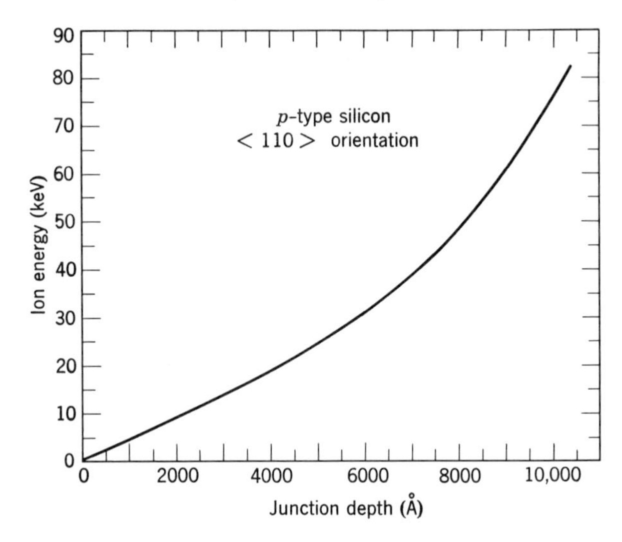

Fig. 7.1 Junction depth vs. kinetic energy of phosphorus ions in an oriented crystal of silicon [7, adapted].

6. The introduction of donor or acceptor atoms can be made quantitative and precise, for either junction formation or bulk compensation.
7. Mass- and energy-resolved ion beams provide the ideal approach for device fabrication and automation.

The above list must admittedly be viewed with some reservations. Unanswered questions remain relating to crystal damage by ion impact and "channeling" phenomena.

Implantation of ions has been achieved by using large mass spectrometers or by isotope separators. Of particular interest have been the implantation of the III–V groups, using atoms of boron or phosphorus. Although the larger electromagnets are usually employed, the required ion-beam currents are easily attainable in a mass spectrometer—provided that a high-voltage source can be used. A relatively small number of atoms is needed to produce a single junction. For example, consider a thin junction region of 10,000 Å (1.0 μ) that is to be uniformly doped with $\sim 10^{15}$ atoms/cm^3. For a 1-cm^2 cross section, this is only 10^{11} atoms. This number of atoms requires an ion beam of only 1.6×10^{-8} A for 1 sec. If the accelerating voltage is 50 keV, the power generated at the surface will be about 5×10^{-4} W, so the specimen temperature rise will be small.

RANGE OF IONS IN MATERIALS

A determination of the range or penetration depth of highly energetic particles is relatively easy; experimental range measurements of low-energy heavy ions present considerably greater difficulty. Furthermore, we cannot extrapolate from data that exist for high-energy ions to the low-energy range, which is of greater interest in problems in ion doping.

The penetration depth of heavy ions at low energies (<50 keV) is so small that it is very difficult to prepare self-supporting films to measure stopping power. An alternative method successfully applied by Davies et al. [8] is that of accelerating radioactive atoms into a thick target. Uniform layers of accurately known thickness are then gradually removed by an electrochemical process. The residual radioactivity after successive "peeling" of layers then permits a computation of the penetration depth of the radioactive ions in the original thick target. A review of range measurements made by this method has been prepared by Domeig [9] for several ion specie in aluminum and tungsten.

Other range measurements by Ferber [5] for boron, phosphorus, and other elements in silicon are shown in Figure 7.2. While these measurements clearly indicated that ions could be implanted at distances comparable to the depletion depths in junctions, the *straggling* in ion range was considerable. For the most part, however, these measurements were made in specimens without regard to specific crystal orientation.

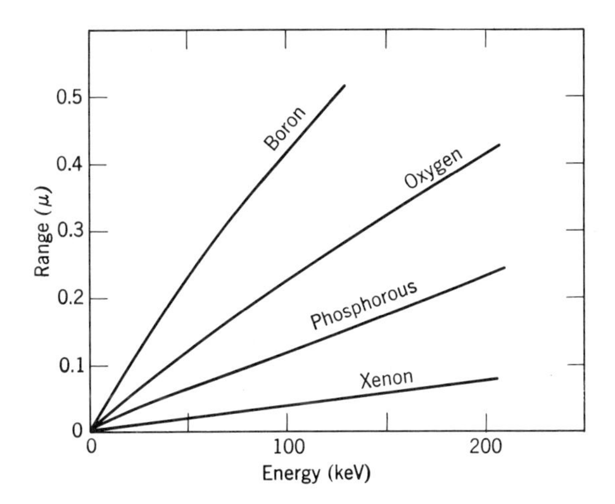

Fig. 7.2 Approximate range of positive ions in silicon as a function of ion energy [5, adapted].

CHANNELING PHENOMENA IN SINGLE CRYSTALS

The recently discovered phenomenon of "channeling" in single crystals is, in a large measure, responsible for the increased optimism for ion doping in semiconducting materials. Channeling can be loosely defined as the enhanced penetration of ions along preferred directions in a crystal structure. A somewhat analogous phenomenon has been observed with respect to interstitial diffusion, where large directional differences in diffusivity have been noted for atoms in a host lattice. An example of an easy channeling path, the ⟨110⟩ direction in a diamond lattice, can be noted in Figure 7.3. Channeling has been observed in crystals for protons, alpha particles, and heavy ions of very high energy. More recently, it has also been explored for atoms of intermediate energies (i.e., in the 0.1–1.5-MeV range). The work of Davies, Eriksson, and Jespersgaard [10] is illustrative of the fact that the crystalline nature of a solid has a marked effect upon its stopping power for ions.

A tungsten crystal was aligned by x-ray diffraction techniques and used as a target for ions in a heavy ion accelerator. Typical crystal orientations were within 1°. The ions were radioactive isotopes that penetrated into the crystal (fewer than 10^{13} atoms per square centimeter were sufficient to give accurate statistical data). The range or penetration depth of the embedded radioactive atoms was then determined by remov-

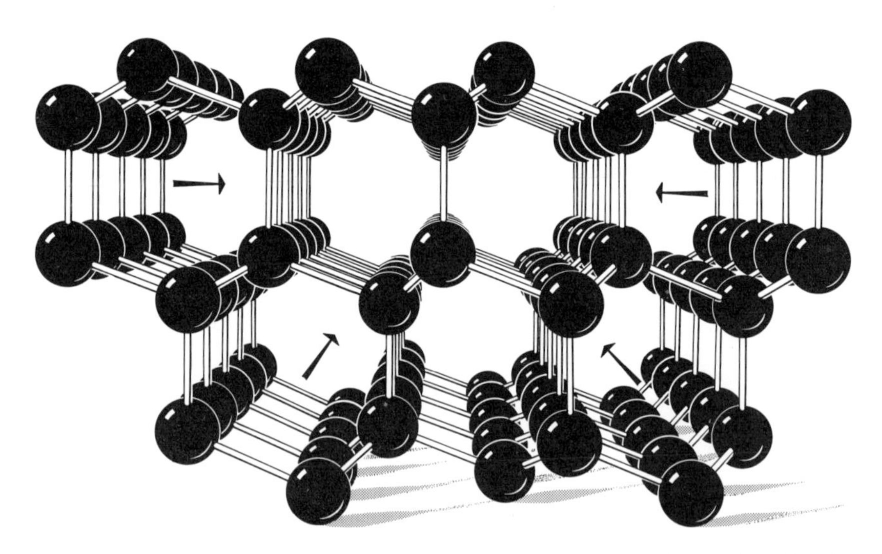

Fig. 7.3 View of ⟨110⟩ direction in a diamond lattice, showing paths of easiest penetration.

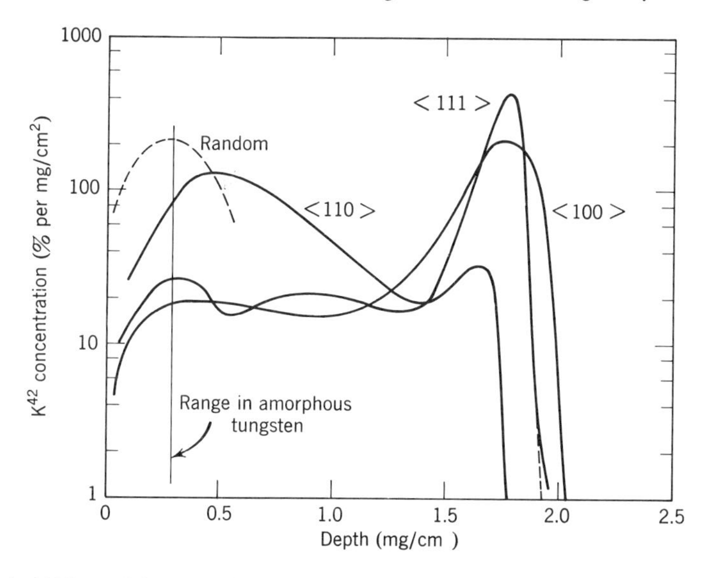

Fig. 7.4 Differential range distributions of 500 keV K^{42} ions along the ⟨100⟩, ⟨111⟩, and ⟨110⟩ directions in tungsten [10].

ing thin, uniform layers of tungsten metal by the electrochemical "peeling" procedure mentioned in the preceding section. The results of this type of ion implantation as a function of crystal orientation are graphically portrayed in Figures 7.4 and 7.5.

It will be noted that the range of K^{42} atoms in amorphous tungsten is short and has an appreciable straggling or tailing. Channeled atoms exhibit both an increased range and a sharper cut-off. Somewhat similar results are displayed for Xe^{133}.

Qualitatively this effect can be explained by assuming nuclear scattering is a more important mechanism for ion energy loss in amorphous material. For an atom travelling between ion centers, such a "channeled" ion would interact primarily with valence electrons—and a perfectly channeled ion would interact with a lower electron density. If only "electronic stopping" is involved, we should also expect that ions would have a better defined range, as evidenced in Figures 7.4 and 7.5.

These results are of great significance. First, if the range of ions in crystals under ideal conditions is found to be sharply defined, ion doping for junction formation becomes exceedingly attractive. Second, channeling (meaning interaction with electrons only) will produce considerably less damage in the doped crystal. Clearly the beams for ion doping will have to be tightly collimated to prevent nuclear scattering.

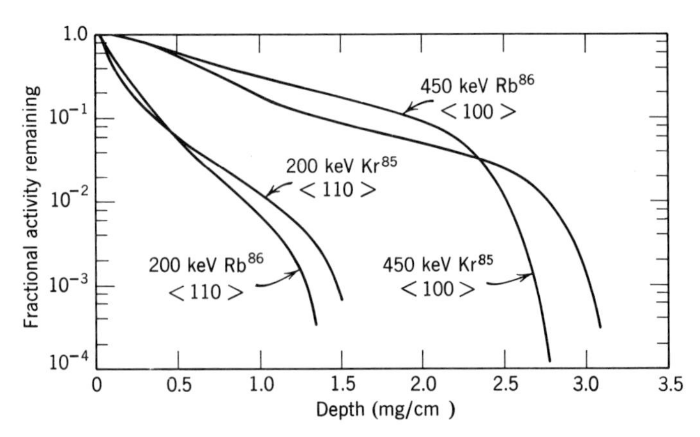

Fig. 7.5 Integral range distributions of 200 keV and 450 keV Kr^{85} and Rb^{86} ions along the ⟨100⟩ and ⟨110⟩ directions in tungsten, observed by radioactive counting [10].

Third, ion implantation in preferred directions should require considerably lower ion-accelerating voltages to achieve depths comparable to those in randomly oriented crystals.

Matzke and Davies [11] have also stated that channeling can be used to detect the location of foreign atoms in a crystal. The reason given is that interactions between a channeled beam and foreign atoms in a crystal will not show any attenuation in yield unless the foreign atoms are located within $\sim 10^{-9}$ cm around each aligned lattice row. Thus by simultaneously observing the interaction of an ion beam with both lattice and foreign atoms, these investigators suggest that a quantitative measurement can be made of the fraction of foreign atoms that are located along a particular set of lattice rows.

ION-IMPLANTATION PROFILES IN SILICON

There have been several recent investigations on the ion implantation of silicon. Bower, Baron, Mayer, and Marsh [12] have reported on the deep (1–10 μ) penetration of donors resulting from 20-keV Sb ion implantations into high-resistivity crystals of ⟨110⟩ and ⟨111⟩ orientation. Density profiles were measured by a capacitance-voltage method. Their results indicated a deep penetrating tail, substantially independent of temperature ($T \geq 300°C$), orientation, annealing, and surface condition. An empirical relationship for the tail of the ion implant is of the form $N \propto (x + B)^{-2.2}$ for more than four orders of magnitude

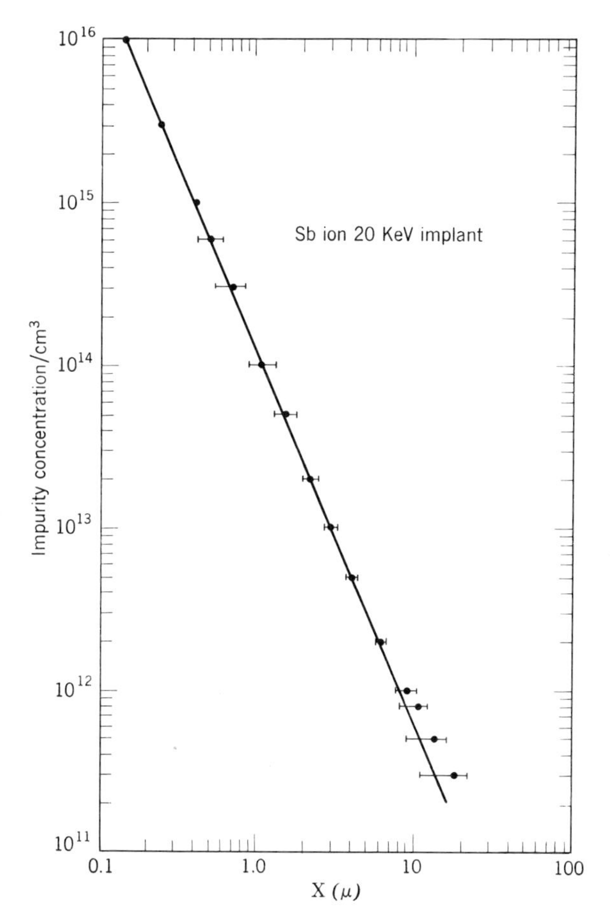

Fig. 7.6 Empirical ion-implant relationship for the penetrating dopant component of Sb ions in silicon [12, adapted].

$(B \approx 0.15\,\mu)$. The experimental data are shown in Figure 7.6. Bower et al. also report that this phenomenon was observed up to carrier concentrations 10^{-3} of the surface concentration and reflect a penetration mechanism basically different from normal channeling.

In contrast to the above findings, Gibbons, El-Hoshy, Manchester, and Vogel [13] found temperature to be an important parameter in ion-implantation profiles of 40-keV phosphorus ions in silicon. They also concluded that the "channeling" properties of crystalline solids is a function of previous implant history and may approach amorphous behavior, apparently caused by (a) the role of previously implanted ions that have come to rest as interstitials in the lattice or (b) the formation of an amorphous surface phase caused by an ion-lattice interaction. The latter is substantially a damage state induced by ion bombardment.

NEUTRON TRANSMUTATION DOPING

The technique of neutron transmutation doping is mentioned as this process may be used in conjunction with ion-beam implantation methods at some future date. In the usual neutron-transmutation method a "radiation die" or shield encloses the semiconductor specimen in a manner so as to shield a portion of the sample but expose the potential junction region to a beam of thermal neutrons. Figure 7.7 is a simple schematic diagram of the irradiation assembly, which comprises (a) a reactor irradiation capsule, (b) bulk neutron shielding, and (c) collimating slits that limit the neutron radiation to a local region of the semiconductor sample. If the semiconductor is silicon, phosphorus atoms will be formed by the reaction

$$Si^{30} + n \rightarrow Si^{31} \xrightarrow[2.6\text{ hr}]{\beta^-} P^{31}. \tag{7.1}$$

Si^{30} constitutes only about 3% of the naturally occurring element; however, this is sufficient in many applications to form a sizable number of *n*-type phosphorus impurities. In this manner a *p-n* junction can be formed in a region that was originally *p*-type. Using this approach, Klahr and Cohen [14] have reported on the production of junctions

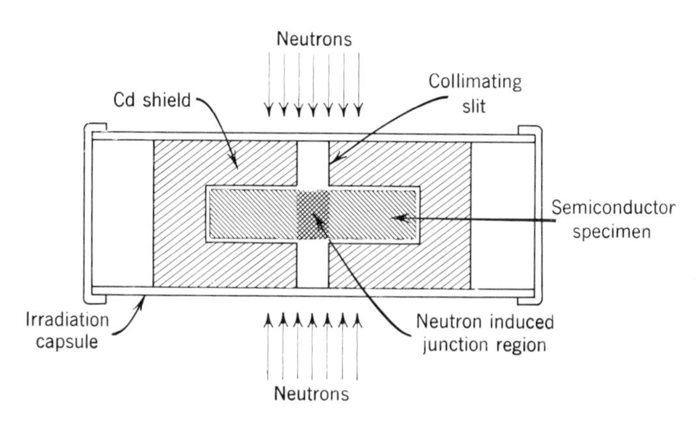

Fig. 7.7 Schematic of neutron irradiation capsule for transmutation doping [14, adapted].

having properties comparable to those fabricated by conventional methods.

But it will now be noted that if isotopically pure Si^{30} is made available, the neutron flux to produce an equivalent number of P^{31} atoms would be reduced by a factor of about 30. In addition, it should be possible to combine general ion-implantation techniques with neutron-transmutation doping to generate new impurity profiles, transistors, and microcircuits that are difficult, if not impossible, to produce by pure diffusion.

SUPERSATURATION

Ion beams can provide doping of great chemical purity. They can also provide chemical compositions that cannot be achieved by the equilibrium conditions that limit other methods, such as diffusion. Ion injection can easily overcome the surface barrier (of the order of magnitude of 1 eV) to form a high concentration of solute atoms in a crystal. A somewhat analogous phenomenon also occurs in electrochemistry—where electrode potentials are known to force hydrogen ions into a metal electrode.

Several specific experiments have conclusively shown that considerably higher impurity concentrations can be achieved by ion injection than by thermal means. For example, a comparison was made of the hydrogen solubility in ZnO caused by ion bombardment (at 500 eV) of a hydrogen ion beam at 1.3×10^{-6} A/cm^2, and the equivalent *thermal* bombardment of hydrogen at 8×10^{-9} torr [15]. The number of hydrogen molecules striking the surface of the ZnO crystal per second were thus equivalent in the two situations. At a crystal temperature of 400°C, the ion-injected hydrogen concentration in atoms per cubic centimeter was greater by a factor of 100.

Supersaturation has also been observed for alkali metals in silicon. Solubilities for sodium have been obtained that are several orders of magnitude greater than by thermal means. Also, large donor concentrations have been observed for lithium in silicon. In addition these high concentrations can be maintained for a considerable period at elevated temperatures. Indeed, complete crystals have been brought to a state of supersaturation, a result that suggests that potentially useful new materials may be available for semiconductor electronics.

In an excellent review paper McCaldin [16] remarks on two other important considerations in ion-doped materials. First, supersaturation doping would be attractive for the Group II–VI compounds like CdS. A current problem relates to the difficulty of achieving heavy doping in special applications. Diamond doping by ion beams has also been considered. Second, supersaturation doping is attractive for producing materials of low work function. By proper technique it may be possible to fabricate crystals with work functions appreciably lower than the ~ 1 eV now available. If the work function or electron affinity can be changed by simple ion injection, some ingenious technological applications should undoubtedly make their appearance.

DIFFUSION ENHANCED BY ION IMPACT

An additional phenomenon has been observed for materials that have served as targets for ion beams. Diffusion is generally enhanced in a manner similar to the radiation-enhanced diffusion that has been observed in studies with gamma rays, protons, and high-energy electrons. For the most part the observations have been made without respect to the crystal orientation. Diffusivities have been enhanced considerably over those in which the specimen is subjected simply to thermal cycling. Strack [17] has reported that boron and phosphorus diffuse as much as 10^4 times faster in Si when the specimen surface is under ion bombardment. It is assumed that this marked increase results from the production of a substantial number of vacancies at the surface. When the vacancies subsequently diffuse into the interior of the crystal, a vacancy concentration is formed in excess of the normally present thermodynamic concentration. Thus if an excess vacancy concentration is produced by ions, an enhancement of the diffusion of substitutional impurities can be expected.

The effect of ion bombardment has also been observed for the case of amphoteric impurities (i.e., impurities that can exist in either interstitial or substitutional sites). Injection of Na^+ ions into *p*-type silicon has been reported [16] to yield *n*-type material (with thermal heating during ion injection). A comparison of the Na^+-implanted material with Na diffusion in a conventional fashion has indicated a substantial enhancement of diffusion rates.

TECHNOLOGICAL IMPORTANCE OF ION BEAMS IN DEVICE FABRICATION

Although some specific advantages of ion-beam doping have been previously cited, it is appropriate to appraise this new tool for its general technological importance.

In many respects semiconductor technology is unique. It has demanded the most detailed researches in the physics of the solid state. It has also placed the most rigorous prerequisites on the purification and handling of material. The tongue-in-cheek suggestion has also been made that only persons of the highest character can be entrusted to processing semiconductors; unless it has been removed since 1962, a large sign is posted at the entrance to an air-conditioned semiconductor room at the Oak Ridge National Laboratory, which in bold 3-in. letters warns: "THINK CLEAN!"

But chemical purity is only one salient advantage of the ion-beam method in device production. Control is another: control with respect to the precise number of donor or acceptor atoms that may be placed in a host lattice and control with respect to the spatial distribution of atoms in this lattice. For as more detailed information becomes available with respect to the interaction of ions in crystals, ion beams can be used to place highly desirable boundary conditions on atom species, the integrated flux and energy, and the relative orientation of the ion beams to the target crystal planes.

In addition to the above, ion beams appear to present us with the sole artifice for producing solid-state devices by atomic doping *through* a passivating surface oxide layer. The array of devices successfully produced to date include (a) large-area solar cells of high efficiency, (b) passivated single counters as radiation detectors, (c) unipolar field-effect and bipolar transistors, and (d) a wide variety of experimental p- and n-type junctions. The use of ion beams for compensation or changing specimens from p to n materials has already been mentioned. Since high temperatures are not required as in diffused-type devices, it appears possible to make a multiplicity of solid-state elements (e.g., p-n-p and n-p-n) in a single-target specimen. King [6] has reported that sections of double implanted layers (P^{31} followed by B^{11}) show two junctions separated by extremely uniform spacings as small as 0.4 μ. B^{11} implants are also being studied for buried conductors in integrated circuits and for making differential and integral (dE/dx, E) particle detectors that are important in nuclear physics.

Some workers are already predicting that junctions can best be made by ion-beam methods, that ion-bombardment damage presents no real problem, and that the electrical parameters of ion-implanted devices are superior to those characterizing conventional fabrication. A few experts also look to the time when ion beams can be used for the wiring of a matrix on a semiconductor chip. There exists, indeed, some evidence for this growing optimism. Crystal damage by ion bombardment takes place, but a relatively low-temperature annealing seems adequate to anneal-out such damage. Acceptable values have also been reported for another important semiconductor parameter, namely, carrier lifetime. In a study of the current-voltage characteristics of heavily doped Si junctions formed by alkali ion bombardment, lifetimes were reported as large as 0.25 μsec [18]. This is comparable to the lifetimes of the best diffused Si junctions.

It is also interesting to note that ion-implanted atoms (up to $10^{18}/cm^3$) have been reported with an apparent one-to-one correspondence between the number of implanted ions and the final number of substitutional atoms in silicon (when these specimens have been heated after ion impact to move interstitial atoms into substitutional sites). Thus a combination of using high-kinetic-energy particles together with thermal cycling seems to be indicated in many instances. The important point is that lower annealing temperatures can probably be employed.

But the greatest enthusiasm among industrialists seems to relate to the unusual doping profiles that may be realized and the possibilities of automation. Burrill et al. [19] report that solar cells are already being made by implantation, with efficiencies equivalent to those of the best production-line diffusion cells. Also, conventional diffusion procedures can, at best, yield only abrupt profiles or exponential shapes. Ion-beam injection affords an opportunity for achieving gaussian and special profiles—with three-dimensional effects being limited only by the sophistication of the ion gun, the beam analyzer, and the physical properties of crystals. The amount of dopant material required is also so small that automated assembly lines could eventually be economically attractive in device production.

If the above is true, the mass spectrometer (or its production counterpart) may find itself becoming the most valuable single tool in all semiconductor work, both for research and device fabrication.

THE NEED FOR RESEARCH

Additional research becomes imperative for assessing the true potential of solid-state-device fabrication by ion doping. Detailed information of the interaction of ion beams with both amorphous and crystalline solids is far from complete—especially in the low-energy range that might be of greatest commercial interest. Special mass spectrometers are needed that can (a) resolve ions of 10 to 100 keV in energy, (b) provide a high degree of angular collimation ($<0.05°$ for channeling), (c) identify the secondary spectrum of ions passing through crystals, and (d) provide quantitative doping.

One such instrument that is specifically designed to accommodate this type of research is currently under construction in our laboratory at the Rensselaer Polytechnic Institute. The frontispiece is a photo of the completed tandem magnet system. These magnets have a mean radius of curvature of 122 cm, a magnet gap of 5.8 cm, and a maximum magnetic field strength of 7300 gauss. With this field, singly charged ions of B^{11}, P^{31}, and Sb^{121} could, in principle, be analyzed at energies of 3.48, 1.24, and 0.32 MeV, respectively. In practice research will probably be limited to 100-keV ion energies, because of limitations of suitable ion sources and the potentials that are feasible for the electrostatic lenses. What is of greater research significance, however, is the fact that the completed instrument will resemble the four-stage spectrometer described in Chapter 2. Thus the apparatus will comprise two independently programmable, double-focusing systems. With these and auxiliary facilities, it is expected that experiments can be performed relating to (a) ion ranges; (b) channeling and Rutherford scattering; (c) charge exchange; (d) "in situ" doping and semiconductor measurements of Hall voltage, conductivity, and lifetime; (e) supersaturation; and (f) experimental fabrication of integrated circuits.

In particular, highly informative measurements can be made on ultra-thin single crystal specimens that are placed at the focal point between the magnets. The primary ion beam can be well collimated, and the initial ion-energy and charge state can be unambiguously determined. As a function of crystal orientation, the momentum spectrum of channeled atoms can then be obtained by programming the second magnetic analyzer. Hopefully, effects relating to ion damage, saturation, diffusion, and other effects will also be amenable to interpretation.

REFERENCES

[1] W. Shockley, U.S. Patent No. 2666814, issued Jan. 19, 1954.
[2] K. Lark-Horowitz, et al. U.S. Patent No. 2588254, issued Mar. 4, 1952.
[3] F. M. Rourke, J. C. Sheffield, and F. A. White, *Rev. Sci. Instr.,* **32,** 455 (1961).
[4] J. Alvager and N. J. Hansen, *Rev. Sci. Instr.,* **33,** 567 (1962).
[5] R. R. Ferber, *IEEE Trans. Nucl. Sci.,* **10,** 15 (1963).
[6] W. J. King, J. T. Burrill, S. Harrison, and F. Martin, *Nucl. Instr. Methods,* **38,** 178 (1965).
[7] K. E. Manchester, C. B. Sibley, and G. Alton, *Nucl. Instr. Methods,* **38,** 169 (1965).
[8] J. A. Davies, J. Friesen, and J. D. McIntyre, *Can. J. Chem.,* **38,** 1526 (1960).
[9] B. Domeig, Presented at the 4th Scandinavian Isotope Separator Symposium; Stockholm, June, 1963.
[10] J. A. Davies, L. Eriksson, and P. Jespersgaard, *Phys. Rev.,* **161,** 219 (1967).
[11] H. J. Matzke and J. A. Davies, *J. Appl. Phys.,* **38,** 805 (1967).
[12] R. W. Bower, R. Baron, J. W. Mayer, and O. J. Marsh, *Appl. Phys. Letters,* **9,** 203 (1966).

[13] J. F. Gibbons, A. El-Hoshy, K. E. Manchester, and F. L. Vogel, *Appl. Phys. Letters,* **8,** 46 (1966).
[14] C. N. Klahr and M. S. Cohen, *Nucleonics,* **22,** 63 (1964).
[15] J. J. Lander, *J. Phys. Chem. Solids,* **3,** 87 (1957).
[16] J. O. McCaldin, *Nucl. Instr. Methods,* **38,** 153 (1965).
[17] H. Strack, *J. Appl. Phys.,* **34,** 2405 (1963).
[18] M. Waldner and D. E. McQuaid, *Solid State Electronics,* **7,** 925 (1964).
[19] J. T. Burrill, W. J. King, P. McNally, and S. Harrison, *IEEE Trans. Electr. Devices,* **ED–14,** 10 (1967).

Chapter 8

Materials Research

Studies relating to materials occupy the center of the stage in modern engineering. Materials are the concern of designers of ballistic missiles that will last 25 min, a satellite that will last a few days, or a Mach 3 aircraft having a useful life of a decade. In other instances materials may be designed for device use when the lifetime is only a microsecond.

Contrasted with the pre-World War II era, classical tests of metals now comprise but a small portion of the total spectrum of materials research. The structures of metals are being examined by x-ray and neutron diffraction, and by electron microscopy. The thermal, electrical, optical, and even acoustical properties of matter are being exploited as transducers in power and guidance systems. Thermoelectric generators, fiber optics, semiconductor diodes, and radiation-treated plastics now comprise the material-sensitive elements of a new generation of components and systems. The gaseous impurities in refractory metals, the vapor pressure of solids, and the concentration of impurities at grain boundaries are but a few of the topics demanding detailed analysis to relate microscopic phenomena to macroscopic characteristics. We can, therefore, expect every possible analytical technique to be exploited in materials-oriented research. A conservative prediction would also be that some type of mass spectrometric assay will be needed in an increasing number of investigations.

SINGLE CRYSTALS

With respect to this selected class of materials, the statement is often made that "there is no such thing as a 'negligible impurity.'" In polycrystalline solids the effect of minute impurities (1 part per million) can rarely be measured. Small changes in much higher concentrations of carbon in steel, for example, defy measurements in material properties. In contrast, single crystals of mercury, which can be obtained in high purity, show a marked change in critical shear and stress, at impurity levels of one atom in one hundred million [1]. The successful production of diamonds also involves the use of impurities as catalysts (e.g., Ni, Co, and Ca) that affect reaction rates and carbon phase change. Further, both carbon and graphite are normally quite inert to water solutions and organic chemicals. Their strong resistance to chemical attack is attributed to the high carbon-carbon bonds ($\sim$10.5 eV). Nevertheless, some reactions can occur, and they appear to be catalyzed by the presence of impurities. In semiconductors, of course, foreign atoms give rise to easily observed changes in conductivity.

If the impurity atom is soluble, it may reside in the lattice at either a substitutional or interstitial site. An insoluble foreign atom (e.g., SiO_2 in iron), however, has no lattice bonding: such inclusions may produce local stresses and inhomogeneities that lead to material failure.

Of course, the presence of foreign atoms in single crystals represents only one type of "defect" in the "ideal" crystal. Other common types of defects are so-called Frenkel and Schottky disorders. The former is the interstitial displacement in the crystal lattice of an ion with a hole. The latter is simply a hole or vacancy caused by an ion deficiency. Such deviations from ideal crystals give rise to color centers in halides, and in certain instances these color centers are caused by an excess of ions. Properties of phosphors are especially affected by interstitial ion excesses. Clearly the mass spectrometer cannot (at least in bulk material) provide detailed information on the above type of defects, but it can provide average impurity concentration data for impurity sensitive systems. Such information is basic to a theoretical understanding of solid state phenomena and the determination of quantized energy levels.

GRAIN-BOUNDARY IMPURITIES

The presence of internal surfaces, or grain boundaries, can greatly affect a number of properties of a solid. These regions between polycrystalline aggregates especially influence the structure-sensitive characteristics of solids, such as yield and tensile strength, ductility, and others. In some instances, yield strength changes of as much as 1000% may result from changes in grain size and the number and magnitude of the associated grain-boundary regions. Information concerning these regions is obtained by an examination of photomicrographs, by crystallographic techniques, and even through electrical measurements. Electron-beam probe methods are also now being employed to determine the chemical composition of the boundaries, which are usually markedly different from those of the polycrystalline matrix. Even in cases where the bulk purity of the solid may be quite high, the impurity content of the grain-boundary region may significantly influence a specific structural parameter. A graphic example is presented in Figure 8.1.

In the work reported by Westbrook and Aust [2], the hardness of a specimen is plotted as a function of distance from the grain boundary. It should be noted that in this instance the average impurity concentration is only of the order of one part per million. This situation points up the need for obtaining much more detailed impurity information. Not only are average impurity values needed, but one would like to identify and resolve the *spatial distribution* of all chemical constituents at either high or exceedingly low impurity levels. To date such information is difficult to obtain.

Electron-beam probe methods are very useful as high resolution and sensitivity can often be achieved. There are, however, two specific limitations. One is that the electron-beam probe is not good for detecting elements of very low atomic number. A second relates to the "range" of electrons in materials. Even if the cross-sectional area of an electron beam is very small, the penetration depth is not negligible (e.g., in silicon, even a 10-keV electron will penetrate $\sim 1.75\,\mu$). Hence it is impossible to obtain a real "surface scan" of a polycrystalline specimen.

A technique warranting serious development is that of utilizing the sputtering method suggested in Chapter 2. An ion-beam probe can have a "depth range" as shallow as only a 100 Å, which should prove to

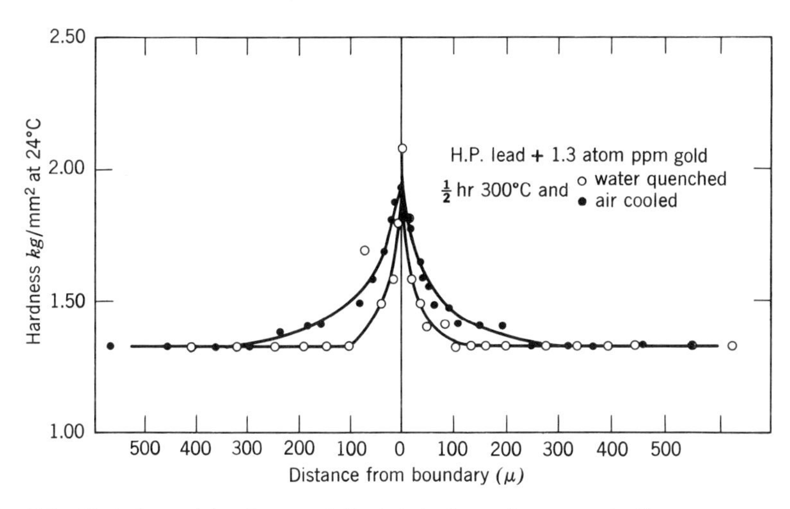

Fig. 8.1 Variation of hardness, attributed to impurity concentrations near a grain boundary [2, adapted].

offer a significant advantage. It should also provide great sensitivity. Although it is true that the sputtered atoms may have only 0.1% of their number emitted as positive ions, their initial kinetic energy is so low ($\sim$10 eV) that each secondary ion can be focused, accelerated, mass analyzed, and detected. Thus the arrival of only 1000 atoms at a suitable detector would provide specific information relating to foreign impurities. The electron-beam method, on the other hand, has a detector that can accept characteristic x-rays only from a very small solid angle. Hence the sensitivity of a mass spectral method may finally exceed all other approaches. At the very least it presents an important complementary technique.

There exist formidable development problems in obtaining a primary sputtering ion beam of 100 Å or even 1 μ (10,000 Å) diameter. Diffraction effects are negligible (e.g., helium ions of 1000 eV have an associated de Broglie wavelength of only 10^{-11} cm), but the problem of producing a sharply focused primary ion beam is exceedingly difficult. Probably the most sophisticated attempt to obtain a semiquantitative chemical analysis having a high degree of spatial resolution is the method used by Castaing and Slodzian [3]. In their apparatus, characteristic secondary ions emitted from a sample by primary ion bombardment are focused to give an image of the bombarded surface by means of suitable ion optics. Figure 8.2 is a schematic diagram of their apparatus. The secondary ions leaving the sample are accelerated by a nearly uniform electric field existing between the sample (biased several kilovolts positive) and the first electrode, which is at ground potential. Auxiliary electrodes provide focusing action to produce an ionic image of the material that constitutes the sample. Without an analyzing magnet, this image would be a superposition of several images, each resulting from the focusing of a particular type of ion. With the magnet, however, the several ion images are mass separated, with the net result that as many images are formed as there are elements. Each of these images (which can be brought into focus by the magnet) gives the distribution of the element to which it corresponds, with a resolving power of the order of a micron. Because the secondary or sputtered ions have a small distribution in their initial velocities, there will be some aberrations, but most of these can be effectively cancelled. The final ion image is post-accelerated to produce secondary electrons, which are then projected to form a final electronic image (with a magnification) on a fluorescent screen.

Although the method is only semiquantitative, it has the distinct advantage of displaying a direct distribution image in terms of a particular ion specie. It also has an advantage over the electron beam probe for light elements. A somewhat simpler approach might be to resort to col-

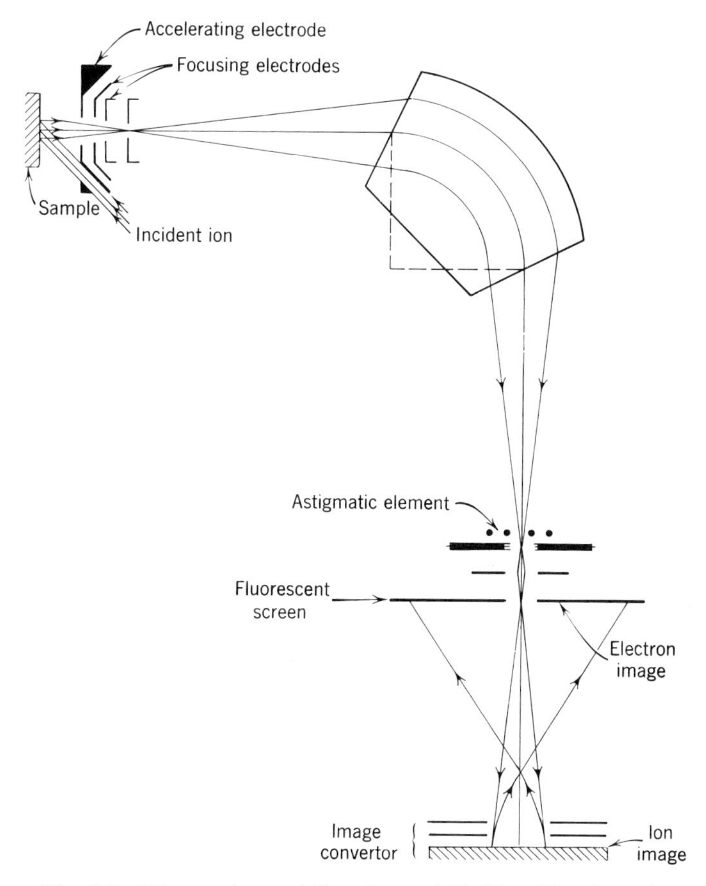

Fig. 8.2 Microanalyzer of Castaing and Slodzian [3, adapted].

limation (after some focusing) and to employ apertures of very small dimensions. One recent method for obtaining small "apertures" is to use thin mica and allow a fission fragment ($\sim$50 MeV energy) to penetrate the thin sheet. The intense ionization track can be subsequently "etched," resulting in a small hole that is impossible to produce by mechanical means. The hole diameter can, in fact, be controlled. In this manner, a primary ion beam (even if poorly focused) could be restricted in cross section to less than a micron and provide a "scanning beam" of dimensions comparable to grain dimensions. If the sample specimen were then translated with respect to the highly collimated ion beam and a cascade-type spectrometer were employed (see Chapter 2), it might be possible to display a complete sputtered mass spectrum as a function of any localized region.

To date, however, the vacuum spark microprobe is perhaps the most useful mass spectrometric tool. In this case a single spark from a 0.010-in. diameter wire will produce a small crater in a sample specimen, reported to be only 25 μ in diameter and only a few microns in depth [4]. The quantity of material removed is of the order of 10^{14} atoms. By using very sophisticated techniques, this same general method has been reported by Hickam and Sweeney [5] to provide a mass spectrometric sampling area of only 2 μ^2.

There exists a great need, however, for improved techniques in this particular phase of mass spectrometry. The embrittlement of steels, for example, which sometimes renders heavy forgings and castings liable to catastrophic failure, can probably be related to grain anomalies [6]. Hence information relating to the composition of such boundaries has a high priority.

DIFFUSION

The phenomenon of diffusion is another important parameter relating to materials. It will assume even greater importance in succeeding decades as attention is focused on components that must perform at elevated temperatures and for extended lifetimes. For impurity-sensitive systems (semiconductors, superconductors, and the like) diffusion effects may result in dramatic device failure. But the problem is also a general one. Jet aircraft will develop surface temperatures and stresses for which their exists no real engineering precedent. Nuclear reactors will be pushed to 2000°F or greater in order that higher thermodynamic efficiencies may be attained. It is true that we have both theoretical and experimental data on diffusion for pure metals and highly ordered systems. But as one leading metallurgist has expressed the situation, "We have exceedingly little high temperature diffusion data on the *practical alloys* that are envisioned for future use." A check in any materials handbook will substantiate this appraisal.

Furthermore, it is generally agreed that the mass transport of atoms for alloys cannot be calculated or extrapolated from existing data available for simpler systems. The kinetics of atomic motion in liquids and solids is complex and cannot be generalized as in the case of gases. Particles do not move independently, nor can it be assumed that the various transport properties—diffusion, thermal conductivity, viscosity, and so forth—are uniquely related. The mobility of an atom in a solid will depend not only on the temperature but also on the specific crystal structure, impurities, grain-boundary regions, and even the formation of virtual chemical compounds. The rates of atomic transport vary over phenomenal limits. The interstitial diffusion for carbon in bcc iron is such that an atom migrates about 10^{-5} cm in 10^7 sec ($\sim$4 months) at room temperature [7]. This corresponds to a diffusion coefficient, D, of approximately 10^{-17} cm^2/sec. In contrast, gold will diffuse through a $\frac{1}{16}$-in. silicon wafer at 1000°C in about 20 min [8], corresponding to a diffusion coefficient of about 2×10^{-6} cm^2/sec.

The classical basis for computing the number of diffusing atoms in a volume element between two parallel planes, x and $x + dx$, is given by

$$\frac{\partial c}{\partial t} = D \frac{\partial^2 c}{\partial x^2}. \tag{8.1}$$

This is Fick's second law for the case when the diffusion coefficient does not depend upon the concentration, c. In the general case a variation of D with temperature should be expected to follow the exponential relationship $D = Ae^{-(E/RT)}$, where A is a constant and E is the "activation energy," namely, the energy required to allow atoms to migrate over potential barriers that impede their motion. If the diffusion sample is semi-infinite (with a thickness $\gg$ penetration depth), a solution to the above equation will have the form

$$c = \frac{Q}{2\sqrt{\pi D t}} \exp\left(-\frac{x^2}{4Dt}\right). \tag{8.2}$$

Here c represents the concentration of diffusant atoms at a distance, x, and Q is the surface density of original diffusant atoms deposited upon the sample. If a thin sample (not semi-infinite) is used, a somewhat more complex expression is required.

The experimental objective is then to ascertain the spatial distribution factor in terms of the time, t. The information can be obtained in three ways:

1. Use of radioactive diffusant atoms as tracers.
2. Use of nonradioactive diffusant atoms, with postdiffusion neutron activation.
3. Direct mass spectrometric methods.

Advantages and specific techniques relating to the first two methods have been widely reported in the literature. Quite erroneously, however, several authors have stated that the use of radiotracer methods provides the only means for measuring self-diffusion in metals. The point that has been missed is that separated stable isotopes also provide (in favorable cases) an equally unambiguous method of tracing chemically identical atoms in a host matrix. When it is further recalled that two-thirds of the chemical elements possess more than one isotope, this stable tracer method can indeed be developed as a general one.

The mass spectrometric method is especially attractive for high-temperature measurements of metals having high work functions, and when the diffusant atoms have first ionization potentials below 6 eV. McCracken and Love [9] reported one of the earlier mass spectrometric methods by measuring the diffusion coefficients of Li^6 and Li^7 in tungsten. Their method consisted in coating a film of lithium on a ribbon filament of tungsten (0.001-in. thickness). This tungsten ribbon was, in fact, a simple surface-ionization source mounted in conventional fashion. The lithium, however, was placed on the back rather than on the front tungsten surface, such that the lithium atoms had to penetrate the 0.001-in. specimen before entering the accelerating section of the mass spectrometer. A reasonable number of lithium atoms were ionized as they left the tungsten ribbon, thus permitting relative intensity measurements of atom transport as a function of time. The results confirmed previous measurements and displayed discrete diffusion times for the two isotopes. By operating the ribbon filament at different temperatures, adequate data were obtained to compute the corresponding activation energies.

A somewhat different approach has been used by Schwegler [10], in our laboratory to measure uranium diffusing in tungsten at temperatures above 2000°C. A tungsten ribbon filament is initially out-gassed and uranium is subsequently allowed to diffuse into the polycrystalline specimen for several hours. This uranium-doped filament is then placed in the mass spectrometer and heated to a temperature above the boiling point of uranium for the ambient pressure. A small fraction of the uranium atoms leaving the filament will be thermally ionized and thus can be measured by the mass spectrometer. In order to relate the number of detected ions in the spectrometer to a diffusion rate the uranium concentration is assumed to be an arbitrary function, $f(x)$, of the distance into the filament. At 2000°C, it is also assumed that the uranium atoms leave the surface immediately, so that the uranium concentration at the filament surface is always zero. Under these conditions, the uranium concentration, as a function of time and distance into the filament, can be expressed as [11]

$$c(x,t) = \frac{2}{l} \sum_{n=1}^{\infty} \sin \frac{n\pi x}{l} \, e^{-\frac{Dn^2\pi^2 t}{l^2}} \int_0^l f(x') \sin \frac{n\pi x'}{l} \, dx'. \tag{8.3}$$

Here c is the uranium concentration, l is the plate thickness, D is the diffusion coefficient, and the integrals in $f(x')$ are the coefficients of the Fourier expansion representing the initial distribution. For long times, only the first term in (8.3) becomes significant, and the current of uranium atoms leaving the filament is related by the general expression:

$$J(t) = -D \left.\frac{\partial c}{\partial x}\right|_{x=0,l} = \frac{2}{\pi}\left[\int_0^l f(x') \sin \frac{\pi x'}{l}\, dx'\right] \exp\left[-\frac{D\pi^2 t}{l^2}\right], \quad (8.4)$$

$$J(t) = A \exp\left[-\frac{D\pi^2 t}{l^2}\right]. \quad (8.5)$$

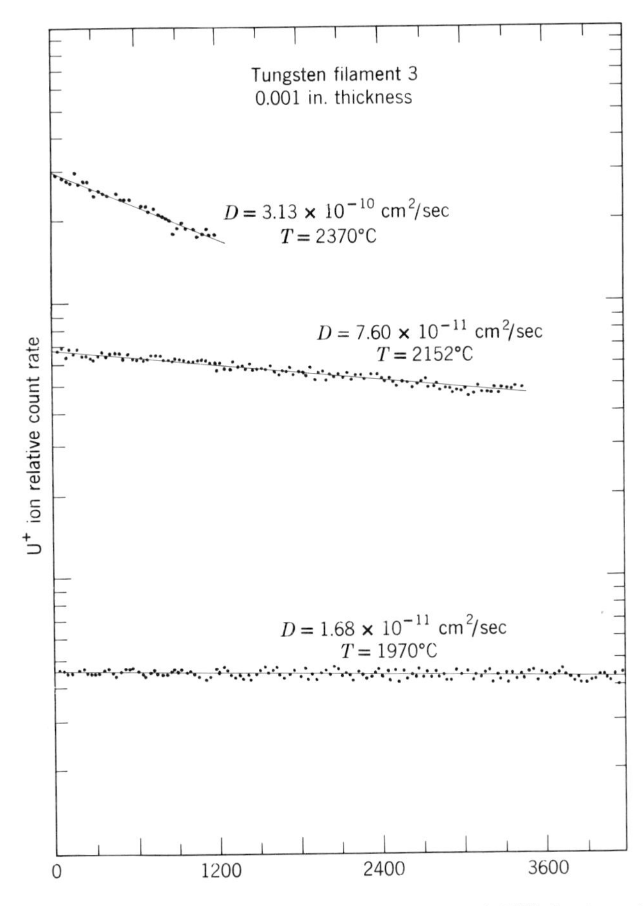

Fig. 8.3 Mass spectrometrically observed diffusion of U^{235} in tungsten at high temperatures [10].

With small uranium concentrations it has been noted that the work function of the tungsten ribbon is substantially unchanged (at a given temperature) by the decreasing uranium content. Hence (8.5) represents the analytical relationship for uranium arriving at the mass spectrometer detector. The result of plotting the logarithm of concentration vs. time for several temperatures is shown in Figure 8.3. Figure 8.4 is a plot of the logarithm of the diffusion coefficient of uranium in tungsten as an inverse function of temperature, the slope of which gives the activation energy. The validity of this method has also been checked by observing the depletion of uranium in tungsten, using reversed biased *p-n* junctions that monitored the uranium alpha activity.

Other diffusion studies have focused on the isotopic mass dependence of the diffusion process. Johnson and Krouse [12] have reported on

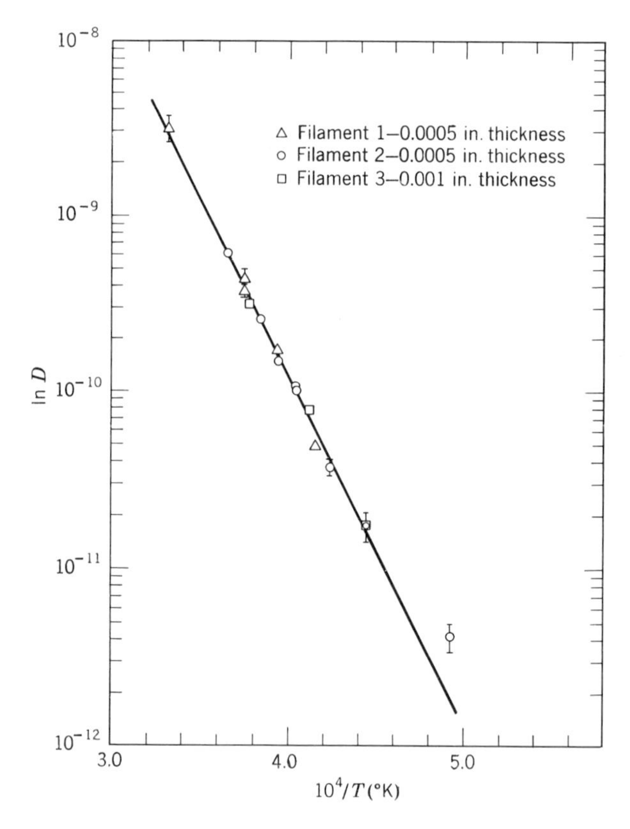

Fig. 8.4 Diffusion coefficient of U^{235} in tungsten. The slope of the graph indicates an activation energy [10].

the interstitial diffusion of Li in TiO_2 (rutile), using mass spectrometric methods, and Lazarus and Okkerse [13] have noted isotopic diffusion anomalies of Fe^{55} and Fe^{59} in silver, using radiotracer methods. In the latter investigation differences in diffusion coefficients were noted that were approximately four times larger than would be predicted by rate theory. These and other studies point up the desirability for developing new mass spectrometric methods. If surface-ionization methods are not applicable, recourse must be taken to sputtering or to other schemes for converting the emitted neutral atoms to ions for mass analysis and detection. Diffusion measurements are also needed to answer questions relating to the influence of impurities of grain-boundary diffusion, self-diffusion in impure metals, the transport of metalloids, and so on, as these phenomena may lead to an acceleration or decrease in corrosion rates. There also exist metallurgical procedures per se—carbonization, homogenization, segregation sintering, and isotopic exchange—for which mass spectrometric measurements may prove useful.

Many spectrometric studies have been made of gaseous diffusion phenomena. Pronko and Krouse [14] have also reported the use of mass spectrometry to study the directional dependence of gaseous diffusion in connection with observations of the effect of electric fields on gas diffusion in solids and a simultaneous measurement on electrical conductivity. They conclude that such studies may yield information about the energy states occupied by rare gas impurity atoms in ionic crystals.

DETECTION OF ALLOTROPIC PHASE CHANGE

In principle mass spectrometers should be able to detect the phase changes that occur in bulk material at high temperatures. Iron, for example, will change its lattice from body-centered to face-centered cubic above 900°C, and zirconium is known to have an allotropic phase change at ~860°C. The detection of this latter phase change was recently made in our laboratory [15], using a mass spectrometric method. A polycrystalline zirconium ribbon that contained approximately 50 ppm rubidium as an impurity was used as a thermal ion source. This ribbon filament was cycled in temperature from 770°C to 1030°C and subsequently down to 785°C. The diffusion coefficients of rubidium in zirconium at several experimental temperatures were obtained in a manner described above. The presence of this allotropic phase change is clearly discernible from the discontinuity in the ln D versus inverse temperature plot of Figure 8.5. The activation energy for rubidium diffusion is considerably higher in the zirconium α-phase than in the β-phase. Also, this α-phase activation energy seems to be larger before the filament has been subjected to

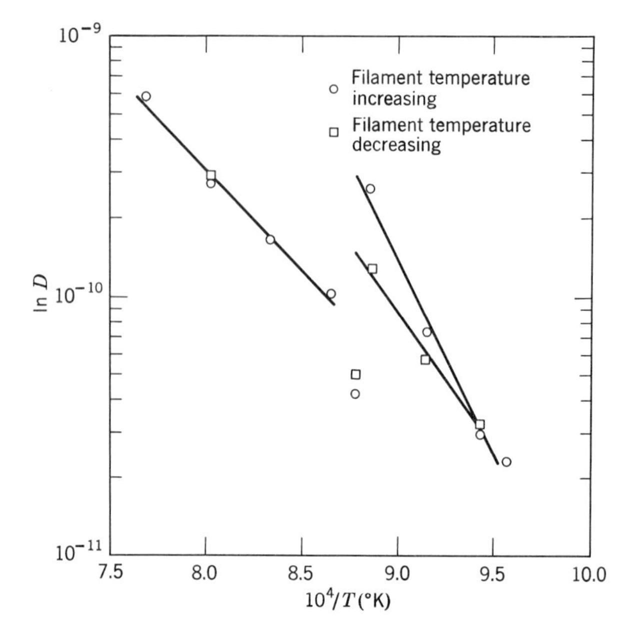

Fig. 8.5 Mass spectrometric detection of the allotropic phase change in zirconium [15].

the phase change than it is after cycling through the phase transition. The work suggests that the mass spectrometer may be developed further for examining the atomic structure of metals.

COMPOSITION OF METALS AND INSULATORS

Previous mention has been made of the importance of the mass spectrometer in the analysis of semiconductors. However, the instrument has also been frequently employed in the analysis of steels, nonferrous metals, and insulating materials. Most analyses have been made with the spark ion source in a double-focusing instrument. Gorman, Jones, and Hipple [16] were among the first investigators to analyze stainless steel samples, and they were able to determine chromium and nickel concentrations ranging from 0.25 to 25% with considerable precision. The composition of stainless and alloy steels can now be determined with a sensitivity in the parts-per-million range for the usual elements that are of interest.

Analyses have also been made for aluminum and aluminum alloys, and the analysis of Craig, Errock, and Waldron [17] (Table 8.1) is

Table 8.1

Impurity Determinations in Aluminum [17]

Element	*Composition (known, in ppm)*	*Spectrometer analysis (av. values)*
Silicon	1975	1370
Titanium	34	36
Manganese	110	93
Nickel	46	49
Copper	83	90
Zinc	33	30

typical of the impurity data that can be obtained without elaborate calibrations. Greater accuracy will usually require fairly specific information on relative sensitivity coefficients for ion production, densitometry measurements of the photographic detector, the use of accurate standard samples, etc.

Mass spectral data can also be obtained for insulating materials such as steatite and quartz [18], and powdered ceramics such as magnesium oxide [19]. In the case of the latter, samples were prepared from powdered MgO by hydrostatic or hot and cold die pressing. Considerable care is required in preparing such samples to prevent the introduction of impurities. By using careful techniques, however, it has been possible to measure impurities in the range of 1 to 5000 ppm for a large number of elements. This type of analysis is important in the ceramics industry for monitoring the impurities that may be introduced as a result of various fabricating and processing methods.

VACUUM FUSION ANALYSIS

Vacuum fusion analysis is an important method for determining oxygen, nitrogen, and hydrogen in metals. In high-grade vacuum melted metals, the gaseous content may be as low as 1 part in 10^7. Hence a sensitive method is required for detecting the relatively small concentrations of gases in test specimens. Procedures and advantages of the mass spectrometric method for this type of analysis have been outlined by Aspinal [20]. The mass spectrometer provides:

1. High sensitivity.
2. The ability to measure gaseous mixtures without prior chemical separation.

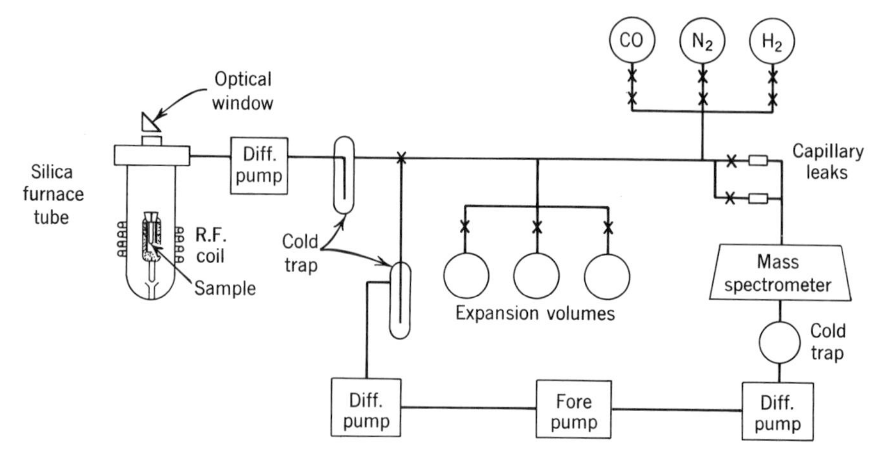

Fig. 8.6 Schematic diagram of apparatus for vacuum fusion analysis [20, adapted].

3. The ability to follow the gas evolution of a particular gaseous constituent during fusion.
4. The ability to measure other gases not normally required in vacuum (e.g., argon).

The basic apparatus required for fusion analysis consists of (a) a furnace section and associated pumps, (b) a gas handling section consisting of known expansion volumes for pressure adjustment and standard gases for calibration, and (c) a mass spectrometer that is coupled to the gas handling section by means of capillary leaks. A schematic diagram is shown in Figure 8.6. In this apparatus a silicon furnace tube supports a graphite crucible that contains the specimen. The sample is heated by a 3-kW, 2-Mc induction heater, and the temperature of the metal sample is monitored by means of an optical port and a pyrometer. According to Aspinal, who used an MS-10 spectrometer, gas concentrations can be monitored over ranges from 1000 to 0.01 parts per million. Limits of detection are approximately 0.1 μg for O_2, 0.1 μg for N_2, and 0.01 μg for H_2. The method has been used to analyze the gaseous content of iron, stainless steels, molybdenum, copper, and zirconium.

TESTS OF CLADDING AND PLATING

The cladding and plating of metals and complete components is of increasing importance in contemporary technology. Cladding of fuel elements for nuclear reactors is important to optimize heat transfer and provide for the containment of uranium fission products. The economic advantages to be gained if products can be protected against corrosion defy realistic estimates, and the protection of surfaces against wear in engines is a continuing engineering objective.

Plating thicknesses vary at least two orders of magnitude, even for electrolytic coatings. The chromium plating of a cylinder wall of a diesel locomotive engine may be 0.010 in. in thickness. Such a thick coating permits reciprocating engines to provide hundreds of thousands of miles of service. In contrast, chromium plating of decorative automotive trim may be "a few atom layers" (0.0001 in.); although this is macroscopically thin, it is still 25,400 Å, a dimension that is large compared to the "range" of low-kinetic-energy ions that are used in mass spectrometry. It is for this reason that ion beams provide the ultimate as a research tool for detecting discontinuities in plating even thinner than the thinnest metallic coatings used by the canning or automobile industries.

Consider the proposed experimental situation depicted schematically in Figure 8.7. Let a substrate material (Cu) be coated with a thin layer of chromium. In this hypothetical example of plating, let the chromium surface be restricted to 1/100,000 of an inch. Let us now take a random sample area for testing with an ion beam (with a cascade ion-beam analyzer or microprobe—see Chapter 2). If the plating is perfect, only the Cr mass spectrum will appear. If, however, even the slightest flaw is present (which may be beyond the resolving power of an optical microscope), eventually sputtered copper atoms will appear in the secondary mass spectrum. This type of test is feasible because even

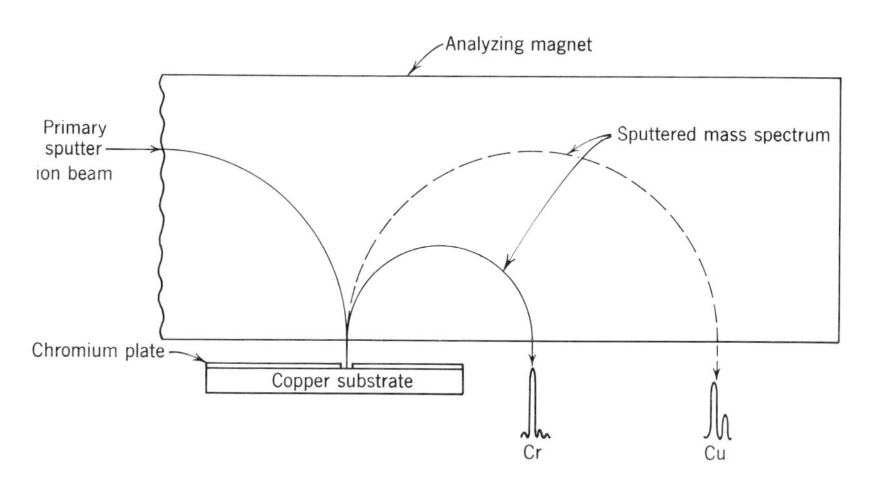

Fig. 8.7 Schematic diagram of an ion-beam probe for detecting defects in thin films.

1/1,000,000 in. of chromium is sufficient to completely stop all the 5-keV argon atoms that might be used as an "ion-beam probe."

To date, there exists no comparable method for making a positive identification of microflaws; I would predict some adaptation of this technique for developing thin film technology in which the integrity of the cladding is an important specification. It might also be noted that multilayer films (of separated isotopes) could be monitored and provide useful information with respect to plating techniques.

MASS SPECTROMETRY AND ACTIVATION ANALYSIS

Because of the wide-spread publication of papers and books that relate "activation" analysis to materials research, the applications of this technique are reasonably well known. Neutron activation is indeed a powerful analytical tool. But it does seem appropriate to make a general statement concerning the potential and real advantages and disadvantages of stable vs. radioactive isotopic assays. Both mass spectral measurements and radiotracers will yield important data on chemical composition and phenomena relating to mass transport. Radiotracers can also furnish macroscopic information relating to thickness and density. The latter have been used to improve methods of zone-refining, determine vapor pressures, examine the distribution of alloying components, locate phase boundaries, and detect the origin of unwanted metallurgical inclusions. Radiotracers are attractive candidates in materials research because of (a) simplicity, (b) sensitivity, and (c) specificity.

The instrumentation required is neither very expensive nor complex. The advent of reversed biased *p-n* junctions as detectors of nuclear radiations has provided greatly increased energy resolution for the detection of fission fragments, alphas, betas, and protons. Geiger counters, ionization chambers, scintillators, and the like have also become more useful as multichannel analyzers have been added to these primary sensors. In specific instances, the sensitivity is impressive. If a β^- emitter of 45-day half-life (e.g., Fe-59) provides only 100 disintegrations per minute, it can be shown that this decay rate is generated from slightly less than 10^{-15} gm of the radionuclide. With good instrumentation and a reasonably "clean" decay spectrum, a radiotracer is also specific and unambiguous. Transuranic elements can be identified by the spontaneous fission of nuclei or high alpha activities. Other excited nuclei have characteristic beta or gamma spectra. If the decay schemes and half-lives are known (and the sample is thin), the material can be identified.

The "activation" of stable nuclides is also a powerful method. The atom to be identified may be transmuted to a radioactive nucleus by thermal or fast neutrons, energetic charged particles, or even via photonuclear (γ,n) reactions. Thermal neutrons have been used to detect parts-per-million impurities in aluminum, iron, copper, nickel, and many metals. The detection of certain rare earths is especially easy because of their relatively high thermal neutron cross section. Organic materials are likewise amenable to similar analysis. By combining neutron activation with quantitative radiography, elemental distributions are observed.

Table 8.2

Neutron Activation and Mass Spectrometer Sensitivities [21]

			Sensitivity (in gm)	
Element	*Activated product*	*Half-life*	*Radioactivity counting*	*Mass spectrometry (estimates only)*
Aluminum	Al^{28}	2.3 min	5×10^{-9}	
Arsenic	As^{76}	26.8 hr	5×10^{-11}	
Barium	Ba^{139}	83 min	3×10^{-9}	
Beryllium	Be^{10}	2.7×10^{6} y		1×10^{-10}
Bismuth	Bi^{210}	5.0 d	5×10^{-8}	
Calcium	Ca^{49}	8.8 min	1.5×10^{-6}	
Cesium	Cs^{134m}	2.9 hr	5×10^{-10}	1×10^{-14}
Cobalt	Co^{60}	5.2 y	5×10^{-10}	
Copper	Cu	12.8 hr	1×10^{-10}	
Erbium	Er^{168}	Stable		5×10^{-11}
Gadolinium	Gd^{158}	Stable		5×10^{-13}
Gold	Au^{198}	2.7 d	5×10^{-12}	
Iodine	I^{128}	25 min	8×10^{-11}	
Iron	Fe^{59}	45.1 d	1×10^{-7}	
Lead	Pb^{290}	3.3 hr	5×10^{-6}	1×10^{-10}
Magnesium	Mg^{27}	9.5 min	1.4×10^{-7}	
Manganese	Mn^{56}	2.58 hr	1×10^{-11}	
Nickel	Ni^{65}	2.56 hr	1×10^{-8}	1×10^{-12}
Potassium	K^{42}	12.5 hr	1×10^{-7}	1×10^{-15}
Rubidium	Rb^{86}	18.7 d		1×10^{-14}
Samarium	Sm^{148}	1.2×10^{13} y		3×10^{-13}
Silicon	Si^{31}	2.62 hr	3×10^{-5}	
Silver	Ag^{108}	2.3 min	3.4×10^{-9}	
Sodium	Na^{24}	15 hr	3×10^{-9}	
Strontium	Sr^{87m}	2.8 hr	1.3×10^{-9}	
Sulfur	S^{37}	5.1 min	7×10^{-5}	
Uranium	U^{236}	2.39×10^{7} y		1×10^{-11}
Vanadium	V^{52}	3.76 min	3.8×10^{-10}	1×10^{-11}
Zinc	Zn^{69m}	13.8 hr	8.3×10^{-8}	

Relatively few fast neutron radiations are in vogue—oxygen in beryllium, and nitrogen in hydrocarbons, N^{14} $(n,2n)$ N^{13}.

Table 8.2 contains a selected list of sensitivities that have been reported [21] for a thermal neutron flux of 4×10^{12} $n/cm^2/sec$, an irradiation time of 1 hr or less, and a counting geometry of 30%. In contrast, estimates are also listed for several elements and the use of the mass spectrometer. With respect to the latter, a prime difficulty is achieving an appropriate ion source. The advantage, however, is that the ionization time can be made exceedingly short, so that even a thousand atoms ($\sim 10^{-19}$ gm) *if simultaneously collected* could yield isotopic data that would be unambiguous as to chemical identity. To date mass spectral assays (for very small samples) is not routine, and experience is usually a prerequisite. But if the development and extensive use of the electron microscope serves as an analogy, technical need rather than cost will condition the real increase in mass spectrometric instrumentation in materials analysis.

ELECTRICAL AND MAGNETIC PROPERTIES

In order to obtain the desired structural, electrical, or magnetic characteristics of materials it has been suggested that the question "How do I avoid defects?" be superseded by "How do I get a particular structure to yield optimum properties?" This is in recognition that not only pure metals but also a wide variety of new alloys and intermetallic compounds provide completely new engineering possibilities. For example, the compound Nb_3Sn remains superconducting to 18.3°K, and at magnetic fields of almost 200,000 gauss—a value which is many times the saturation value of iron [22]. Another example is the development of "composites," which utilize fibrous or particulate strengtheners in a metallic or plastic matrix. With respect to steels, mention has already been made of the use of the mass spectrometer for monitoring steels that are subjected to high vacuum for removing large quantities of dissolved gases. Mass spectrometric analyses have also been made on various grades of copper, to obtain a correlation between cuprous-oxide rectifier performance (reverse leakage current) and carbon, oxygen, and sulphur content [23].

Spectrometry has also been employed for research on electrochemical cells and to measure the gaseous decomposition products from pyrolysis of metal acetylacetonates as a function of pyrolysis temperature [24]. It is potentially useful in research relating to ferromagnetic films and magneto-optical effects, and in thin-film technology generally. It is also reasonable to assume that mass spectrometry will be an adjunct tool in research that is focused on low-coercive force materials and magnetoresistors.

Research in thermoelectricity is another area in which increased attention will be given to the identification of trace elements. This is true of both metals (including alloys) and semiconductors. According to MacDonald [25] there are at least two impurity effects in thermoelectric materials. The first relates to the scattering phenomenon of conduction electrons: if sufficient impurities are added, there will be some alteration in the basic features of the parent lattice itself and in the band structure of the conduction electrons. The second relates to resistivity. Any metal with less than 0.0001% impurity will behave substantially as a pure metal, but with thermoelectric power this will not necessarily be the case. Specifically, small amounts of transition elements (particularly Fe) in relation to other impurities present, can cause a drastic change of thermoelectric power at low temperatures.

An example of recent work is that of Malm and Woods [26], who have investigated low-temperature measurements of electrical resistivity, thermal conductivity, and thermoelectric power for silver alloys containing 0.005, 0.067, 0.11 and 0.31 atomic percent of manganese. The large differences in thermoelectric power, S, is shown as a function of temperature for the several alloys in Fig. 8.8. These investigators' data show that the thermoelectric power of pure silver is positive, while the alloys show a large negative thermoelectric power characteristic of a number of metals that have resistance anomalies at low temperatures.

A final phenomenon might be mentioned that relates more to nuclear and solid-state physics than to the electrical properties of matter. This is the Mössbauer effect, in which some of the energies associated with nuclear events are not necessarily of greater magnitude than the chemical binding of atoms in a crystal. These energies are those associated with the recoil that is imparted to a nucleus by the emission of a low-energy gamma ray (for a gamma ray of 100 keV and a nucleus of mass number 100, the fraction of available energy that is imparted to the recoiling nucleus is only 0.5 parts per million). This implies that information concerning the atoms in a crystal lattice can be obtained from the emitted gamma-ray spectrum. The only point to be made in this monograph, however, is the observation of Wertheim [27]. He has suggested that a rigorous correlation of theory to emitted gamma-ray energies applies only to an atom in a monatomic lattice of identical atoms, and that an impurity atom acts as though the host lattice atoms have the same mass only if the binding forces of all atoms in the system is the same. This suggests the possible use of isotopically pure specimens in research.

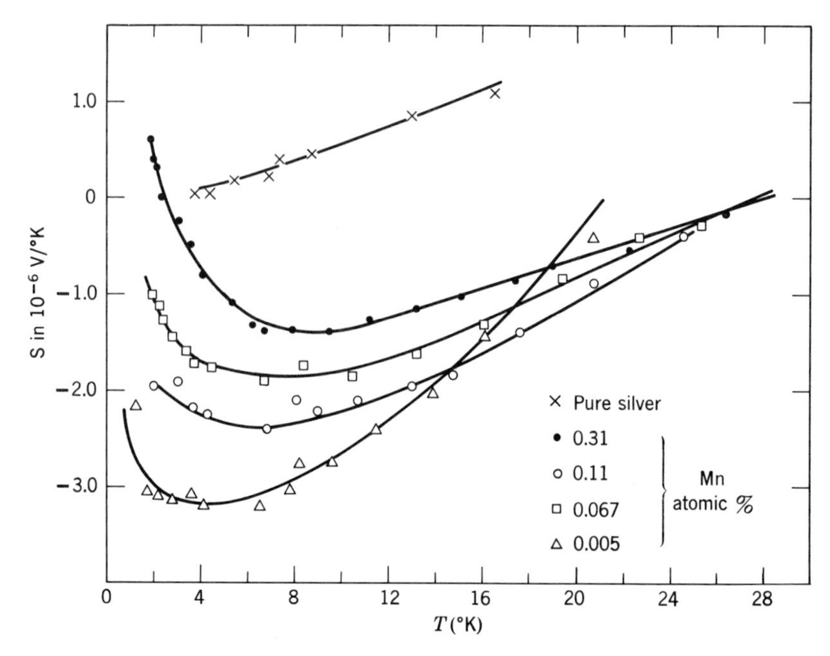

Fig. 8.8 Thermoelectric power of pure silver and of specimens containing small amounts of manganese [26, adapted].

SUPERCONDUCTORS: ISOTOPE EFFECT

The superconducting properties of metals and alloys continue to attract the attention of the physicist and the device engineer. In the area of cryogenic computer research alone, considerable interest has been generated because the sharp transition between the normal and superconducting state is potentially useful in switching circuits. Those working in the mass spectrometer field also have a related interest for two reasons. The first arises from the known effect of impurities on the value of the characteristic transition temperature, T_c, below which certain elements and compounds become superconducting. The materials which are superconducting include some of the common metals: aluminum, cadmium, gallium, indium, iridium, lanthanum, lead, mercury, molybdenum, niobium, osmium, rhenium, ruthenium, thallium, tin, titanium, uranium, vanadium, zinc, and zirconium. For dilute alloys or various solutes such as tin, indium, and aluminum, a definite correlation of T_c has been noted. For low impurity concentrations, T_c decreases linearly with the reciprocal electronic mean free path, which is impurity sensitive.

It has also been noted that for thin films (vanadium) [28], the superconducting transition temperature is increased when electron donors are deposited on the film surface and decreased when electron acceptors are deposited, the magnitude of the change being $\sim 0.1°$K.

Of equal interest is the rather dramatic dependence of T_c on the isotopic mass. According to Lynton [29], the relationship between the onset of superconductivity, an electronic process, and the isotopic mass, which affects only the phonon spectrum of the lattice, implies that superconductivity is very largely due to a strong interaction between the electrons and the lattice. An effect that was independently observed by Maxwell [30] and by Reynolds et al. [31] was that the critical temperature of mercury isotopes depends on the isotopic mass by the relation

$$T_c M^a = \text{constant}, \tag{8.6}$$

where M is the isotopic mass, and $a \approx \frac{1}{2}$. Experimentally, this relationship was observed by using an external magnetic field, for which there is a critical or threshold H_c, which can quench the superconducting phenomenon according to the relation [32]

$$H_c \approx H_o \left[1 - \left(\frac{T}{T_c}\right)^2\right], \tag{8.7}$$

where $H_o = H_c$ at $T = 0°$K. This threshold magnetic field, H_c, plotted against T_c for several enriched isotopic samples of mercury is shown in Fig. 8.9.

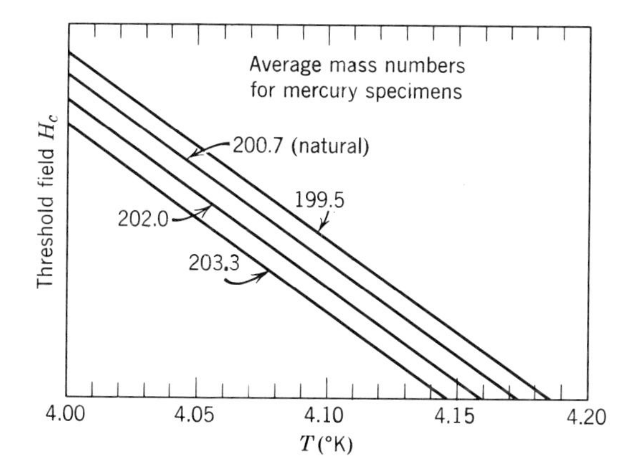

Fig. 8.9 Dependence of the superconducting transition temperature on isotopic mass [31, adapted].

Isotope effect studies have also been reported for enriched isotopic compounds of molybdenum and tungsten [33] and point up the potentially useful role for separated isotopes in basic solid-state studies.

OPTICAL AND PHOTOGRAPHIC MATERIALS

Only recently has mass spectrometry become an important tool in the industrial research laboratories of the optical and photographic industries. Of course the mass spectrometer is not so widely utilized as the optical spectrograph or infrared instrumentation. But today mass spectrometry is being applied to both research and product analysis.

In the area of glass technology the need for impurity monitoring is analogous in several respects to the semiconductor industry. In fact glasses may become one of the most important semiconductor products. Primary uses for glass continue to include precision optical elements, color filters, tube envelopes, protective coatings, insulators, and a vast array of other commercial products. New uses also include light-sensitive glasses whose transmission varies inversely with light intensity, and radiation-sensitive glasses for use in nuclear-radiation dosimetry. The development of quarter-wavelength antireflection coatings and fiber optics has also had wide application in science and commerce. However, according to Mackenzie [34], some of the most important future uses for glasses will be as semiconductors, photoconductors, magnets, transducers, optical switches, and memory materials. Such products are possible, if manufactured from glasses that will conduct either ionically or electronically and which possess a room-temperature resistivity that may range from 10^2 to 10^{20} Ω-cm. Two of the many new developments in glass science relate to (a) the invention of phototropic or photochromic glasses and (b) nonoxide glasses for infrared transmission. In the former materials are observed to change color when irradiated by light of one wavelength, but revert to their original color upon exposure to another wavelength. According to Corning Glass Works scientists [35], some silicate glasses containing about 0.5% silver halide crystals retain their phototropy for more than 300,000 cycles. Among the nonoxide glasses are elemental glassy systems containing various amounts of sulphur, selenium, tellurium, arsenic, antimony, phosphorus, silicon, germanium, chlorine, bromine, and iodine. Some of these glasses are reported to transmit to wavelengths of 20 μ.

The above are isolated examples that point up the general desirability of a total analytical capability, including mass spectrometry, in materials research in optics and glass technology. The control of activator elements in phosphors and electroluminescent materials is even more important.

With respect to photographic products, there is both direct and indirect coupling. Photographic emulsions are one of the primary detectors for mass spectral assay, so there is a continuing interest in the development of improved photoplates for this use. In the nuclear physics field, isotopically enriched impregnations of boron or lithium have been used for many years and spectrometry is used to determine such enrichments. According to Anderson [36] of Eastman Kodak, the mass spectrometer also joins the many other analytical tools required to monitor the development and manufacture of photographic film and similar materials—which is essentially a fine-chemicals industry. As a chemical tool capable of detecting elements at submicrogram levels, there is no superior instrumentation.

PRODUCT CODING

The identification and cataloging of materials, products, and even people seem to be assuming greater importance in every phase of human activity, technology, and commerce. With an exploding world population, the reader may be philosophically disturbed at events that seem to forecast that serial numbers and computer codes will provide the only "instant identification" for people, places, surgical operations, food flavors, colors, music, and human emotions. In any event information-processing and analytical-identification procedures are probably in their embryonic stage. I was surprised to learn only a few years ago that railroads were still interested in more efficient methods for determining the precise location of all freight cars distributed throughout the nation. Presumably the inaccuracies and delays occasioned by usual reporting methods have serious economic consequences. These few paragraphs are not concerned, however, with the general problems of information transmittal or retrieval. Rather a question is raised with respect to a specialized coding problem.

Specifically, what permanent identifying property can accompany materials or products regardless of their form, structure, or ultimate subdivision into microscopic specimens? The question has more than academic interest, even for relatively large objects. For example, there exists growing concern over tracing the circulation of coins from the several mints, or in identifying bullets from an ordnance manufacturer. In criminological laboratories, of course, the objective is literally to trace everything!

The answer to the above is that there are indeed few analytical methods that qualify. Small differences in chemical composition can be purposely introduced, provided such differences are not masked by the possible subsequent introduction of impurities. Radiotracers, if very long lived, can be considered "permanent"—but sensitivity decreases with increasing half-life. There are also very practical limits in the introduction of radioactive atoms. Use of neutron or other radiation "activation" methods provides an excellent means of identification provided that a standard sample can be made available for comparison.

But variations from normally occurring isotopic ratios can be employed even though the chemical composition of multiple products is *identical.* In addition, if a prerequisite exists that a product *normally* be "tracerless" until specially analyzed, such a condition is satisfied with stable isotopes. Stable isotopes do not present a health hazard, trigger Geiger counters, or require labeling to alert a consumer. The use of different isotopic ratios may be currently questioned on the basis of economics and need, but these factors are variable. Experience has shown that uniqueness and positive identification have no fixed price tags, and the same may apply for isotopic spikes.

As an example, let us examine what a mass spectrum of an alloy steel would look like when only *one* isotope of *one* chemical element was enriched (see Figure 8.10). Even if the chemistry is identical to thousands of other ingots, these steels would be unique as initial ingots, structural members, or minute critical parts in an airplane or jet engine. If failure occurred, a mass spectral assay could be traced to a specific heat number and manufacturer. The example is a poor one, both from the standpoint of economics and application, despite the fact that certain specialty steels cost up to $3.50/lb. But it is a fact that if aircraft failure did occur, and there existed the slightest doubt about material history, the cost of product analysis would hardly be brought into question.

Isotopic product coding can be introduced in two ways: (a) by a simple change in the isotopic ratios of chemical elements normally existing in the material, or (b) by the introduction of anomalous isotopic ratios of one or more chemical additives. The number of possible codes (for steels alone) exceeds the millions—and could increase as the sensitivity and precision of mass spectral assays improve. The desirability of isotopic coding in biology and pharmaceutical products seems clear. The reader is left to envision possible exploitations of isotopic coding in his own field of interest—possibly in conjunction with other analytical methods.

Nature has provided a rich variety of mass numbers for the elements that comprise our most useful metals and alloys. In those instances in

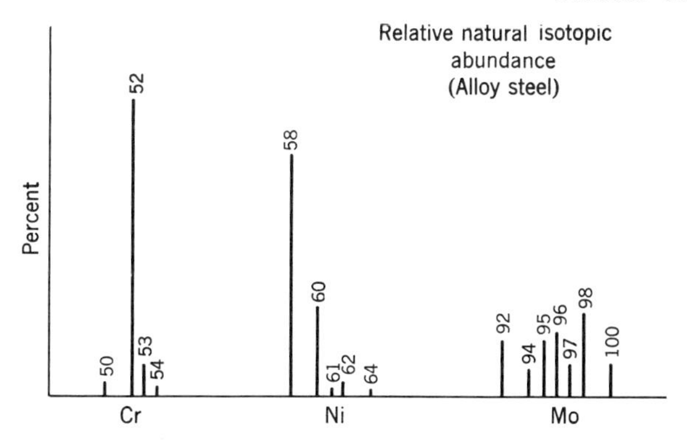

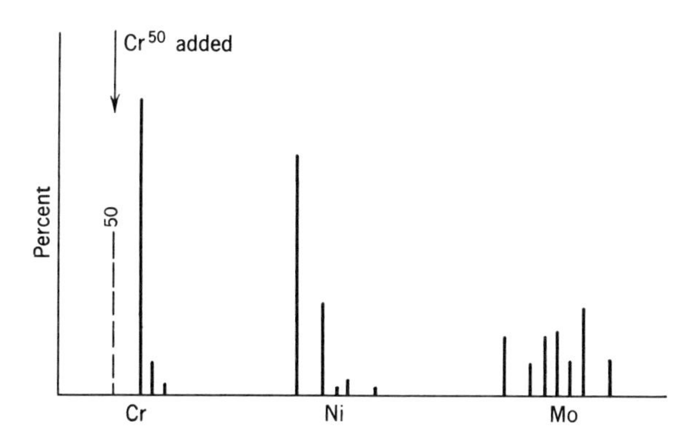

Fig. 8.10 Possible change of an isotopic ratio in an alloy steel (for identification purposes) with Cr^{50} as the tracer.

which isotopic coding is feasible, mass spectrometry will be required to assume a new role in product monitoring.

Finally, there is little doubt that stable isotopic tagging can be adapted for testing purposes on a wide scale. By fabricating isotopically coded specimens having different isotopic contents, one could monitor the wear of cutting tools, gears, or bearings; test the durability of paving asphalts; or examine the properties of lubricants. The same statement might be made for rubber products, abrasives, and manufacturing operations where wear or material transfer requires investigation. It is true that all these areas have to date utilized radioisotopes, but it requires little imagination to predict that stable tracers will surely find their special place in prototype manufacturing, pilot plant operations, and materials investigations of both a basic and applied nature.

REFERENCES

[1] K. M. Greenland, *Proc. Roy. Soc.*, **52,** A163, (1937).
[2] J. H. Westbrook and K. T. Aust, *Acta Met.*, **11,** 1151 (1963), also private communication.
[3] R. Castaing and G. Slodzian, *J. Microscopie*, **1,** 395 (1962).
[4] W. M. Hickam and G. G. Sweeney, in *Mass Spectrometric Analysis of Solids*, ed. by A. J. Ahearn, Elsevier, Amsterdam, 1966, p. 143.
[5] *Ibid.*, p. 152.
[6] J. R. Low, Jr., private communication.
[7] A. H. Cottrell, *The Mechanical Properties of Matter*, Wiley, New York, 1964, p. 201.
[8] G. J. Sprokel and J. M. Fairfield, *J. Electrochem. Soc.*, **112,** 200 (1965).
[9] G. M. McCracken and H. M. Love, *Phys. Rev. Letters*, **5,** 201 (1960).
[10] E. C. Schwegler, Doctoral Thesis, Rensselaer Polytechnic Institute, 1967.
[11] J. Crank, *The Mathematics of Diffusion*, Oxford, London, 1957, p. 45.
[12] O. W. Johnson and H. R. Krouse, *J. Appl. Phys.* **37,** 668 (1966).
[13] D. Lazarus and B. Okkerse, *Phys. Rev.*, **105,** 1677 (1957).
[14] P. P. Pronko and H. R. Krouse, *Rev. Sci. Instr.*, **38,** 871 (1967).
[15] E. C. Schwegler and F. A. White, to be published, *Int. J. Spectry, Ion Phys.*
[16] J. E. Gorman, E. J. Jones, and J. A. Hipple, *Anal. Chem.*, **23,** 438 (1951).
[17] R. D. Craig, G. A. Errock, J. D. Waldron, in *Advances in Mass Spectrometry*, Pergamon, London, 1959, p. 146.
[18] A. J. Ahearn, *J. Appl. Phys.*, **32,** 1195 (1961).
[19] A. J. Socha and M. H. Leipold, *J. Amer. Ceramic Soc.*, **48,** 464 (1965).
[20] M. L. Aspinal, presented at the Mass Spectrometer Symposium, Associated Electrical Industries, Ltd., University of London, April 13, 1965.
[21] W. H. Blahd, ed., *Nuclear Medicine*, McGraw-Hill, New York, 1965, p. 610.
[22] D. W. Lillie, *Industr. Res.* **77,** (November 1966).
[23] C. C. Hein and W. M. Hickam, *J. Appl. Phys.*, **22,** 1192 (1951).
[24] J. von Hoene, R. G. Charles, and W. M. Hickam, *J. Phys. Chem.*, **62,** 1098 (1958).
[25] D. K. C. MacDonald, *Thermoelectricity*, Wiley, New York, 1962, p. 68.
[26] H. L. Malm and S. B. Woods, *Can. J. Physics*, **44,** 2293 (1966).
[27] G. K. Wertheim, *Mössbauer Effect*, Academic, New York, 1964, p. 43.
[28] B. M. Hoffman, F. R. Gamble, and H. M. McConnell, *J. Amer. Chem. Soc.*, **89,** 1 (1967).
[29] E. A. Lynton, *Superconductivity*, Wiley, New York, 1962, p. 132.
[30] E. Maxwell, *Phys. Rev.*, **78,** 477 (1950).
[31] C. A. Reynolds, B. Serin, W. H. Wright, and L. B. Nesbitt, *Phys. Rev.*, **78,** 487 (1950).
[32] E. A. Lynton, *op. cit.*, p. 4, p. 74.
[33] J. J. Engelhardt, G. W. Webb, and B. T. Matthias, *Science*, **155,** 191 (1967).
[34] J. D. Mackenzie, *Electronics*, **1,** 129 (September, 1966).
[35] W. H. Armistead and S. D. Stookey, *Science*, **144,** 150 (1964).
[36] D. Anderson, private communication.

Chapter 9

The Physics and Chemistry of Surfaces

In 1939, when the classic motion picture *Gone With the Wind* was ready for release, the technicolor film was observed to be quite dense. A serious question arose as to the ability of theater projectors to transmit sufficient light onto the screen. The problem was solved by utilizing a phenomenon in surface physics, when the Bausch and Lomb Optical Company provided super cinephor projection lenses, coated by the firm's newly developed antireflection coatings. The intensity in screen illumination increased about 30%. This incident is of some historic interest, but it is no more dramatic than many surface-physics phenomena that are being investigated today.

There are at least three reasons for taking an optimistic view about the future of surface-physics research. The first relates to the rapid advance of vacuum technology, and the fact that 10^{-10} torr can actually be achieved for experimental work. This means that an investigator can examine a surface before a buildup of gaseous atoms takes place at the gas-solid interface. Second, the use of mass-resolved ion beams appears to be an ideal method for either depositing monolayers, introducing surface "transients," or probing the surface for its true chemical composition. At low energies, ions interact with a very small number of surface layers and they can furnish information concerning the first 100 Å with resolution and high sensitivity. An ion microscope can also be coupled to a mass spectrometer in special situations. Third, the design of advanced devices is often specifically dependent upon surface properties. This latter factor is perhaps the greatest single stimulus.

MACROSCOPIC CHARACTERISTICS OF SURFACES

Brittle solids such as glass are significantly affected by the quality of the surface. The tensile strength of rocksalt is orders of magnitudes lower than theoretically predicted, essentially due to surface imperfections and defects. The adhesion of solids and the magnitude of frictional forces are largely, but not completely, the result of surface conditions of solid-solid surfaces. The physical absorption or chemisorption of gases, catalysis, the absorption of solutes from solution, wettability, evaporation or distillation, and surface tension are phenomena reflecting the nature of the solid-gas, solid-liquid, and liquid-gas interfaces. The uptake of water in certain clays, important in such diverse fields as agriculture, civil engineering, and ceramics, has long been recognized; but only recently has the basic "surface" nature of these phenomena been even partially understood.

The importance of surface conditions has also been graphically demonstrated in research outside the scope of the usual physical sciences. With respect to water conservation, it is known that monomolecular films on water reservoirs could, in principle, conserve billions of gallons of water that would otherwise be lost by evaporation.

In entomological studies it is known that insects in dry environments may die by surface coverage of powders which are chemically inert. Admixtures of inert powders such as alumina have been used in insecticides (such as DDT) to increase their effectiveness. Death is at least partly attributed to desiccation resulting from discontinuities produced in the epicuticular wax layer and the subsequent abnormal transfer rate of water from the insect [1]. In these areas as well as in more obvious applications, separated isotopes of stable tracers may well augment other classical and radiotracer-type techniques.

A single example with respect to physical macroscopic studies should point up the possible utilization of separated isotopes in investigating mass transfer and mechanisms of boundary lubrication. A thin film of lubricant between two polished surfaces will (a) reduce metal pickup and (b) cause a decrease in frictional forces. Most lubricants serve the dual function. But while friction may be reduced by one or two orders of magnitude, mass transfer of solid to solid may be reduced ten-thousand-fold. Consider two copper specimens that are utilized for conducting friction and mass transfer tests. If the opposing surfaces consist of very pure films of Cu^{63} and Cu^{65} respectively, the mass transfer can be monitored by suitable mass spectral assay. Bearings comprising alloys and special metals could be similarly tested. Such an experimental approach may be attractive for exceedingly long life-time tests in which a sufficiently long-lived radioactive specie is not available with adequate intensity or when a radiotracer might present other disadvantages.

ELECTRICAL PROPERTIES: WORK FUNCTION

The practical importance of the electrical characteristics of the solid-gas interface can hardly be overemphasized. In particular, the measurement of the work function of a metallic surface has been a continuing objective of many investigations. The work function ϕ and changes of this parameter are crucial in devices that relate to (a) the photoelectric effect and (b) the thermionic emission of electrons.

The Einstein equation, relating the maximum kinetic energy of an electron $(E)_{max}$ leaving a surface (at $T = 0°K$) resulting from photon absorption of frequency ν is

$$E_{max} = h\nu - (e\phi)_{T=0}. \tag{9.1}$$

Thus the escape of electrons is possible only if $h\nu > e\phi$ ($\nu_0 = e\phi/h$ is termed the threshold photon frequency or "red limit.") Equation 9.1 can then be written as

$$E_{max} = h(\nu - \nu_0). \tag{9.2}$$

This relationship has been useful in determining the work function with a variety of experimental techniques. One method of observing photoelectric emission, which may also indicate the cathode temperature, T, is to plot photoemission vs. wavelength, λ. Figures 9.1*a* and 9.1*b* indicate the substantial differences in spectral responses and sensitivities of pure metals and complex surfaces [2]. Figure 9.1*a* is a characteristic curve that indicates the usual increase in photocurrent with decrease in photon wavelength for a "clean" metal. Cathodes that are coated with compounds of alkali metals tend to display maxima in a particular frequency range, as seen in Figure 9.1*b*. The attainment of a particular spectral response is, of course, a technical objective in photocathode design.

The work function plays an equally important role in the thermal emission of electrons. As in photoemission, thermal emission measurements provide a mechanism for determining relative values of the thermionic work function, ϕ. For a surface that is not space-charge limited, Richardson's equation is

$$J = AT^2e^{-\frac{e\phi}{KT}} \quad \text{or} \quad J = AT^2e^{-\frac{11,600\phi}{T}}, \tag{9.3}$$

where J is the thermionic emission current density (amp/cm^2), T is the absolute temperature (°K), k is the Boltzmann constant, A is a semiempirical constant, and ϕ is the work function (in volts). A plot of $ln\ (J/T^2)$ vs. $1/T$ yields the Richardson correspondence, where the slope is a measure of the work function of the emitting surface.

The wide variation of thermionic emission as a function of work function and temperature of several cathode materials has been reported

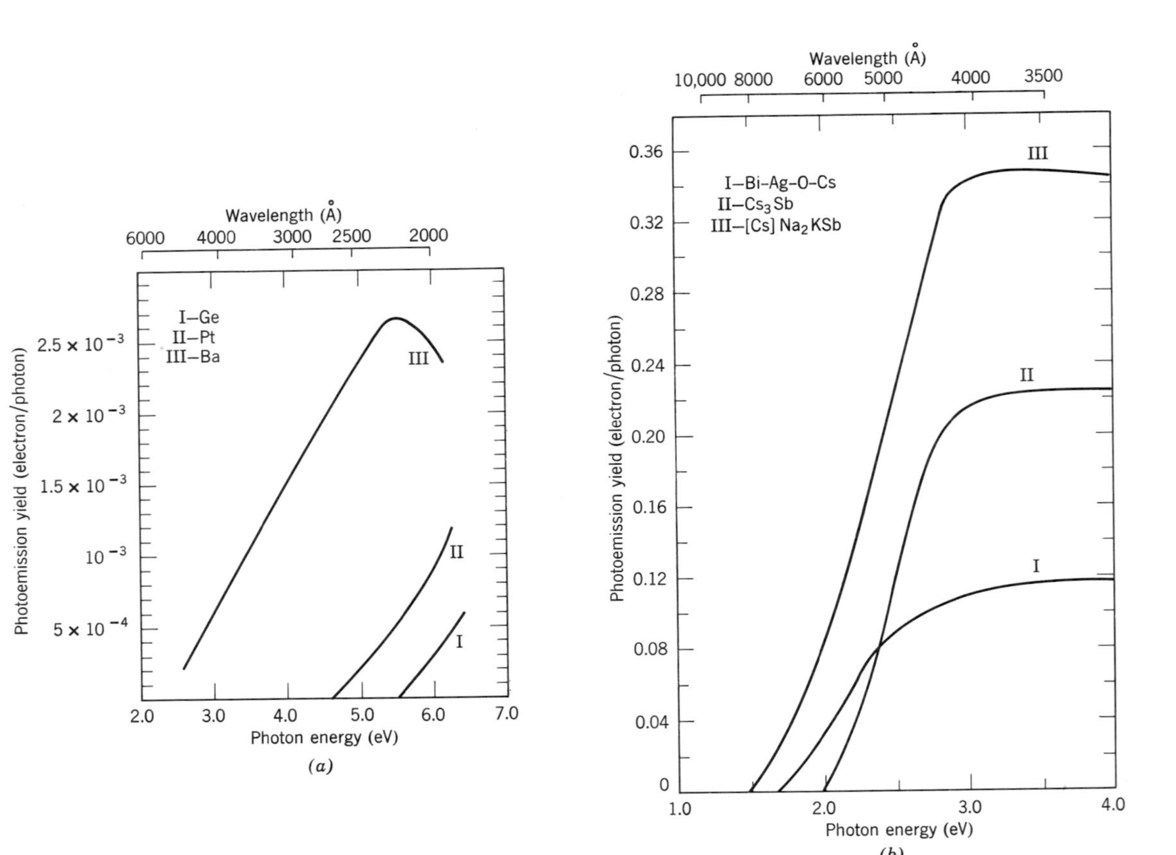

Fig. 9.1 Ratio of the photoelectron current to the photon flux as a function of photon frequency, for (*a*) a "clean" metal and for (*b*) typical composite films [**2**, adapted]. (By permission of Rheinhold Book Corp.)

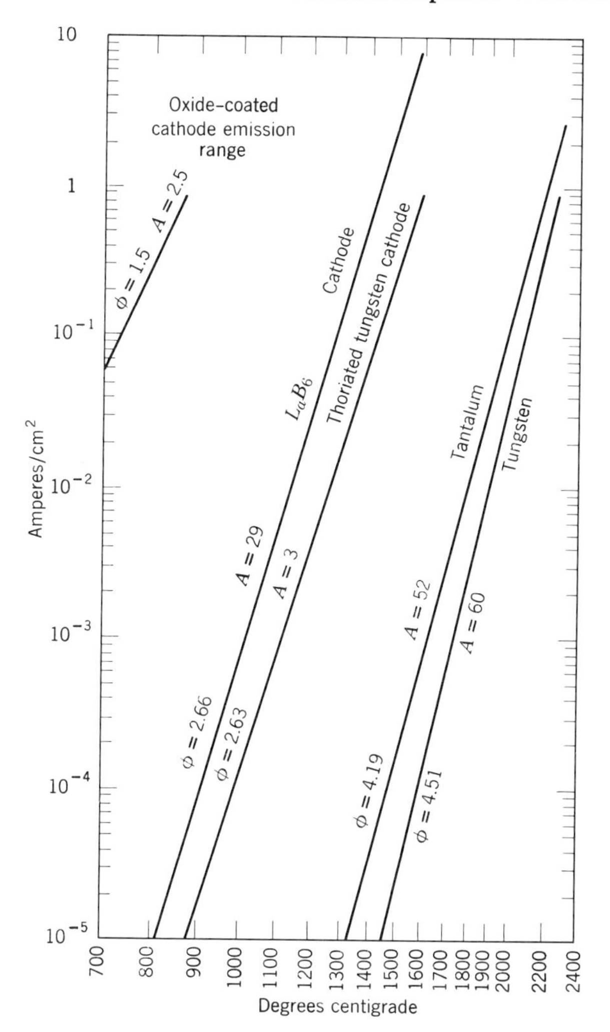

Fig. 9.2 Thermionic emission and work functions of several cathode materials [3, adapted]. (By permission of Rheinhold Book Corp.)

by Lafferty [3], and results are shown in Figure 9.2. (The work function values are for polycrystalline surfaces.) Tungsten and tantalum are widely used as cathodes because of their low evaporation rate. It will be noted that the more efficient thermionic emitters are the thoriated or oxide surfaces. Cesium, barium, and lanthanum, as well as thorium, have been used to enhance emission. If approximately 0.7 of a monolayer of these metals is absorbed on a base metal such as tungsten, the atoms are believed to be polarized, producing a dipole moment that substantially reduces the work function. In an actual electron tube some mechanism must also exist to replenish the atoms that are removed by evaporation.

In recent photoelectric and thermionic emission measurements, increased attention is being focused on the microscopic physical and chemical heterogeneity of the surface. The polycrystalline surface will yield a work function that represents an average of the work functions of all the crystal planes that are present on the surface. Rather large differences in the work function have been reported for specific crystal orientations. For example, Smith [4] has reported work functions of 4.38, 4.56, 5.26, and 4.29 for the $\langle 111 \rangle$, $\langle 001 \rangle$, $\langle 011 \rangle$, and $\langle 110 \rangle$ respective crystal planes of tungsten.

Fine et al. [5] have also employed a mass spectrometric method to measure the work function and temperature dependency for a $\langle 110 \rangle$ oriented tungsten surface in the temperature range of 1800 to 2200°K. Their work is unique in that their technique consisted of measuring the ratio of positive to negative surface ionization of an incident NaCl beam. Mean values for the work function ϕ and the temperature coefficient α in the equation $\phi = \phi_0 + \alpha T$ (for the 110 surface) are $\phi = 5.03 \pm 0.15$ eV and $\alpha = -(1.4 \pm 0.2) \times 10^{-4}$ eV/°K.

The mass spectrometer cannot usually furnish information on crystalline structure per se, but it can be exceedingly helpful in identifying "imperfections" with respect to the presence of foreign atoms in surface layers of a host lattice. Using mass spectrometric techniques, Datz and Taylor [6] have made several investigations to determine local concentrations of impurities. One of their conclusions was that these impurities were heavily concentrated in microcracks (large compared to atomic spacings) between single crystal domains. Ahearn [7] and others have also used the mass spectrometer for general assay of surface impurities. Using a spark source instrument, surface impurities have been monitored in concentrations less than 10^7 foreign atoms per 10^{17} lattice atoms. Many workers are now engaged in similar studies.

THERMAL PRODUCTION OF POSITIVE AND NEGATIVE IONS

Mention has already been made of thermal positive ion generation (Chapter 3) in connection with mass spectrometer ion sources. However, interest in thermal ionization phenomena arises from many basic and applied areas. The Saha-Langmuir [8] theory has been exceedingly useful in predicting ionization processes, yet experimental agreement with theory is generally qualitative. Only in rare instances can quantitative predictions be made, namely for a few of the alkali metals. Hughes, Levinstein, and Kaplan [9] have studied the thermal production of sodium ions from a tungsten surface. For the case of a composite superstructure of ⟨112⟩ surfaces, they obtained a work function of 5.20 eV, and from a microstructure of ⟨110⟩ faces they obtained a work function of 5.26 eV. From the slope of the "Saha-Langmuir line" they then determined the "effective" work function to be 5.25 ± 0.05 eV. This value is in substantial agreement with corresponding data from thermionic emission studies.

Deviations from the theory are generally attributed to other mechanisms that are barely understood. Effects of adsorbed layers of oxygen, diffusion, chemical reactions between ionizing species and impurities, and failure of the theory to account for extra valence electrons in cases other than the alkali metals are only a few of the effects that are not understood in detail. Perhaps the point should again be made that several interrelated phenomena, photoionization, thermionic emission, and positive or negative ion production, can probably be best further explored in a spectrometer. For example, Lichtman et al. [10] used the technique of bombarding a surface with slow electrons, coupled with mass analysis of the desorbed species, to study the system CO on polycrystalline molybdenum in the temperature range 300 to 1500°K.

There is also an increasing appreciation for the practical implications of surface ionization phenomena. Effects of ion sputtering in electron tubes, ion bombardment of electrodes and secondary electron production, and production of spurious voltages on insulators are well-known effects that, on occasion, can be deleterious. On the other hand, the production of copious quantities of positive ions is a prerequisite to acceptable performance of advanced engineering devices such as the thermionic converter and ion engines for space propulsion (Chapter 10). The fact that about two-thirds of the elements in the periodic table can yield positive ions from high-temperature, high-work-function surfaces emphasizes the technological importance of this process. Negative ions can also be produced by surface ionization, but relatively few elements have an "electron affinity" sufficiently high to generate large negative ion currents.

THE SOLID SURFACE IN CHEMICAL REACTIONS

Surfaces play a predominant role in chemical reactions and considerable literature exists on this subject. Some mass spectrometric work has been done with time-of-flight techniques, but the importance of this field appears to warrant many new investigations.

Chemisorption

Chemisorption is an effect that can be loosely defined as the formation of a chemical compound at a gas-solid interface, involving an exchange or sharing of electrons between the absorbent and absorbate molecules. Additional criteria, however, must sometimes be considered to distinguish completely between chemical and physical effects. Among the more important factors that separate the chemical from purely physical reactions are (a) reaction heat, (b) the *rate* at which adsorption proceeds, (c) the thickness of the adsorbed layer, and (d) selectivity.

In general the heats of the physical adsorption mechanisms are considerably less than the heats of chemisorption. For example, heats of chemisorption of carbon monoxide on a variety of adsorbents approximate 20 kcal./mole, whereas the heat of physical adsorption of this gas is usually less than 30% of this value [11]. As might be anticipated, the rate of chemisorption is temperature dependent and increases significantly with rising temperatures. Barrer and Rideal [12] cite the adsorption of hydrogen on graphite. At −78°C, only a physical adsorption is observed, whereas at 300°C a weak chemisorption effect was detected; above 700°C the rate was fast enough to permit the determination of an adsorption isotherm. The rate will also be dependent, of course, on other factors—such as surface porosity, which may significantly affect both the chemical and physical reactions rates.

A further distinguishing characteristic of chemisorption is the limitation in the depth of the adsorbed surface layer. Generally single atomic or molecular layers are involved in chemisorption, whereas multimolecular layers are frequently related to physical adsorption processes. Lastly, a solid surface is highly selective in that the chemisorption of gases by metals can vary over exceedingly wide limits. Although hydrogen is chemisorbed by nickel and tungsten, it does not react with aluminum or copper; oxygen is chemisorbed by charcoal but not by certain metals [13].

Modern mass spectrometry provides the basic instrumentation for exploring most of these processes. Independent control of adsorbent gas and the surface temperature of a specimen are possible. The most difficult assignment is that of ion production without dissociation in the case of secondary molecular spectra. The quantitative deposition of surface atoms (and of a particular specie) appears to be well within the capability of specialized instruments. The arrival of primary ions or neutral atoms on a "target" can also be made high compared with the arrival of "background" molecules by employing high-vacuum techniques (i.e., at $<10^{-10}$ torr). Oxygen chemisorption on tungsten [14] has been studied directly by mass spectrometry at low pressures and time-of-flight spectrometry has been used to investigate desorption phenomena of tungsten oxide surfaces in some detail [15].

Contact Catalysis

Contact catalysis can be viewed as including chemisorption processes or vice versa. It is difficult to visualize mass spectrometric studies contributing to catalysis studies of gaseous and liquid reactions or substances in solution. But for basic studies involving reaction rates of gases on surfaces of solids, mass spectrometry provides a powerful technique. Mass-resolved ion beams can be programmed to interact with diverse metallic surfaces. Auxiliary apparatus will generally be needed, however, to provide a detailed picture of catalytic phenomena, chemical intermediates, and reaction rates. In addition to classical problems such as hydrogenation it is possible that ion-beam analyzers may be used to evaluate other factors influencing catalytic activity, such as surface texture, crystal concentration, and the effect of adsorbed molecules having the role of "poisons or promoters" [16].

SURFACE COLLISIONS OF IONS AND NEUTRAL ATOMS

Inelastic Collisions

Consider the case of a gas atom or molecule having an energy, E_0, corresponding to a temperature T_0. Upon striking a solid surface having a temperature T_1 (with a surface molecular energy E_1), an energy exchange may occur between surface atoms and the incident atom or molecule. If the incident atom eventually leaves the surface in accordance with some mean absorption lifetime, it may have some intermediate energy $kT_2(E_2)$. Accordingly, a thermal accommodation coefficient, α, is defined that serves to relate the magnitude of energy change [17]

$$\alpha = \lim_{E_1 \to E_0} \frac{E_2 - E_0}{E_1 - E_0}, \quad \text{or} \quad \lim_{T_1 \to T_0} \frac{T_2 - T_0}{T_1 - T_0}. \tag{9.4}$$

It will be noted that this parameter depends not only on the temperature of an incident atom and the surface but also on a particular temperature interval. Theoretical attempts have been made to relate the accommodation coefficient to vibrational frequencies of the atoms in the solid surface, amplitudes of lattice vibrations, potential functions, and transition probabilities. An extensive bibliography relating to both theory and experimental measurements is given by Kaminsky [18]. Because of the importance of obtaining precise values of accommodation coefficients, many approaches have been employed: (a) collimated molecular beams with velocity distribution determinations of the reflected atoms; (b) thermal conductivity measurements of gases at low pressures; (c) momentum transfer studies using delicate torsion balances; and (d) detailed analyses of recoil spectra by special methods.

These experiments have dealt primarily with surfaces of tungsten, platinum, nickel, and iron, with varying conditions of the surface and at temperatures as high as 2000°K. Incident monatomic gas atoms have usually been helium, potassium, neon and argon. In general the measurements call attention to the important role of foreign atoms or surface states in the exchange of energy between a gas and a metal surface.

Elastic Collisions

Another important class of interactions arises when the impinging atom or molecule immediately leaves the surface without energy exchange (i.e., $\alpha = 0$). For the special case in which incident and "reflected" atoms make equal angles with the normal to the surface we have specular reflection. This situation is quite analogous to optics in that for a small glancing angle, θ, the projection of the height, h, of surface irregularities must be small compared to the wavelength, λ, of an incident photon, namely

$$h \sin \theta < \lambda. \tag{9.5}$$

If we recall that the De Broglie wavelength for hydrogen molecules or helium atoms at $\frac{1}{40}$ eV (room temperature) is of the order of 1 Å (10^{-8} cm), little specular reflection can be expected from most surfaces. Highly polished polycrystalline surfaces have a surface "roughness" approximating 500 Å, without special treatment. Only cleavage planes of ionic crystals provide a surface finish approximating the 1 Å value. Specular reflection for hydrogen, with a glancing angle of ~0.001°, of approximately 5%, has been observed for specially prepared specular metal alloys [19]. Diffuse reflection, however, is the predominant phenomenon that is usually encountered. Hagstrum [20] and Bradley et al. [21] have studied the reflection of noble gas ions incident normally to clean metal surface. These investigators have found that the number of reflected ions is substantially independent of energy in the 100–1000-eV region. Representative data of Hagstrum is shown in Figure 9.3, which indicates a reflection coefficient, ρ, of 10^{-3} to 10^{-4} for clean tungsten.

Although these data may be regarded as "typical" of this phenomenon for noble gas atoms, a considerable departure from the above values has been observed under varying experimental conditions. Comparison of experimental data is often difficult because of the simultaneous occurrence of several factors such as surface etching by ion impact, desorbed

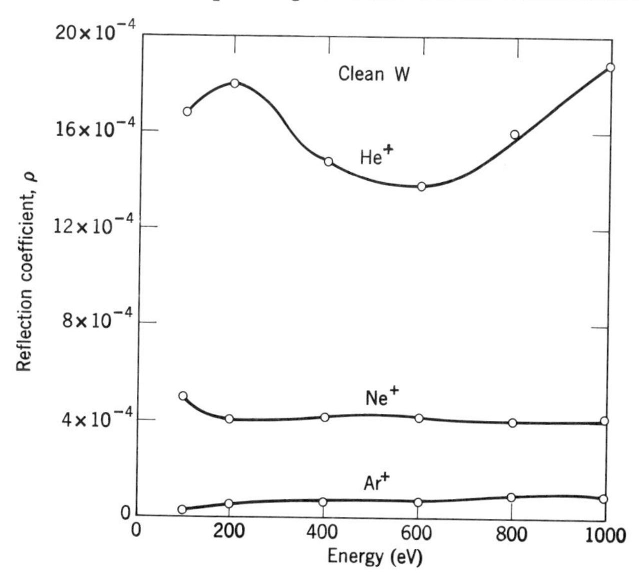

Fig. 9.3 Ion reflection coefficient, ρ, as a function of the kinetic energy of He^+, Ne^+, and Ar^+ for a clean polycrystalline surface [20, adapted].

ions, and formation of surface layers. In contrast to a pure reflection, Hagstrum [22] has also shown that He^+, Ne^+, and Ar^+, when incident on clean tungsten, may transform into metastable ions. At low energies, (100–1000 eV) this transformation increases with energy and is about an order of magnitude greater than the reflection coefficient ($\sim 10^{-2}$).

SPUTTERING OF METALS FROM ION BOMBARDMENT

As distinguished from the reflection or re-emission of incident ions, sputtering is usually defined as the emission of *target* atoms from a metal surface. It plays a role in specialized vacuum-tube devices and gaseous discharges, and it is given serious consideration in large systems such as ion engines and high-voltage generators. Sputtering is also usefully employed in the production of thin uniform films or coatings. This same phenomenon is used for achieving high vacuum in "ion-getter" pumps. Other engineering applications include surface etching, out-gassing, and cleaning. It is also probably responsible for "cathode poisoning" and shortened lifetimes of countless vacuum tubes.

A survey of experimental results up to 1955 has been reported by Wehner [23]. Much of this earlier work provided qualitative data only. In most instances the sputtered species were not mass analyzed. The relative degree of sputtering was often determined by either detecting weight differences of the targets or by optical transmission measurements of thin films. These methods were supplemented by techniques utilizing radiotracer atoms in the target. This latter method is limited, however, to target materials having convenient half lives. During the last decade mass spectrometers and electromagnetic isotope separators have provided the distinct advantage of providing detailed information relative to the charge, mass, and kinetic energy of sputtered atoms. Such instruments usually surpass all other methods in sensitivity, although the detection of neutral sputtered atoms still poses many problems.

Honig [24] and Bradley [25] conducted some of the earlier mass spectrometric measurements at low energies. Almen and Bruce [26] have extended these measurements in an electromagnetic separator in Gothenburg, Sweden. Interesting results of some of their work is shown in Figures 9.4 and 9.5. These experimental measurements were obtained using a mass-resolved primary beam from an isotope separator in which the ion current density was 10–300 $\mu A/cm^2$. Targets were commercial metal foils, and the sputtering ratio (in atoms per bombarding ion) was determined by measuring (a) the loss of weight of the bombarded target, (b) the primary ion current, and (c) the time of bombardment. Figure 9.4 shows the atom-per-ion yield of Ne^+, Ar^+, Kr^+, and Xe^+ for a copper target. The data fit the general analytical expression of Rol [27], who assumes that the sputtering ratio is proportional to the energy dissipated by the impinging particle in the first atomic layers.

Rol suggests that the chance for a collision occurring near the surface is inversely proportional to a calculated mean free path, λ, and that

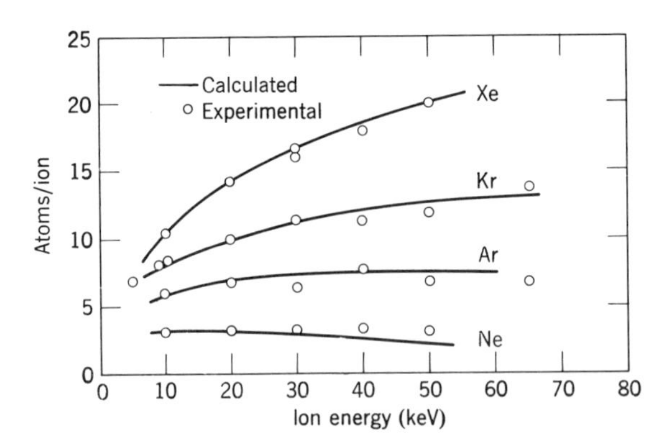

Fig. 9.4 Sputtering yield for copper as a function of ion energy of several noble gases [26, adapted].

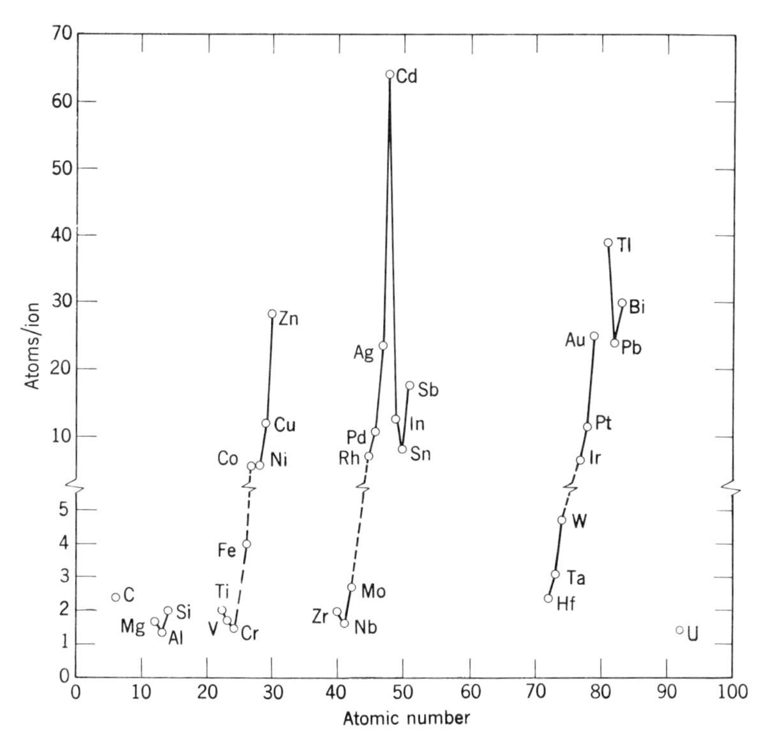

Fig. 9.5 Sputtering yields for 45-keV Kr^+ ions for different target materials [26, adapted].

the sputtering yield, S, has the form

$$S = K\frac{E}{\lambda}\frac{M_1 M_2}{(M_1 + M_2)^2} \qquad \text{atoms/ion,} \tag{9.6}$$

where E = energy of the impinging particle, M_1 and M_2 are the mass numbers of ion and target atoms, respectively, and K is a constant for a certain combination of ion and target material. By properly adjusting K values, Almen and Bruce [26] found rather good agreement with this expression over a large energy range. They also observed the interesting relative sputtering yields of many metallic targets shown in Figure 9.5.

At low ion energies (100–10,000 eV), specialized mass spectrometers provide a means for analyzing both primary and sputtered ion energies. A rough threshold value for the occurrence of sputtering is about 25 eV for the incident ion, regardless of mass. From this value to 10 keV, the number of sputtered atoms per ion is usually observed to be a monotonically increasing function of energy. The number of sputtered atoms that leave the surface as charged particles is small compared to the number of neutral atoms or molecules (10^{-2} to 10^{-4}), and the actual fraction will be highly dependent upon the particular surface. However, it is convenient to examine the sputtered *ion* spectrum and assume that this spectrum is somewhat typical of the total number of sputtered atoms or molecules. (This assumption may not always be justified.)

A spectrometer reported by White, Sheffield, and Rourke [28], which has been used for a variety of sputtering experiments at low ion energies, is similar to that of Figure 2.23. One experiment demonstrated the feasibility of obtaining reasonably precise isotopic abundance ratios from a target [29]. As indicated in Figure 2.23, a surface-ionization source produces primary ions that are accelerated through a fixed potential, E_1. These ions are analyzed in a 180° sector (12-in. radius) and they bombard the target specimen, which is biased at any desired potential, E_2. Ions originating from the target are then accelerated by the positive voltage, E_2, and follow 180° trajectories the radii of which correspond to their mass and kinetic energy. Primary and sputtered ion beams are analyzed at the same magnetic field strength in a common pole piece, but the large dimensions of the magnet and range of bombardment energies ($E_1 - E_2$) allow considerable choice in the mass of the primary ion and its energy.

Singly charged Cs^{133} ions were employed as the primary bombarding particles. E_1 was maintained at 10 kV and E_2 was varied during this investigation between 9700 and 7000 V, corresponding to Cs^+ bombarding energies between 300 to 3000 eV. The sputtered copper ions were thus accelerated by the same potentials that served as retarding potentials for the beam of primary ions. The resolution of the Cu^+ ions was complete, indicating little energy spread in the energy of sputtered ions (< 10 eV; see Figure 9.6). Measurements of the isotopic ratio were made by alternately moving the magnetic multiplier to the required mass positions. Pulses from single positive ions were fed into appropriate amplifiers, discriminators, and scaling circuits, and counted with the aid of a timer that "gated" these circuits in 10-sec intervals.

The bombarding Cs^+ ion beam was of the order of 10^{-12} A, hence there was relatively little "doping" of the sample. Cu^+ intensities at the multiplier were $\sim 10^{-16}$ to 10^{-15} A. It was necessary to "clean" the copper ribbon by heating it to $\sim$800°C to minimize the hydrocarbon background, but after such heating, ion-beam intensity ratios were found to be independent of temperature from 800°C to room temperature. Also, no change in isotopic ratio was observed as a function of the energy

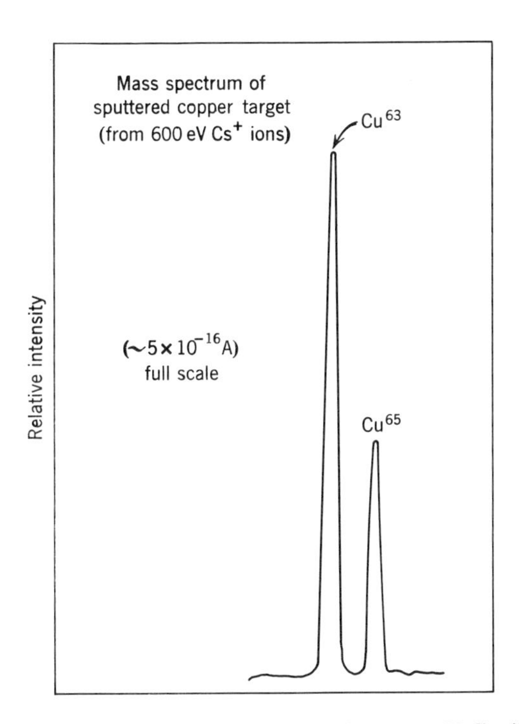

Fig. 9.6 Sputtered spectrum of copper from 600-eV Cs^+ ions [28].

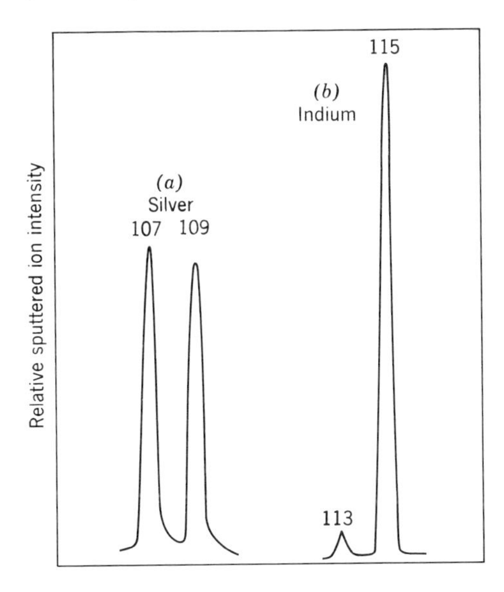

Fig. 9.7 Silver and indium sputtered spectrum from 3000-eV Rb^+ ions incident on an Ag-Mg-In-Al alloy [28].

of the bombarding Cs^+ ions, although the ionic sputtering yield increased with energy.

The mean value obtained for the isotopic ratio of Cu^{63}/Cu^{65} in this work was 2.25 ± 0.02. This value compares favorably with the adopted value of 2.24 obtained from electron bombardment of $CuCl_2$ vapor by Brown and Inghram [30]. A partial mass spectrum is also shown in Figure 9.7, of sputtered atoms of an alloy of Ag, Cd, In, and Al; the silver and indium spectrum was produced by 3000 eV Rb^+ ions.

A quite different experiment has been reported by McHugh and Sheffield [31], who used the same apparatus. The secondary ion emission from a tantalum surface was observed, using Hg^{202} ions for the bombarding beam. Eighteen different sputtered species were noted together with their relative intensities, as listed in Table 9.1.

The data of Table 9.1 were obtained with 7.4-keV Hg^{202+} ions at a beam intensity of 2×10^{-9} A, and with the tantalum target maintained at 750°C. The pressure near the tantalum specimen was 9×10^{-7} torr, so that the primary ion beam was substantially smaller than the arrival rate of foreign atoms. It was inferred, however, that at elevated temperatures the metallic ions were produced from the bulk solid rather than

Table 9.1

Secondary Ions from a Tantalum Surface [31]

Mass No.	*Specie*	*Relative intensity*
181	Ta^+	9250
197	TaO^+	10000
213	TaO_2^+	5600
362	Ta_2^+	790
378	Ta_2O^+	1000
394	$Ta_2O_2^+$	2100
410	$Ta_2O_2^+$	500
426	$Ta_2O_4^+$	380
442	$Ta_2O_5^+$	290
543	Ta_3^+	51
559	Ta_3O^+	48
575	$Ta_3O_2^+$	170
591	$Ta_3O_3^+$	80
607	$Ta_3O_4^+$	75
623	$Ta_3O_5^+$	20
639	$Ta_3O_6^+$	12
655	$Ta_3O_7^+$	3
$\approx$780	$Ta_4O_{2-6}{}^+$	1

from fragmentation of surface oxides. In this work the reflection of normally incident gas ions from the Ta surface was not observed; the specular reflection coefficient of ions to ions was estimated to be less than 10^{-6}.

Many mass spectrometric studies relating to sputtering are currently in progress, and a unified sputtering theory has been recently outlined by Brandt and Laubert [32]. Data have been obtained by Comas and Cooper [33] on the sputtering yield of single crystals of SiC, InSb, GaAs, GaP, and PbTe under argon ion bombardment. Also using argon as a bombarding gas, the energy distribution of sputtered ions from various metal targets has been carefully examined by Herzog, Poschenrieder, and Satkiewicz [34]. They have noted sputtered ions of high kinetic energy (several hundred volts), although the number of such ions is very small compared with the maximum at about 10 eV.

SECONDARY ELECTRON EMISSION BY ION IMPACT

Large numbers of secondary electrons (>10) are ejected from metal surfaces by ions of high energy (e.g., alpha particles of 5 MeV, and fission fragments of 50 MeV). The secondary electron yield from ions of low and intermediate energy is determined by many factors, and, as in the case of sputtering, the mass spectrometer proves to be an indispensable tool for obtaining reliable data. Secondary electron emission has been shown to be a function of ion energy, velocity, charge, and molecular state. It is also believed that at very low ion energies an ion approaching a metal surface may be neutralized and undergo Auger de-excitation [35] from its higher energy state. It has been difficult, however, clearly to distinguish between secondary electrons ejected by metastable atoms and those directly emitted by incident ions. Surface parameters that influence secondary yields include the specific metal, ion angle of incidence, the work-function and oxidation state, temperature, and coverage of adsorbed atoms. There is also a large effect (as is the case in photoelectric emission) if the target is covered by alkali atoms or composite layers at the surface [36].

The theory of secondary electron emission is still incomplete. Hagstrum [35] and others have discussed emission by an Auger neutralization process at low energies. This mode is suggested for systems where $eI > 2e\phi$, where eI is the ionization energy and ϕ is the work function of the metal. At incident ion energies in the kilovolt range, a "kinetic" mechanism is proposed. Several theoretical models have included (a) a microscopic zone of high temperature, (b) the interaction of free metal electrons and the electromagnetic field generated by the decelerated ion, and (c) an electric stripping of the bombarding particle. A somewhat simpler theory is suggested by Sternglass [37], who assumes that observed secondary yields can be explained by (a) considering the number of secondary electrons, n_e, produced in a metal by the incident ion and (b) finding an expression for the probability $P(x)$ for electron escape at depth x from the surface. He thus states that the secondary yield $d\gamma$ from an ion of velocity V will be related by

$$d\gamma = n_e(V,x)P(x)\,dx. \tag{9.7}$$

This equation, however, has not been transformed into a closed expression involving atomic parameters, stopping power, and the like that would allow an accurate quantitative prediction of secondary electron yields.

For low-energy ions, the energy distribution of secondary electrons has been measured by many investigators. The secondary energy spectrum reported by Hagstrum [35] for 200-eV krypton ions in different charge states is shown in Figure 9.8. The mean energy of secondary

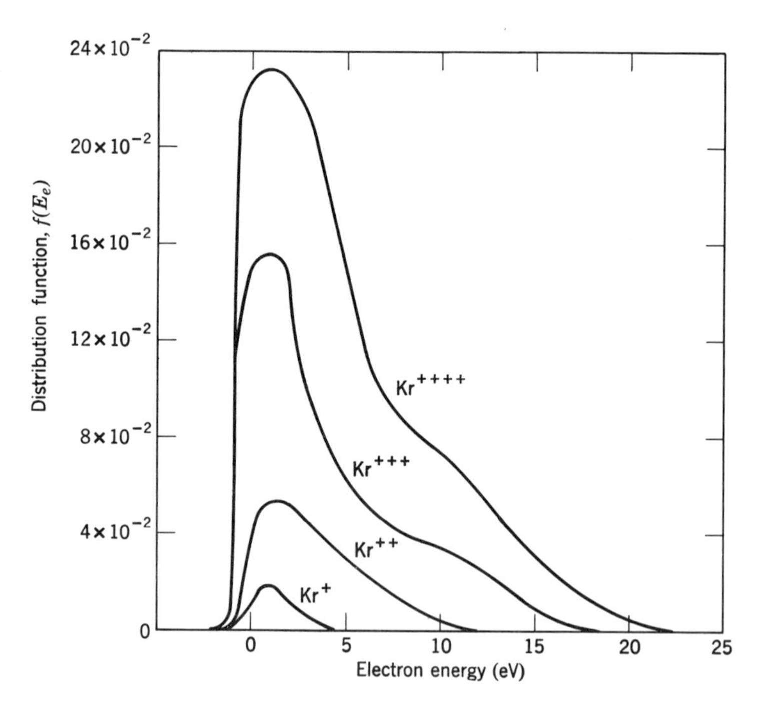

Fig. 9.8 Energy distribution of secondary electrons emitted from a molybdenum surface by krypton ions of several charge states [35, adapted].

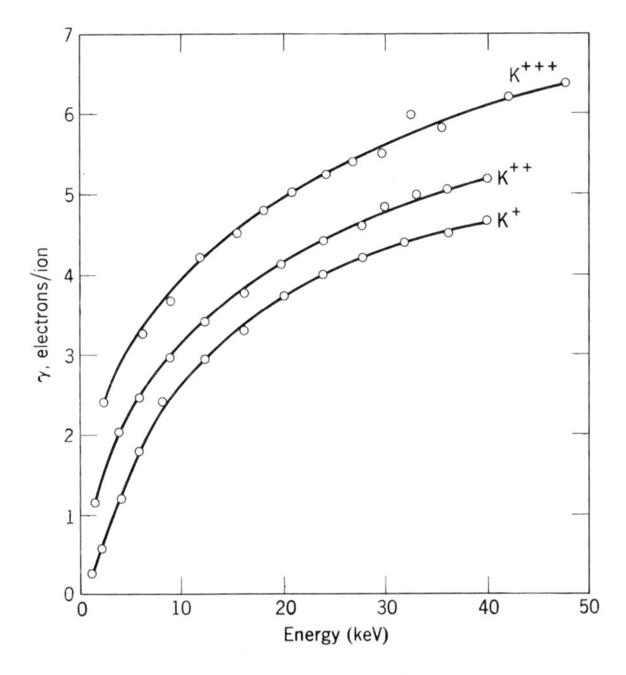

Fig. 9.9 Secondary electron yield as a function of primary ion energy and charge state for a Pt surface [38, adapted].

electrons is quite independent of the charge state, but there exists a substantial increase in relative yield with increased ionic charge. (The apparent negative energies indicated in the figure result from an experimental bias.)

Somewhat typical of the monotonically increasing yield of secondaries with primary ion energy are the data of Flahs [38]. Some of his results are shown in Figure 9.9 for singly, doubly, and triply charged potassium ions. In this case the surface is Pt, but gas-covered rather than clean. It will be noted that the difference in secondary yield is nearly independent of the kinetic energy of the ion. For a detailed review of kinetic emission and other experimental measurements, the reader is referred to the excellent survey by Kaminsky [39].

REFERENCES

[1] P. Alexander, A. J. Kitchener, and H. V. A. Briscoe, *Ann. Appl. Biol.*, **31**, 43 (1944).

[2] W. F. Spicer, in *Encyclopedia of Electronics*, ed. by C. Susskind, Rheinhold, New York, 1962, p. 600.

[3] J. M. Lafferty, in *Encyclopedia of Electronics,* ed. by C. Susskind, Rheinhold, New York, 1962, p. 859.
[4] G. F. Smith, *Phys. Rev.,* **94,** 295 (1954).
[5] J. Fine, T. E. Madey, and M. D. Scheer, *Surface Science,* **3,** 227 (1965).
[6] S. Datz and E. H. Taylor, *J. Chem. Physics,* **25,** 389 (1956), also **25,** 395 (1956).
[7] A. J. Ahearn, in *Transactions of the 6th National Symposium on Vacuum Technology,* Pergamon, New York, 1959, p. 1.
[8] M. Kaminsky, *Atomic and Ionic Impact Phenomena on Metal Surfaces,* Academic, New York, 1965, p. 101.
[9] F. L. Hughes, H. Levinstein, and R. Kaplan, *Phys. Rev.,* **113,** 1023 (1959).
[10] D. Lichtman, R. B. McQuistan, and T. R. Kirst, *Surface Science,* **5,** 120 (1966).
[11] S. J. Gregg, *The Surface Chemistry of Solids,* Reinhold, New York, 1961, p. 84.
[12] R. M. Barrer and E. K. Rideal, *Proc. Royal Soc,* **140A,** 231, 253 (1935).
[13] S. J. Gregg, *The Surface Chemistry of Solids,* Rheinhold, New York, 1961, p. 86.
[14] Y. E. Ptushinskii and B. A. Chuikov, *Surface Science,* **6,** 42 (1967).
[15] B. McCarroll, *Surface Science,* **7,** 499 (1967).
[16] G. M. Schwab, trans. by H. S. Taylor and R. Spice, *Catalysis* Macmillan, New York, 1938, p. 272.
[17] M. Kaminsky, *op. cit.,* p. 56.
[18] M. Kaminsky, *op. cit.,* pp. 70–92.
[19] F. Knauer and O. Stern, *Z. Physik,* **58,** 779 (1929).
[20] H. D. Hagstrum, *Phys. Rev.,* **123,** 758 (1961).
[21] R. C. Bradley, A. Arking, and D. S. Beers, *J. Chem. Physics,* **33,** 764 (1960).
[22] H. D. Hagstrum, *op. cit.,* p. 762.
[23] G. K. Wehner, *Adv. Electronics Electron Physics,* **7,** 239 (1955).
[24] R. E. Honig, *J. Appl. Physics,* **29,** 549 (1959).
[25] R. C. Bradley, *J. Appl. Physics,* **30,** 1 (1959).
[26] O. Almen and G. Bruce, in *Transactions of the Eighth National Symposium on Vacuum Technology,* Pergamon, New York, 1962, p. 245.
[27] P. K. Rol, thesis, Amsterdam, 1960.
[28] F. A. White, J. C. Sheffield, and F. M. Rourke, *Appl. Spectry.,* **17,** 39 (1963).
[29] F. A. White, J. C. Sheffield, and F. M. Rourke, *J. Appl. Phys.,* **33,** 2195 (1962).
[30] H. Brown and M. G. Inghram, *Phys. Rev.,* **72,** 347L (1947).
[31] J. A. McHugh and J. C. Sheffield, *J. Appl. Phys.,* **35,** 512 (1964).
[32] W. Brandt and R. Laubert, *Nucl. Instr. Methods,* **17,** 201 (1967).
[33] J. Comas and C. B. Cooper, *J. Appl. Phys.,* **37,** 2820 (1966).
[34] NASA Report CR-683, R. F. K. Herzog, W. P. Poschenrieder and F. E. Satkiewicz, Washington, D.C., 1967.
[35] H. D. Hagstrum, *Phys. Rev.,* **96,** 325 (1954).
[36] H. S. W. Massey and E. H. S. Burhop, *Electronic and Ionic Impact Phenomena,* Oxford, London, 1952, p. 308.
[37] E. J. Sternglass, *Phys. Rev.,* **108,** 1 (1957).
[38] I. P. Flahs, *J. Tech. Phys. (S.S.S.R.),* **25,** 2463 (1965).
[39] M. Kaminsky, *Atomic and Ionic Impact Phenomena on Metal Surfaces,* Academic, New York, 1965.

Chapter 10

The Space Sciences

Many mass spectrometric studies have some point of contact with space exploration. The objective of this chapter is to delineate a few areas to which mass spectrometry has already been applied and to anticipate something of the scope of its future contributions.

Today's over-all goals of the space program relate to (a) surveillance of the earth and its atmosphere; (b) communications; (c) knowledge of radiations, fields, and particles in space; (d) exploration of the moon and the near planets; (e) stellar astronomy; and (f) human space flight. These goals will be achieved with the aid of specially instrumented terrestrial-based laboratories and through the media of sounding rockets, satellites, and lunar, planetary, or deep space probes. Radiation detectors and infrared, optical, microwave, and mass spectrometers are among the more important analytical sensors that will be included in these probes.

MATTER IN SPACE

First consider the general distribution of matter in space. We might well question whether mass spectrometry would ever be called on to monitor interstellar space, for this region is "almost empty." The mean distance between stars is about ten million times that of their diameter, and the mean distance between galaxies is about 100 times that of a mean diameter. The average density of this interstellar void is practically zero, about 10^{-28} gm/cm^3, a density roughly equivalent to one proton per 10 liters. The result is the total silence and darkness of space. All acoustic signals vanish when the medium is rarefied to the extent that the distance between its constituent particles is comparable to the wavelength of sound. Thus space silence begins at about 130 km (50 miles) and above this altitude a Mach number becomes meaningless. The darkness of space also begins at roughly this same altitude, the illumination being equivalent to a moonlit night at sea level. At 160 km one approaches the blackness of deep space.

However, there are "trace" amounts of our planetary matter that are indeed amenable to mass spectrometric assay. These are sometimes subdivided as follows.

1. The planets proper.
2. Planetoids or asteroids.
3. Comets, meteorites, and micrometeorites.
4. Interplanetary particles and dust.

Some of these "particles" are of more than theoretical interest because of their potentially damaging effects on spaceships. A meteorite can be quite large and weigh up to several thousands of tons. Its size is sufficient so that not all of it is vaporized if it passes through the earth's atmosphere. Mass spectral assays of this type of material are noted in Chapter 12. Smaller-sized bodies (0.01 to 300 cm in diameter) will burn up completely in the atmosphere.

Micrometeorites and interplanetary dust have diameters of 0.01 cm to less than $0.5\,\mu$. Their velocities are only about 10 km/sec and their small size and large surface-area-to-mass ratio permits them to radiate away the heat acquired by interaction with the atmosphere; as a consequence, they are not completely vaporized. It is estimated that the earth's surface accumulates several thousands of tons of meteoric material per day. To date, considerable research has been related to measurements of particle densities and particle sizes. But these particles can also be analyzed by mass spectrometry, for both chemical composition and isotopic abundance measurements.

THE EARTH'S ATMOSPHERE

Between the earth's surface and an altitude of about 30 km is the zone that comprises about 99% of the atmosphere of our planet. This is the region of primary interest to ecologists and meteorologists, for in this zone the major forces are in evidence that determine our weather and climate. Parameters of interest at these lower altitudes include wind speed and turbulence, temperature and humidity, density, and chemical composition. Several of these factors are pertinent to modern jet transport. The jet stream itself is of considerable practical interest; at 40,000 ft, wind speeds (west to east) reach 150 mph. In a single day air masses are carried across one-fifth of the globe at latitudes of 40° to 50°.

At lower altitudes, the earth's atmosphere consists of a mixture of a number of gases, a small percentage of ionized species, meteoritic dust, and a suspension of small particulate matter. (In industrial areas the number of dust particles may be greater than $10^5/cm^3$; some of these are below optical microscope visibility.) The chemical composition of dry air at the earth's surface is given in Table 10.1 [1].

Ozone, an allotropic form of oxygen, is highly reactive and exists only in trace quantities at the earth's surface. There are also trace quantities of methane (CH_4) and nitrous oxide (N_2O). The amount of water vapor present in the atmosphere may vary from nearly 0 to 5%; an increase in the mixt[illegible] content will cause a relative decrease in the other constituents.

Atmospheric absorption of electromagnetic radiation not only influences the chemical composition of the atmosphere, but it excludes all radiation having wavelengths less than 2900 Å from the earth's surface. It is also responsible for the electron density distributions, and the temperature inversions [2] that are shown in Figure 10.1. A temperature maximum occurs at about 50 km and a minimum occurs near 85 km. Atmospheric attenuation reduces the solar constant by about one-half, or 1 cal/cm²/min. Much of the infrared spectrum fails to penetrate atmospheric carbon dioxide and water vapor, but substantially all light in the visible region reaches the earth.

Ultraviolet penetrates the atmosphere to about 30 to 50 km, where it dissociates molecular oxygen into atomic oxygen. Substantial amounts of ozone are subsequently formed in a three-body collision and additional ultraviolet interaction dissociates some ozone back into oxygen. The

Table 10.1

Composition of the Earth's Atmosphere

Element or compound	*Percent (by volume)*	*Density (gm/l STP)*
Nitrogen	78.08	1.2506
Oxygen	20.95	1.4390
Argon	0.93	1.7837
Carbon dioxide	0.03	1.9767
Neon	1.8×10^{-3}	0.9004
Helium	5.2×10^{-4}	0.1785
Krypton	1.0×10^{-4}	3.708
Hydrogen	5.0×10^{-5}	0.0899
Xenon	8.0×10^{-6}	5.851
Ozone (O_3)	1.0×10^{-6}	2.22

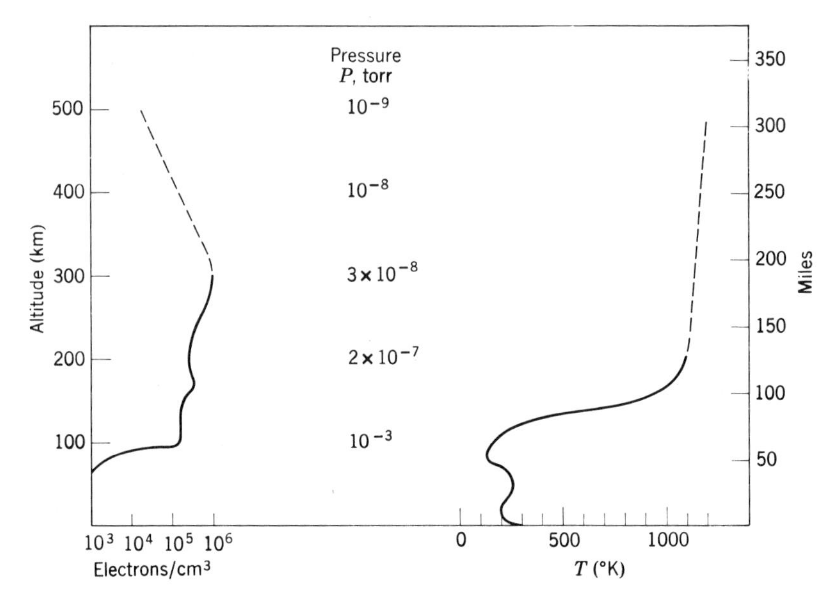

Fig. 10.1 Temperature inversion resulting from the absorption of electromagnetic radiation [2, adapted]. (By permission of McGraw-Hill Book Co.)

net photochemical effect is intense UV absorption giving rise to a temperature increase in this region (the mesosphere).

COMPOSITION AND DENSITY OF THE IONOSPHERE

The ionosphere is the highly interesting region in which the earth's upper atmosphere merges into the vacuum of space. Beginning at an altitude of approximately 80 km, many atoms and molecules become ionized, and the magnetic field of the earth assumes an important influence on the ions and free electrons. At this height a few light atoms may escape the earth's gravitational field, and in this region particulate matter is entrained from space. Long x-ray and short ultraviolet absorption by nitrogen and oxygen are in a large measure responsible for the high degree of ionization and dissociation. Some authors even suggest that there is a sufficient free radical density to serve as a useful source of energy. Oxygen becomes ionized at about 1000 Å, corresponding to a threshold energy of 12.2 eV. Nitrogen requires a threshold radiation of about 800 Å. Thus

$$O_2 + h\nu \rightarrow O_2^+ + e^-$$

$$N_2 + h\nu \rightarrow N_2^+ + e^- \qquad (10.1)$$

Other reactions that can take place at about 100 km include [3]

$$\begin{aligned}
&O + h\nu \rightarrow O^+ + e^- \\
&O^+ + N_2 \rightarrow NO^+ + N \\
&N_2^+ + e^- \rightarrow N + N \\
&N_2^+ + O \rightarrow NO^+ + N \\
&N + O_2 \rightarrow NO + O
\end{aligned} \tag{10.2}$$

At 200 to 300 km the amount of ionized gas is indeed large—about 100 parts per million; at distances greater than 1000 km, nearly all gas atoms exist in an ionized state.

Physical parameters of interest in this near space region include:

1. Chemical composition.
2. Information on the relative densities of neutral molecules, negative and positive ions, radicals, and electrons.
3. Measurements of temperature, diffusion and collision rates, and recombination.
4. Nuclear and electromagnetic radiations and radiation intensities.
5. Magnetic fields.

All of these parameters are of interest as functions of altitude, longitude and latitude, and time. Certain gross measurements can be made to yield general information (e.g., a rough estimate of atmospheric density can be measured by the effective drag encountered by a satellite), but highly sophisticated instrumentation is required to furnish specific analytical data. The mass spectrometer is clearly one of the more effective tools for the measurement of the first three of the above categories.

Mass spectrometric data that have been taken are extensive. Early United States experiments in Aerobee sounding rockets with radiofrequency mass spectrometers began in 1954. The Russians reported on radiofrequency mass spectrometer measurements of ionic compositions in 1957 and an rf mass spectrometer was carried aboard the third Soviet satellite.

The quadrupole mass spectrometer has been especially useful in ionospheric research and extensive data have been reported by Narcisi and Bailey [4]. Their instrument was the small quadrupole mass filter originally developed by Paul et al. [5] and fabricated at the Bell and Howell Research Center, Pasadena. The instrument package was placed in a Nike-Cajun rocket in a configuration as shown in Figure 10.2. The rocket nose tip was programmed for ejection at an altitude of 60 km in order

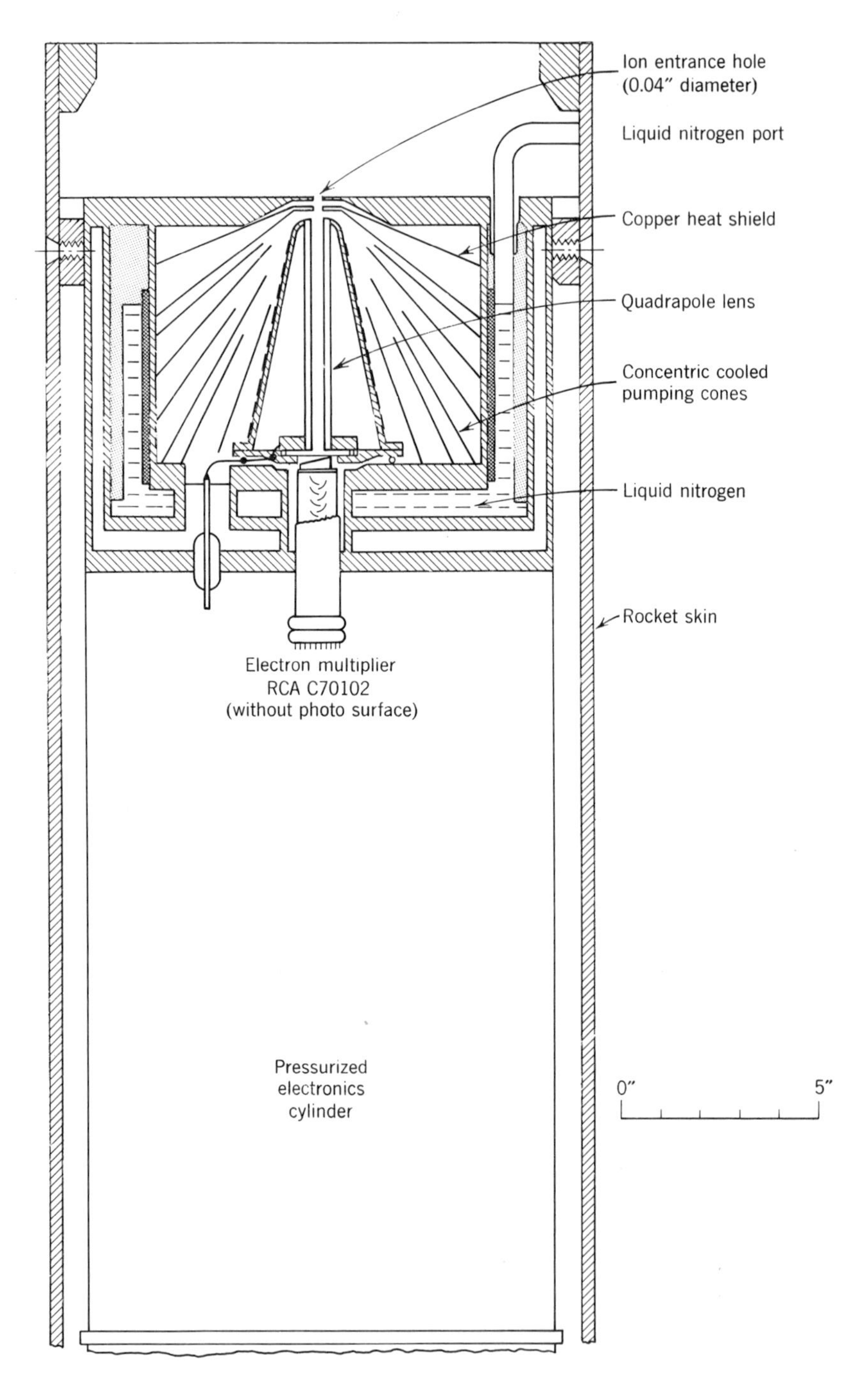

Fig. 10.2 Cross section of the quadrupole mass spectrometer in a Nike-Cajun rocket [4, adapted].

that a survey could be made of the D-region and lower E-region of the ionosphere. Preflight testing of the device in a simulated test (at the low-pressure plasma tunnel facility, University of Toronto, Institute of Aerophysics) had indicated that ion densities of $\sim 10^3$ ions/cm^3 could be monitored at the 60-km altitude. The electron multiplier amplifier system had a current sensitivity of 5×10^{-16} A. A liquid-nitrogen-cooled zeolite absorption pump surrounded the quadrupole structure, capable of 100 l/sec pumping speeds for the flight period.

Owing to the high sensitivity of the system, 20 positive ion peaks were detected in the mass range of 10 to 45 amu within an altitude range of 64 to 112 km. In addition to gaseous ion species, five ion peaks were noted that were attributed to the presence of metallic ions of $(Na^{23})^+$, $(Mg^{24,25,26})^+$, and $(Ca^{40})^+$. The metallic ions all exhibited a similar altitude profile. A partial summary of the ion-concentration profile, as a function of altitude, is displayed in Figure 10.3.

Excellent mass spectrometric data have also been obtained and reported by Nier et al. [6] on the neutral constituents of the upper atmosphere. Measurements were made at an altitude of 100–200 km, using

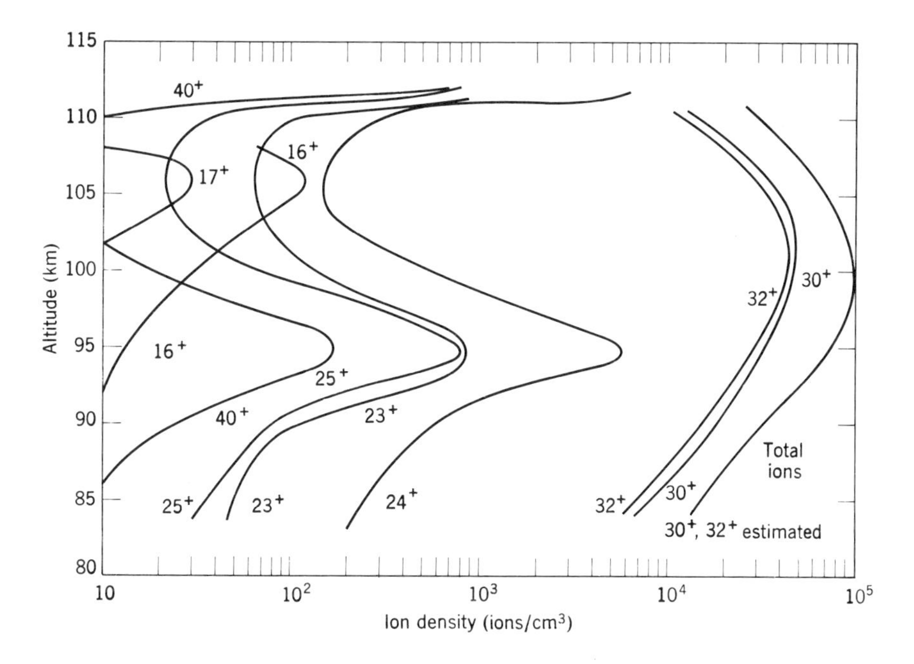

Fig. 10.3 Concentrations of positive ions detected by the quadrupole mass-spectrometer of Narcisi and Bailey within the altitude range of 83 to 112 km [4, adapted].

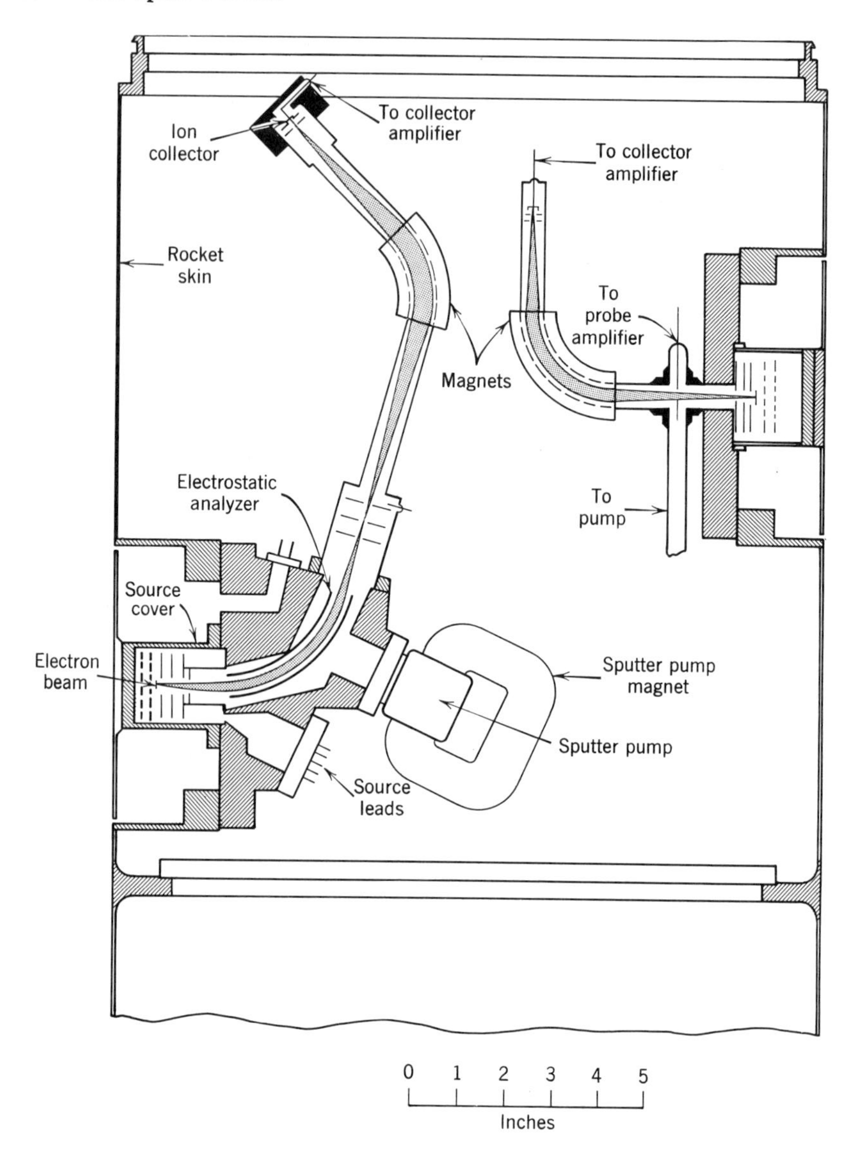

Fig. 10.4 Schematic diagram of Aerobee rocket nose cone with spectrometers [6, adapted].

a small magnetic spectrometer mounted in the nose cone of an Aerobee rocket. A schematic diagram of the rocket section, with a double-focusing and single magnet spectrometer, are shown in Figure 10.4. This instrumentation represents a highly sophisticated package and it was designed to cover the mass range to approximately 50 amu. Mass spectra are obtained by charging a condenser that is connected to the ion-accelerating, high-voltage electrodes. Mass spectra are observed during the exponential condenser decay. The condenser can be charged every 2 sec. The exponential type of voltage decay provides a mass scale that is roughly linear, and information is telemetered to ground. Caps covering the ion source region were removed at an altitude of about 100 km and a total of over 150 mass spectra were obtained during a 300-sec ascent and descent.

According to Nier [7], up to about 100 km there is complete gaseous mixing, so that the chemical composition is essentially that at earth's surface—although the pressure is reduced to about 10^{-4} torr. Above 100 km atomic oxygen formed by dissociation of molecular oxygen from ultraviolet absorption becomes a significant constituent. At 200 km there are approximately equal parts of molecular nitrogen and atomic oxygen. At 600 km, helium, which has an abundance of only 2 parts per million at ground level, becomes the principal specie. Some of the more interesting variations reported by Nier [7] for the ratios of O/O_2, O_2/N_2, and O/N_2 with altitude are shown in Figure 10.5. Other data obtained with

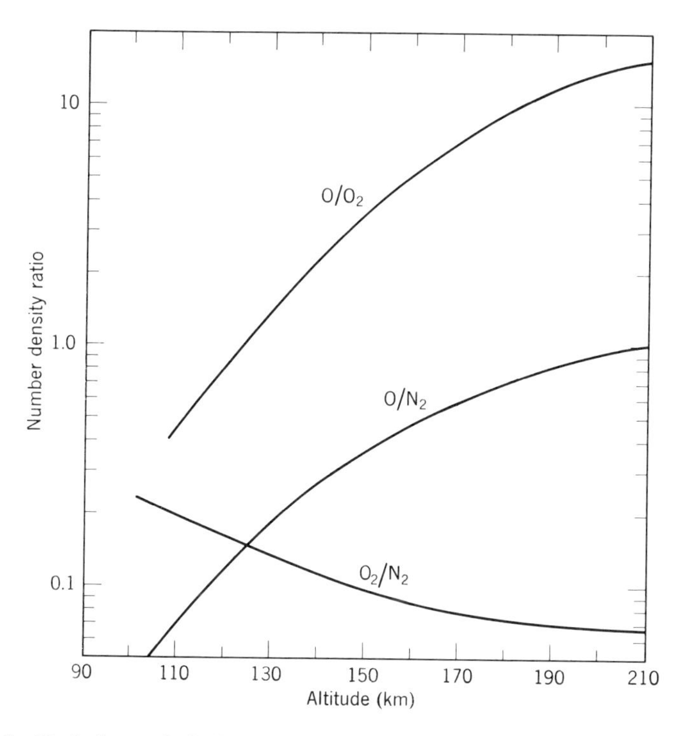

Fig. 10.5 Variations of O/O_2, O_2N_2, and O/N_2 between 100 and 200 km [6, 7, adapted].

this apparatus have been measurements of the argon-nitrogen ratio, which indicates that diffusive separation of Ar and N_2 sets in somewhat below 100 km. The accurate determination of minor constituents in the atmosphere (CO, H_2O, or CO_2) has been somewhat limited by trace impurities associated with mass spectrometer components or the rocket.

The distribution of helium, nitrogen, and argon at altitudes of up to 430 km has also been reported by Pozhunkov [8], using a rocket-mounted radiofrequency spectrometer. No neutral helium was detected in the altitude range of 130 to 370 km, but helium ions were noted in a concentration of $(7 \pm 2) \times 10^3/\text{cm}^3$ at an altitude of 430 km. Direct measurements of the helium and hydrogen ion concentrations and total ion density have also been reported by Taylor et al. [9]. Using a Bennett-type ion mass spectrometer equipped with a cylindrical electrostatic probe, data were recorded up to 940 km; H^+ and He^+ concentrations

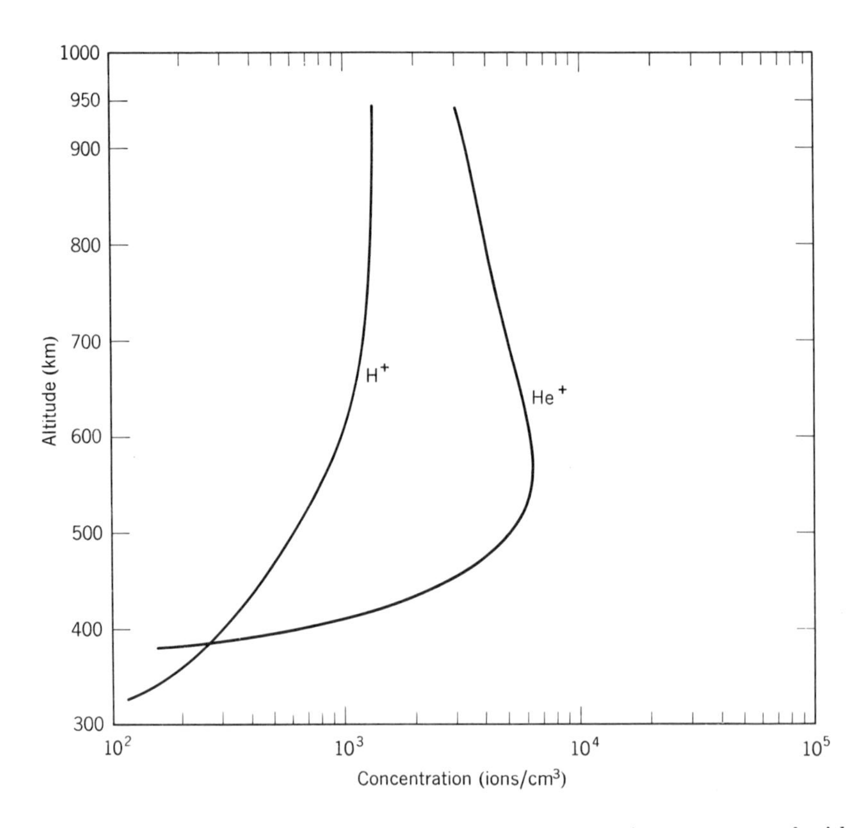

Fig. 10.6 Concentration profiles of helium and hydrogen ions as measured with a Bennett-type mass spectrometer [9, adapted].

were found to rise steeply above 350 km. Results of these measurements are shown in Figure 10.6.

UPPER-ATMOSPHERE TEMPERATURE MEASUREMENTS

Temperature measurements of the upper atmosphere pose considerable experimental hurdles, but such data are of great theoretical and practical interest. At high altitudes the electron temperature may differ considerably from the temperature of heavy particles. Further, the temperature distribution tends to explain certain physical and chemical phenomena at high altitudes. The methods for measuring temperature include (a) the determination of gaseous constituents as a function of altitude and applying classical kinetic theory, and (b) the direct measurement of the energy of upper-atmospheric ions.

A calculation can be made of the temperature, T, from the hydrostatic relationship for a gaseous system, and the perfect gas law, namely

$$\frac{dp}{dh} = -g\rho, \tag{10.3}$$

$$p = nkT, \tag{10.4}$$

where p is the gas pressure, h is altitude, ρ is gas density, g is the gravitational constant, and n is the total number density of gas. Let n_j and m_j also be defined as the concentration and mass, respectively, of the jth gaseous specie. From (10.3) and (10.4) it can then be shown that

$$\frac{dn}{dh} = -\frac{1}{T}\left(\frac{g\Sigma n_j m_j}{k} + n\frac{dT}{dh}\right) \tag{10.5}$$

This relationship suggests two measurements for the determination of T as a function of altitude, h. One is to measure the variations of n_j with altitude for a single gaseous species, and normalize to a known temperature, T, at some specific altitude. A second is to determine dn/dh for two or more gaseous constituents. Mass spectrometry is clearly the tool for this latter approach.

If the temperature is to be measured in terms of ion energies, a simple "ion probe" can be used. These probes are small spherical chambers ($\sim$10-cm radius) aboard rockets or satellites, which allow ions to enter through an orifice. Electrons are excluded by a spherical grid maintained at an appropriate negative potential. If ions alone are permitted to enter the probe, a positive ion current can be monitored as a function of the potential of the probe relative to the surrounding ion-electron plasma.

Massey [10] has shown that the general shape of the positive ion current for three discrete masses, M_1, M_2, M_3 will have the form of Figure 10.7. Successive changes of slope occur at voltages, V_1, V_2, and V_3 in the expression

$$eV_i = (\tfrac{1}{2})M_i v_s^2, \tag{10.6}$$

where v_s is the satellite speed.

Values of $i = 1, 2, 3$ correspond to the successive elimination by an appropriate voltage bias of ion currents contributed by masses M_1, M_2, and M_3, respectively. Now the satellite speed, v_s, is much higher than the average speed of the ions—because of thermal agitation. (For O^+ ions the kT may be $\sim$0.1 eV but $(1/2)Mv_s^2$ may be $\sim$5 eV). Therefore, the ion energy of mass M, relative to the satellite is approximated by $(1/2)Mv_s^2$, hence is proportional to M. Massey [11] has further shown that this assumption leads to a reasonably simple distribution function for the energy of ions, namely

$$f(E)\,dE = \frac{n}{(2\pi MkT)^{1/2}v_s} \exp\left[-\frac{(E - E_s)^2}{4kTE_s}\right] dE, \tag{10.7}$$

where n is the ion density.

This is a Gaussian distribution centered around $E = E_s = 1/2Mv_s^2$, with a half-width $2(kTE_s)^{1/2} = (2kTM)^{1/2}v_s$. For a given temperature, the width of the peak increases as the square root of the ion mass. Thus for

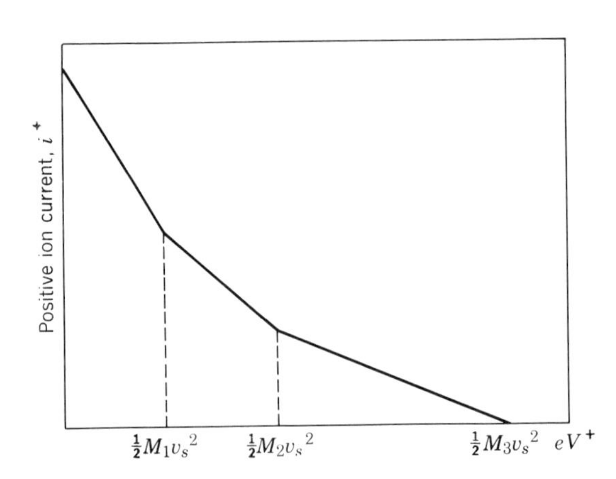

Fig. 10.7 Positive ion current versus probe potential for a screened spherical probe on a satellite moving with a speed v_s. V^+ is the potential of the probe relative to the surrounding plasma [10, adapted].

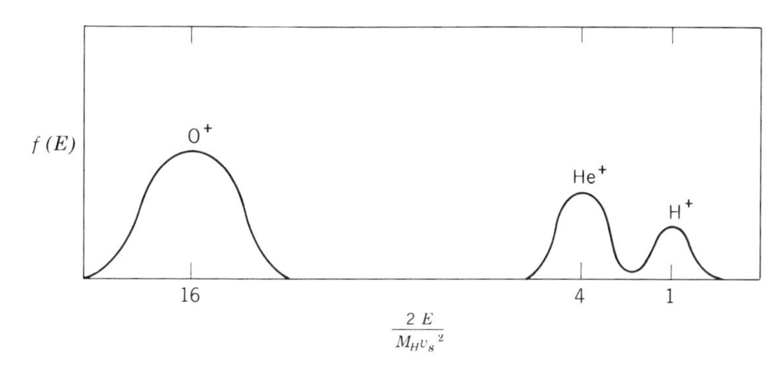

Fig. 10.8 An energy distribution function for ions of several masses detected by an ion probe moving with a satellite velocity, v_s. O^+, He^+, and H^+ are assumed to be present and M_H is the mass of a hydrogen atom [11].

a mixture of ions of different mass number the general distribution curve will be similar to Figure 10.8.

Simple ion probes can yield definitive data regarding the temperature, although this type of instrumentation does not have the resolution of mass spectrometers at high mass numbers. Somewhat more accurate data can also be obtained if a pair of ion probes or mass spectrometers are mounted in orthogonal directions—one directed along the velocity vector of the rocket or satellite (Figure 10.2) and a second perpendicular to the satellite spin axis. In the latter case, the ion source will be alternately "looking" at molecules in a "forward" or "backward" direction, and the resulting modulation in ion beam intensity can be interpreted in terms of the rocket spin rate and the ion energy spectrum.

THE PLANETARY ATMOSPHERES

Considerable speculation is included in contemporary discussions regarding the planetary atmospheres. Urey [12] suggests, however, that if instrumental landings and direct analyses of the planetary atmospheres are ultimately realized, we shall indeed obtain much more precise information than is possible to gain by present methods. Spectroscopic data gathered by numerous investigators have led to the conclusion that the gases listed in Table 10.2 may exist in the planetary atmospheres.

Mercury is too small a planet to retain an appreciable atmosphere; if an atmosphere really exists, the gaseous matter may be escaping as fast as it is being produced by some chemical process. Presumably an equilibrium condition occurs with only the heavy gases being retained.

Table 10.2

Planetary Atmospheres [13]

Planet	*Sun-planet distance (AU)*	*Atmospheric composition (in order of abundance)*				
Mercury	0.387	Ar	Kr	Xe		
Venus	0.721	N_2	Co_2	H_2O		
Earth	1.00	N_2	O_2	(see Table 10.1)		
Mars	1.52	N_2	Ar	CO_2	H_2O	
Jupiter	5.20	H_2	He	CH_4	NH_3	H_2O*
Saturn	9.54	H_2	He	NH_3	CH_4	H_2O*
Uranus	19.19	H_2	He	CH_4	NH_3^*	H_2O*
Neptune	30.07	H_2	He	CH_4	NH_3^*	H_2O*
Pluto	39.46	H_2	He			

* In frozen state.

Venus is of considerable interest, but what has been observed until recently is essentially the outer thick layer of clouds. Space probes of this planet should reveal considerably more information.

Mars has a thin atmosphere (its diameter is one-half that of Earth) and its lower temperature favors the retention of some gases. Observations have been made with both radiometric and spectroscopic instrumentation, and there is evidence that CO_2 is somewhat more abundant than it is on the earth.

From optical spectroscopic measurements that have been made of Jupiter, we know there are substantial amounts of CH_4 and NH_3, the latter being in the form of crystals because of the low temperatures. The outer layer temperatures range to —130°C. In lower layers it is presumed that the pressure is so high that hydrogen must exist in the liquid or solid phases. Saturn has nine known satellites, and the largest of these (Titan) is also presumed to have an atmosphere.

Questions relating to the moon's surface will soon be answered in a dramatic fashion, and arguments as to whether a small amount of atmosphere exists in the lower part of its surface will then be resolved. Even before man lands on the moon, however, it is expected that a considerable amount of instrumental hardware will be soft-landed on the lunar surface. Among many instruments that are expected to be used in the NASA Surveyor program is a robot mass spectrometer. One such type is termed a "nude" spectrometer because it is designed without the usual vacuum pump system. The lack of an appreciable atmosphere on the moon makes such an instrument operational without any vacuum components, thus greatly reducing complexity and weight. Further, the truly remarkable advances in telemetering information to earth seem to indicate that mass spectra of at least the near planets will eventually be recorded.

SPACE BIOLOGY AND LIFE-SUPPORT SYSTEMS

What role, if any, will mass spectrometry have in space-generated biomedical problems and life-support systems? A partial answer to this question can be made by considering the general hazards of human space flight.

A space capsule must contain a livable atmosphere for an astronaut. A normal or near-normal amount of oxygen must be supplied to his vital organs and tissues, and this establishes tolerance limits for both gaseous content and the pressure of the space capsule atmosphere. Noxious gases such as exhaled CO_2 and water vapor must be restricted to acceptable concentrations, thus requiring a life-support system that can absorb specific quantities of these gases at an appropriate rate. In addition a completely sealed space ship may require other controls. Toxic gases may evolve from the inner walls of the capsule, from organic plastics and electronic insulating materials, or from other materials that are stored aboard. This potential hazard is undoubtedly a negligible one for flights of short duration. For extended flights, however, objectionable trace quantities of gases may develop at elevated temperatures, or by the dissociation products of ambient ionizing radiation. These gases must be identified, and they may need to be controlled. Thus the "livable" environment for short periods may be unacceptable for a flight duration of many weeks.

For short-period excursions into space, the chemical air purification systems that absorb CO_2 and water vapor and supply oxygen have already been developed. Life-support systems for astronauts who must live in space for many months are still being studied. These systems must have a regenerative capability for recycling and reinstituting metabolic outputs of the body for reuse. From a weight standpoint alone, it is not feasible to equip a spaceship with non-reusable water, oxygen, or purifying agents. Hence space capsules may eventually contain living plants to complement man's respiration, absorb CO_2, and replenish oxygen. Algae have been evaluated to function not only as an oxygen supply but to accomodate the disposal of wastes. Auxiliary power sources may also be engineered capable of dissociating water and CO_2 at least for the partial fulfillment of metabolic needs.

The specific composition of the "ideal space-capsule atmosphere" is still open to research. Some physiologists are concerned about long-term effects the breathing of pure oxygen have on the lungs and central nervous system. The specific pressure is also a variable. Lower-than-atmospheric pressures have been tested for supplying oxygen with good results, but the general air pollution problem is enhanced at lower operating pressures. Toxic metals such as cadmium and lead can "boil off" from solder at an increased rate at reduced pressure. Thus supplying pure oxygen at 187 mm of pressure (atmospheric pressure at 34,000-ft altitude), which is equivalent to breathing air at sea level, has some potential disadvantages. Obvious engineering advantages of reduced cabin pressure, of course, include the greatly reduced weight of a vehicle necessary to withstand the pressure gradient from cabin pressure to the vacuum of space. The early manned space capsules launched by the United states contained 100% oxygen at 0.34 atm. On January 27, 1967, the tragic flash fire that took the lives of three American astronauts caused a reappraisal to be made of the pure oxygen system, even though the potential fire hazard had been well recognized. But quite apart from such a catastrophe, the possible recurrence of which may be greatly reduced by instituting other precautions, there are many space scientists who believe that atmospheres consisting of gas mixtures are desirable and possibly essential for *prolonged* missions in space [14].

In any event it is clear that partial pressure analyzers and some special types of mass spectrometers will be called upon, with increasing frequency, to furnish basic research data on the atmospheres of space cabin prototypes. Eventually spectrometers may be integrated into the flight instrumentation of all manned vehicles.

EFFECT OF SPACE VACUUM ON MATERIALS

The general effect of a space vacuum on materials cannot be predicted without some qualifications, as superposed upon a high-vacuum environment may be radiation, thermal, or other phenomena. All materials, however, will release matter by one or more of the following processes.

1. Evaporation.
2. Permeation and diffusion.
3. Sorption (absorption, adsorption, desorption).
4. Decomposition.
5. Ion or neutral particle impact.

The pressure in space at an altitude of 120 miles approaches 10^{-6} torr; at 8000 miles and deep space it is less than 10^{-12} torr. At such pressures one must consider the effects on both inorganic and organic materials. With inorganic materials, a rough estimate of rates of decomposition to volatile products can be made by substituting the equilibrium-decomposition pressure in the Langmuir equation [15]. The equation indicates the rate, W, of evaporation or sublimation in gm/cm^2/sec as

$$W = \frac{P}{17.14} \cdot \frac{(M)^{1/2}}{T}. \tag{10.8}$$

P is the vapor pressure of the material in torr, M is the molecular weight of material in the gas phase, and T is the Kelvin temperature. Evaporation rates for some common metals at 10^{-9} torr are given in Table 10.3.

The organic materials used in spacecraft are often long-chain polymers. Sublimation of the surface layers of such material into vacua proceeds by decomposition into smaller fragments and by subsequent escape of the more volatile products. Some polymers can lose up to

Table 10.3

Evaporation of Metals at 10^{-9} Torr [16]

	*Temperature**	
Element	*10^{-5} cm/yr*	*10^{-3} cm/yr*
Cd	100	170
Zn	160	260
Mg	260	350
Li	300	410
Pb	510	630
Ag	890	1090
Sn	1020	1220
Al	1020	1260
Be	1140	1300
Cu	1160	1400
Au	1220	1480
Ge	1220	1480
Fe	1420	1650
Si	1450	1690
Ni	1480	1720
Ti	1690	1960
Mo	2520	2960
W	3400	3900

* In degrees Fahrenheit at given evaporation rate.

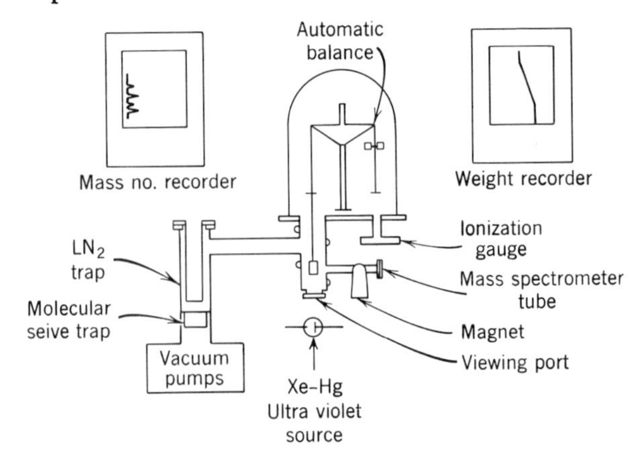

Fig. 10.9 Typical test system with a mass spectrometer for evaluating the sublimation of materials [17, adapted].

10% of their weight per year at temperatures greater than 100°C; others are relatively stable.

The sublimation of both inorganic and organic materials from spacecraft surfaces poses some specific problems for the spacecraft designer. One may be the condensation of out-gassed products on the optical surfaces of instruments, yielding decreased transmission and image distortion of lenses or windows. Another may be the slight change in the thermal radiative properties of the outer skin of the spacecraft. Finally, surfaces in frictional contact have a tendency to "freeze" or weld together when exposed for long periods in space. Thus such surfaces must have a lubricating interface which will not evaporate in a vacuum. Because of these and other considerations, extensive testing of materials has been made of spacecraft materials. Figure 10.9 indicates a typical test configuration for the evaluation of ultraviolet, temperature, and vacuum effects.

The sample under study is placed in a direct line of sight with a mass spectrometer and an ultraviolet source. The gross weight loss of out-gassing products can be monitored by an analytical balance and the mass spectrum of evaporated products is made by periodic scanning over an extended mass scale. The condensation of materials on lenses (which occurs in actual systems) can be simulated at increased rates by maintaining these optical components at ambient to cryogenic temperatures. Paints, silicone rubbers, polyethylene, and polyesters are among the many materials that have been mass spectrometrically analyzed for outgassing, weight loss, and effects of secondary condensation.

ION ENGINES FOR SPACE PROPULSION

The use of ions for space propulsion systems is intriguing to the mass spectrometrist in several respects. The basic ion engine can be considered to be a scaled-up version of a mass spectrometer or an isotope separator ion source, and the accelerating voltages are of the same order of magnitude. In addition, the mass spectrometer is one of the more basic analytical tools required to investigate surface ionization, charge exchange, and other phenomena that are specifically related to advanced ion rocket design.

The basic concept of the ion engine is remarkably simple. The three essential elements are (a) the ion source, or emitter; (b) an accelerating electrode; and (c) a beam neutralizer. A typical ion engine is shown in Figure 10.10. Ions are produced from the "propellant fuel" in the ion source region by surface ionization, electron bombardment, or arc discharge. They are subsequently accelerated by a high-voltage electrode that provides them with an energy of 10 to 30 keV. After emerging from an exit nozzle, these ions are neutralized by recombining with electrons, so that a high-velocity neutral beam is discharged from the engine to produce thrust. If ions emerged from the ion gun into space without neutralization, the immediate build-up of electrical charge on

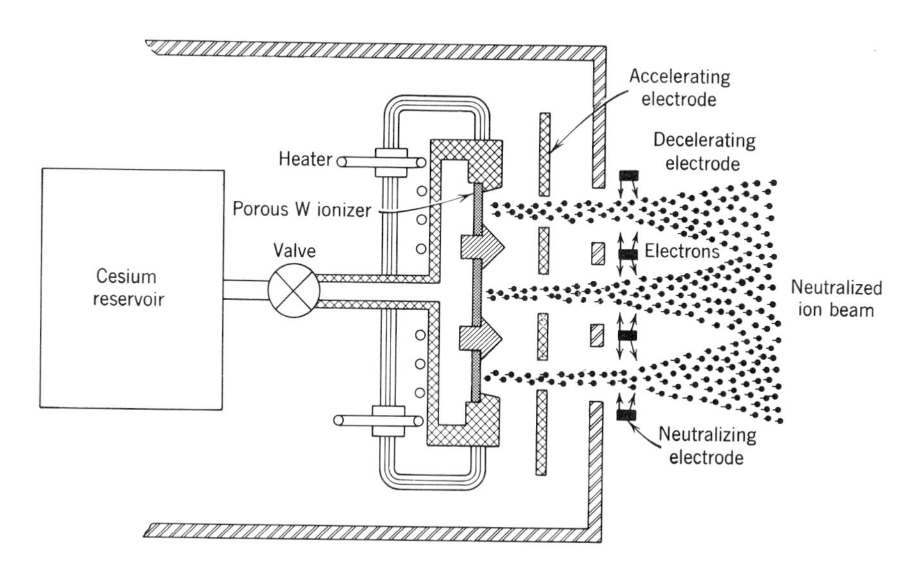

Fig. 10.10 Schematic diagram of an ion engine having a porous tungsten ionizer and utilizing cesium as a propellant.

the engine or associated space vehicle would cause some ions to return. An electrostatic "drag" would thus result.

The thrust developed by the continuous expulsion of a neutralized atomic beam from an ion rocket operating in a vacuum can be computed directly from Newton's second law:

$$F = \frac{d(m\bar{v}_e)}{dt}, \tag{10.9}$$

where $m\bar{v}_e$ is the momentum of the ion exhaust in the frame of reference of the space vehicle. Carrying out the indicated differentiation results in

$$\begin{aligned} F &= \bar{v}_e \frac{dm}{dt} + m \frac{d\bar{v}_e}{dt} \\ &= \dot{m}\bar{v}_e \text{ (for a constant exhaust velocity).} \end{aligned} \tag{10.10}$$

Thus the net thrust of an ion engine is directly proportional to $\dot{m}$, the rate at which propellant mass is discharged from the nozzle, and the terminal velocity of the electrically accelerated ions.

It is the exceedingly large ion velocity that makes the ion engine an attractive candidate for certain applications. In particular the ion engine has a very high *specific impulse,* a parameter whose defining equation is

$$I_{sp} = \frac{F}{\dot{m}g_0} = \frac{\dot{m}\bar{v}_e}{\dot{m}g_0} = \frac{\bar{v}_e}{g_0}, \tag{10.11}$$

where g_0 is the gravitational constant. Specific impulse is thus the thrust generated per rate of propellant discharged (i.e., the number of pounds thrust per pound of fuel per second). Specific impulse then has the dimensions of seconds and its magnitude is a direct measure of propellant economy. In this respect the ion rocket is superior to all other practical space propulsion systems. A rough comparison of the ion engine with other systems, using practical temperatures, is suggested in Table 10.4.

Table 10.4

Comparison of Propulsion Systems [18]

Propulsion system	*Specific impulse*
Chemical rockets	400 sec
Nuclear rockets (with H_2)	800 sec
Ion engines	15,000 sec

A major disadvantage of the ion engine system is its exceedingly low total thrust (of the order of pounds in any system conceived and fractions of this value for test models to date). But the specific impulse far exceeds that of other propulsion methods. In the chemical rocket I_{sp} is limited by (a) the relatively high mass of the combustion products and (b) the energy of the molecular bond. The best structural materials are limited to about 5000 to 6000°F (about 0.25 eV), a value that is small compared with ion energies. The use of hydrogen in a nuclear-reactor heat exchanger somewhat improves the situation, as the lower mass number propellant (at the same temperature) increases the exhaust velocity.

Ion rockets are envisioned for small payloads and deep space missions in which high terminal velocities can be reached after many days of acceleration. They are also being considered as auxiliary thrustors for satellites (e.g., for attitude control), because of their inherent fuel economy and the advantages of electrical programming. Cesium, rubidium, potassium, sodium, lithium, and mercury have all been given some consideration as propellants. The high degree of ionization that can be achieved for the alkali metals at modest temperatures, using a tungsten cathode, have made these elements especially attractive.

Problems inherent to the optimization of the ideal ion engine include:

1. Improved cathodes.
2. Ion optics.
3. Measurement of atomic cross sections.
4. Anode sputtering.
5. Beam neutralization schemes.
6. Minimization of space charge (which limits ion emission).

Quite clearly, the mass spectrometer can contribute substantially to virtually all of these studies.

THE MASS SPECTROMETER IN HIGH-VACUUM TECHNOLOGY

The use of mass spectrometers in the general field of vacuum technology has been widely reported in the literature, and this monograph can provide only a few comments relating to this exceedingly broad area. Applications of high-vacuum technology are, of course, not restricted to space-related investigations, but such studies have strongly influenced the increased use and development of mass spectrometers for high-vacuum applications. No major space-testing chamber is without some type of mass spectrometric analyzer, and it is possible that small mass spectrometers will soon replace conventional ionization gauges on virtually all important high vacuum facilities and apparatus.

The earliest use of a simple spectrometer in vacuum work per se was for leak detection—using helium as a sensing gas. Today, convenient commercial forms of the helium leak detector will be seen wherever vacuum-related systems are being built. Such devices reduce the hours of leak hunting to a minimum, and in many instances the specifications for vacuum products includes the quantitative limits of helium leak rates.

It has only been within the last few years, however, that there has been a general recognition of the need for small analytical spectrometers or residual gas analyzers. These have been developed with a sensitivity that far exceeds the simple ionization gauges. In addition, such instruments furnish the specific data that a general pressure gauge can never provide. A residual gas analyzer can indicate a high nitrogen content (suggestive of an air leak), a high hydrogen content (resulting from the out-gassing of metal components); or a hydrocarbon spectrum that might be indicative of a malfunctioning oil-diffusion pump or cold trap. The residual gas or partial pressure analyzer can also furnish information relating to chemical kinetics. For example, it can reveal whether trace quantities of water vapor exist in amounts sufficient to act as a catalyst. In the area of thin-film technology, such a tool can monitor the composition of gaseous species in evaporating chambers, and materially aid in developing techniques that insure film adhesion, uniform film densities, and desirable electrical characteristics. Even trace quantities of certain gases in evaporating or sputtering apparatus may have more than minor effects upon the electronic, magnetic, or superconducting film properties.

The fact that a mass spectrum can be obtained of a "complete vacuum" (e.g., 10^{-8} torr) seems impressive until one considers the actual particle densities of such a pressure. Even at 10^{-12} torr there are about 3×10^4 molecules per cubic centimeter, so that the ionization and detection of a small fraction of these molecules at least appears plausible. (The mean free path, at this pressure, is about 30,000 miles.) Two of the factors that have permitted these low pressures to be analyzed have been (a) small volume spectrometers through which an ion source could have multiple chances for ionization and (b) high-gain electron multipliers for ion-beam detection.

Reynolds [19] was among early investigators to report a small spectrometer in which static total pressure of 5×10^{-10} torr and partial pressures of 10^{-12} torr were analyzed. Davis and Vanderslice [20] subsequently reported the development of another small spectrometer capable of even lower pressure measurements. The spectrometer comprised a bakeable envelope, a 90° 5-cm radius of curvature magnetic analyzer, a Nier-type hot cathode ion source, and a 10-stage electrostatically fo-

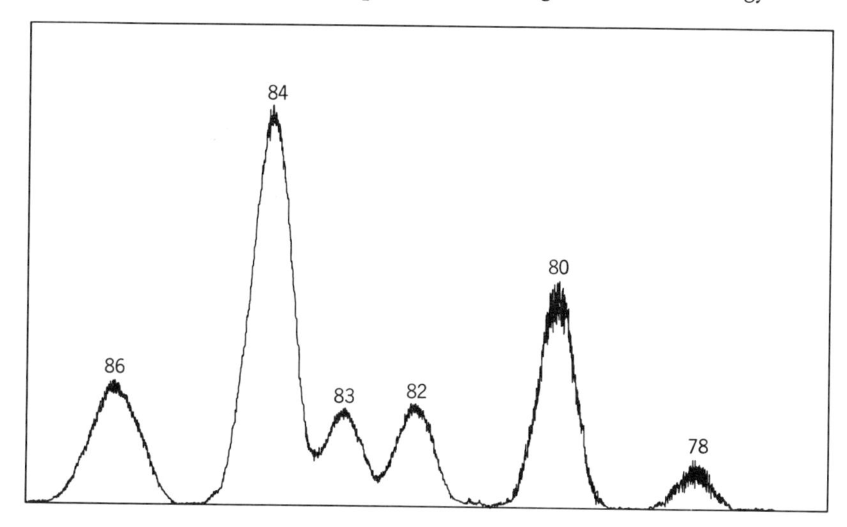

Fig. 10-11 Mass spectrum of the isotopes of krypton at 10^{-8} torr [21, adapted].

cused electron multiplier. An additional feature developed in this system was a fast mass scan with an oscilloscope display. Sweep rates as high as 1.5 μsec/amu were attained. Figure 10.11 is a mass spectrum of krypton obtained with this instrument at a pressure of about 10^{-8} torr [21]. More recent developments by Davis [22] have made possible the measurement of partial pressures down to 10^{-16} torr. Such a pressure corresponds to a density approximating only 1 molecule/cm^3, a pressure presumably comparable to that of outer space. Such pressures can be obtained only for short periods and at cryogenic temperatures.

A problem that is more likely to be encounted in practice is that of measuring very small partial pressures of gases in the presence of other gases. For example, it is sometimes necessary to detect trace quantities of gases at total pressures of 10^{-6} torr. The instrumentation that has been successfully developed to cope with this problem is a partial pressure analyzer consisting of multiple electrostatic and magnetic analyzing sectors, similar in principle to the large mass spectrometers described in Chapter 2. A schematic diagram of a recent Aero Vac Corporation model is shown in Figure 10.12. The combination of electromagnets in tandem arrangement permits the identification of low mass number gaseous species at less than the 1-part-per-million level, with complete discrimination from large adjacent mass peaks. Such an instrument, of course, qualifies more as a general analytical instrument than solely as a partial pressure monitor.

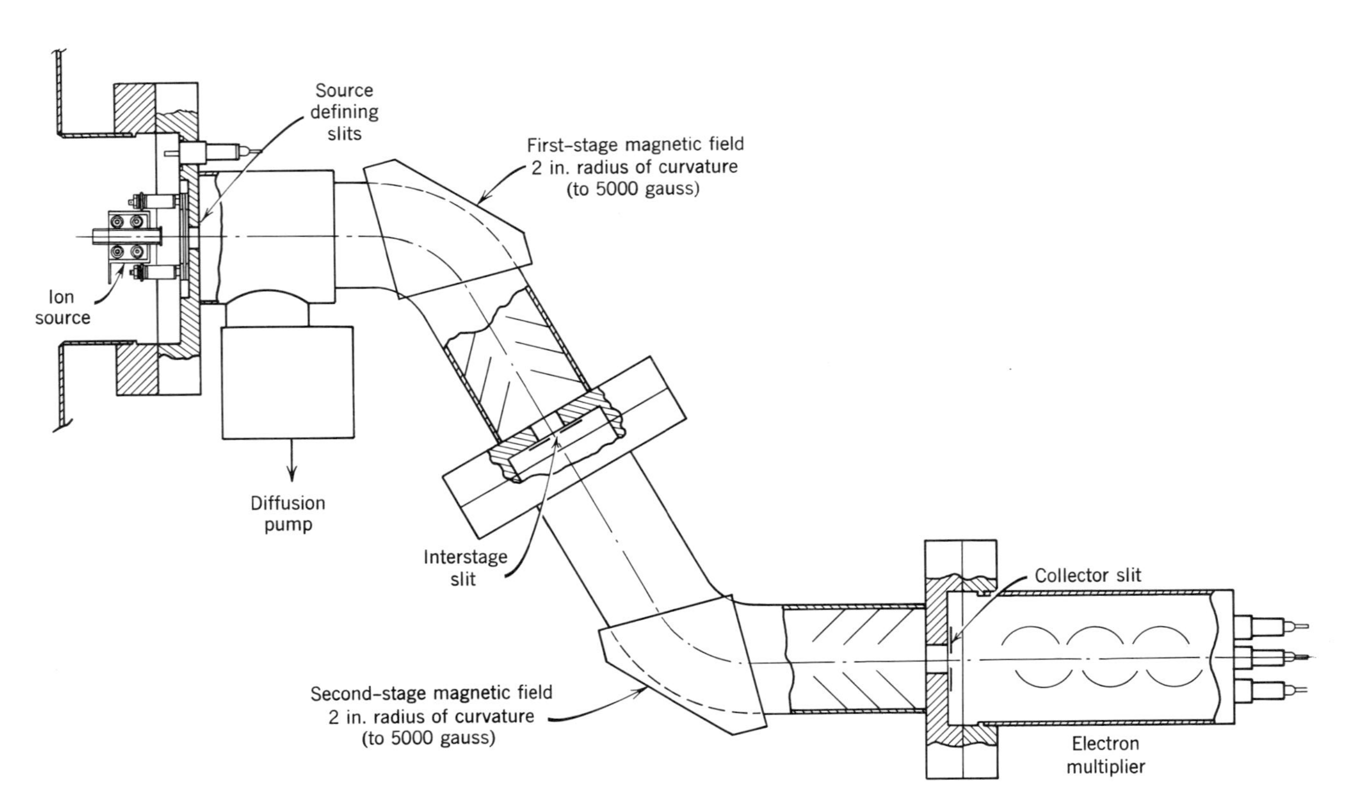

Fig. 10.12 Schematic diagram of partial-pressure spectrometer for analyzing trace gases (courtesy Aero Vac Corp.).

These and other spectrometer types will undoubtedly provide the detailed answers to many of the problems encountered in high-vacuum technology. Only by the quantitative monitoring of particular gaseous species can the most meaningful measurements be made of out-gassing, gettering, or the removal rate of gas molecules by diffusion, ion pump, and cryogenic methods.

REFERENCES

[1] E. Stuhlinger and G. Mesmer, *Space Science and Engineering,* McGraw-Hill, New York, 1965, p. 50.

[2] *Ibid.,* p. 313.

[3] *Ibid.,* p. 88.

[4] R. S. Narcisi and A. D. Bailey, U.S. Air Force Office of Aerospace Research; Experimental Research Paper No. 82; Bedford, Mass., 1965.

[5] W. Paul, H. P. Rheinhard, and U. von Zahn, *Z. Physik,* **152,** 143 (1958).

[6] A. O. Nier, J. H. Hoffman, C. Y. Johnson, and J. C. Holmes, *J. Geophys. Res.,* **69,** 986 (1964); also **69,** 4629 (1964).

[7] A. O. Nier, *Am. Scientist,* **54,** 379 (1966).

[8] A. A. Pozhunkov, "Distribution of He, N, and A in the Earth's Atmosphere at Altitudes of Up to 430 Km," trans. from *Kosmicheskie Issled.,* **1,** 147 (1963).

[9] H. A. Taylor, Jr., L. H. Brace, H. C. Brinton, and C. R. Smith, *J. Geophys. Res.,* **68,** 5339 (1963).

[10] H. Massey, *Space Physics,* Cambridge, London, 1964, p. 147.

[11] *Ibid.,* p. 149.

[12] H. C. Urey, "The Planets," in *Science in Space,* ed. by L. V. Berkner and H. Odishaw, McGraw-Hill, New York, 1961, Ch. 10.

[13] U. T. Slager, *Space Medicine,* Prentice-Hall, Englewood Cliffs, N.J., 1962, p. 364.

[14] R. L. Latterel, *Science,* **153,** 69 (1966).

[15] L. D. Jaffe, *Nucleonics,* **19,** 93 (1961).

[16] E. S. Pederson, *Nuclear Energy in Space,* Prentice-Hall, Englewood Cliffs, N.J., 1964, p. 470.

[17] C. Boebel, N. Mackie, and C. Quaintance, "Materials for Space Vehicle Use," in *Transactions of the 10th National Vacuum Symposium,* Pergamon, New York, 1962, p. 59.

[18] W. R. Michelsen, *Aerospace Eng.,* **19,** 7 (1960).

[19] J. H. Reynolds, *Rev. Sci. Instr.,* **27,** 928 (1956).

[20] W. D. Davis and T. A. Vanderslice, in *Transactions of the 8th National Symposium of Vacuum Technology,* 1960, p. 417.

[21] W. D. Davis, private communication.

[22] W. D. Davis, in *Transactions of the 10th National Symposium of Vacuum Technology,* Pergamon, New York, 1962, p. 363.

Chapter 11

Nuclear Geology and Isotope Cosmology

The term *nuclear geology* is sometimes used to denote the exceedingly close coupling of nuclear science and certain phases of geology. The term is an appropriate one for since Henri Becquerel in 1896 discovered the radioactivity of uranium salts and uranium-bearing minerals, much of our knowledge of the composition and formation of our planet has been gained from nuclear-related phenomena. Even before isotopes were discovered, J. J. Thomson in 1905 detected the radioactivity of long-lived rubidium. In 1906 potassium was also found to possess a trace quantity of radioactivity. This was the same year that Rutherford attempted to measure the absolute geological age of minerals from a helium/uranium ratio. About this same time the first crude analyses were being made of the radium content of meteorites. The weak alpha activity of samarium remained undiscovered until 1934.

In 1935 Nier employed mass spectrometry to show that all the activity of potassium was contributed by the K^{40} isotope [1], and in subsequent studies he demonstrated the vast potential of isotopic measurements in geochronology [2]. During this 1935–1940 period, the mass spectrometer was first considered as a tool in geochemical prospecting. In fact Herbert Hoover, Jr., son of the late president, was among the pioneers who proposed that mass spectrometry might be employed for soil analysis, in connection with gas and petroleum prospecting.

Today, the mass spectrometer is a standard analytical tool in geological research, providing highly specific information relating to the composition and age of the earth. In many instances the mass spectrometer has superceded emission spectroscopy, flame photometry, and other chemical methods. As a single example of sensitivity, a granite specimen was recently analyzed by spark-source mass spectrometry and over 40 "trace" elements were measured—most of these in the part-per-million concentration [3].

TERRESTRIAL ABUNDANCE OF THE ELEMENTS

Terrestrial abundances of the elements in the earth's crust are based on extensive chemical geological data. Abundances of the elements for the earth as a whole are deduced primarily from the analysis of chondrites, and most common meteorites. Figure 11.1 indicates the relative abundance of the more common elements in the earth's crust, and for the earth as a whole [4].

It will be noted that oxygen is the most common element in the crust; iron is the more abundant, however, for the entire earth. Approximately 90% of the earth is comprised of Fe, O, Si, and Mg, and 99.9% is made up of only fifteen of the elements. Thus, there exists a marked differentiation between our planetary body and stellar matter. Helium, an abundant stellar gas, is only a trace element on earth. It is available in quantity primarily in the United States (Kansas, Oklahoma, Texas, and New Mexico), where it is separated from natural gas. There are two naturally occurring isotopes of helium, He^3 and He^4. Perhaps some of the He^4 is primordial, but most of it is assumed to be a product of alpha decay.

"NORMAL" ISOTOPIC ABUNDANCES

If we focus our attention on the isotopic composition of the terrestrial elements, we find that a chemical element will have almost the same

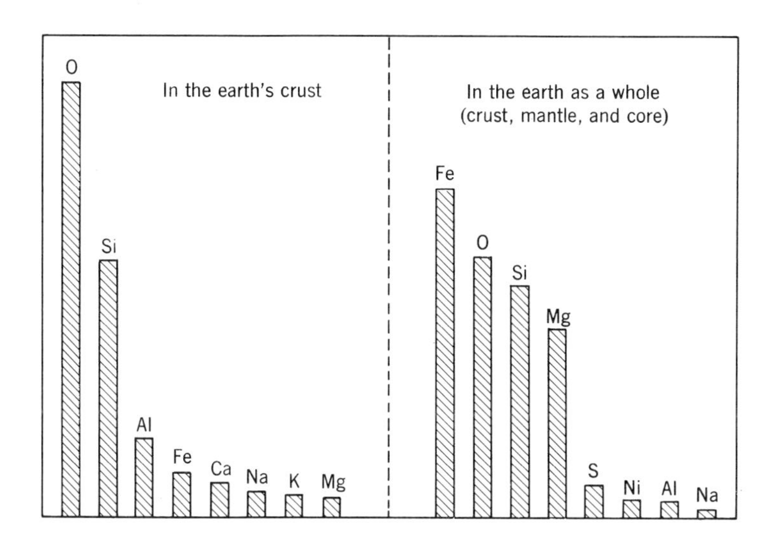

Fig. 11.1 Relative abundances of the common terrestrial elements [4, adapted].

isotopic abundance, regardless of where it is found on the earth. And if radioactive decay products are excluded, the variation in the percent abundance of the isotopes is very small. But the "normal" isotopic abundance values per se are of interest to the nuclear geologist and the nuclear physicist alike. In particular, the relative abundances of the several isotopes of a given chemical element clearly indicate which nuclides are favored from a standpoint of nuclear stability. In other words, one would expect the more abundant isotopes to correspond to the most stable nuclear configurations of protons and neutrons. Consider the isotopic structure of tin (having the largest number of isotopes) and cerium. If we plot the percentage abundance of the isotopes of these elements according to the number of neutrons in the nucleus $(A\text{-}Z)$, we obtain the graph of Figure 11.2. An even number of protons and neutrons are clearly favored over the even-odd proton-neutron species. The neutron number 82 for the Ce^{140} isotope is also one of the "magic" numbers associated with a particularly stable nuclear configuration. Indeed, isotopic spectra of the naturally occurring nuclides have been important sources for developing modern theories of nuclear structure and in establishing limits of geologic age. The isotopic invariance of widely dispersed matter (e.g., in the solar system) also suggests that such material had been exposed to a nuclear history comparable to that of the earth.

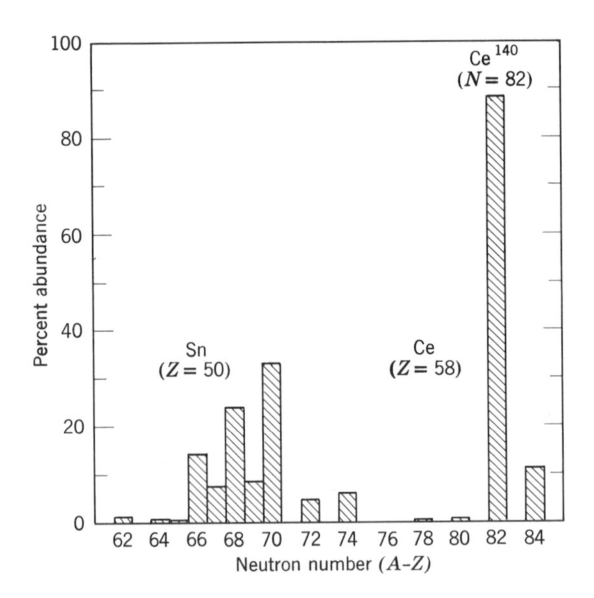

Fig. 11.2 Isotopic abundance of tin and cerium, showing the nuclear stability of "even-even" nuclides.

Thus the general isotopic invariance of matter is an important phenomenon, for it establishes a primary baseline against which changes in isotopic ratios can be compared and evaluated for their significance. And as the accuracy of mass spectrometer measurements is increased, it is quite probable that a closer look will be taken of the "normal" isotopic compositions of all the elements, as these are dispersed throughout the world and in extraterrestrial bodies.

When very large isotopic deviations occur from normal abundances, we can expect that nuclear transmutations have occurred. An analogous situation takes place in a nuclear reactor in which high fluxes of neutrons breed many new nuclides, fission products, and literally hundreds of new isotopes in trace amounts. In nuclear geology, the problem is to extract information produced by the past "natural nuclear environment." We thus inquire as to the "natural" rather than artificially produced species that gives rise to isotopic anomalies.

NATURAL UNSTABLE NUCLIDES

Provided that the rate of disintegration of an unstable nuclide is relatively low, we should expect some residue of that substance in nature. Table 11.1 is a partial listing of some geochemically important nuclides, together with mass number, half-life, decay mode, and product.

With the exception of the neutron, all of the nuclides listed in Table 11.1 can be detected by a mass spectrometer. By stretching a point, we could even consider that the neutron has been observed by mass spectrometry. For, in fact, the half-life of the neutron was determined by the ingenious coincidence detection of its decay products—a beta particle and a proton, with a beta ray spectrometer and a proton spectrometer, respectively [5].

A glance at the half-lives shown in Table 11.1 suggests why the mass spectrometer is the ideal tool in geological studies. On a laboratory time scale most of these nuclides are quite stable, but on a geological time scale many of these nuclides have an appreciable decay rate. We shall therefore consider several of these nuclides that are effective in perturbing isotopic abundances from their normal or equilibrium values.

NUCLEAR INDUCED VARIATIONS IN THE ISOTOPIC RATIOS OF THE NOBLE GASES

From a geological point of view, the noble gases are an especially interesting class of elements. They are chemically inert, so that if initially confined in some geological reservoir, they remain in their initial

Table 11.1

Natural Unstable Nuclides*

Nuclide	*Mass No.*	*Half-life*	*Decay mode*	*Product*
Neutron	1	12.8 min	β^-	H^1
Tritium	3	12.4 y	β^-	He^3
Carbon	14	5568 ± 30 yr	β^-	N^{14}
Potassium	40	1.27×10^9 yr	β^-; K-capture	Ca^{40}; Ar^{40}
Vanadium	50	6×10^{15} yr	β^-; K-capture	Cr^{50}; Ti^{50}
Rubidium	87	6.15×10^{10} yr	β^-	Sr^{87}
Lanthanum	138	1.1×10^{11} yr	β^-; K-capture	Ce^{138}; Ba^{138}
Samarium	147	6.7×10^{11} yr	α	Nd^{143}
Rhenium	187	4×10^{12} yr	β^-	Os^{187}
Lead	204	1.4×10^{17} yr	α	Hg^{200}
Radium	226	1622 yr	α	Rn^{222}†
Thorium	232	1.39×10^{10} yr	α	Ra^{228}‡
Uranium	234	2.69×10^5 yr	α	Th^{230}†
	235	7.07×10^8 yr	α	Th^{231}§
	238	4.5×10^9 yr	α	Th^{234}†

* Partial list.
† Final decay product: Pb^{206}.
‡ Final decay product: Pb^{208}.
§ Final decay product: Pb^{207}.

state. Further, these gases are produced in easily detectable quantities by natural nuclear reactions which have taken place over billions of years. Helium may be found wherever alpha particles are a reaction product, as indicated in Table 11.1. Argon is produced by K-capture in K^{40}, and krypton and xenon are fission product gases. Radon, the only naturally radioactive noble gas, has three isotopic species produced from the decay series of uranium and thorium.

We find two isotopes of helium in the earth's crust and in the atmosphere. The lightest isotope, He^3, a daughter product of H^3, is thought to be generated in rocks by the reactions [6]:

$$Li^6(n,\alpha)H^3,$$

$$H^3(\beta^-)He^3. \tag{11.1}$$

Production in the atmosphere is believed to result from the three reactions [7]:

$$N^{14}(n,C^{12})H^3$$

$$N^{14}(n,3\alpha)H^3$$

$$H^3(\beta^-)He^3 \tag{11.2}$$

The upper atmosphere contains an adequate flux of neutrons from cosmic rays to explain the abundance of atmospheric He^3. Cosmic ray neutrons, however, are highly attenuated below the surface of the earth. Hence the generation of He^3 in minerals is ascribed to neutrons from the spontaneous fission of U^{238}, and an approximately equal number of neutrons arising from various (α,n) reactions on elements of low atomic numbers. An early substantiation of such a reaction was reported by Aldrich and Nier [8], who discovered that spodumene ($LiAlSi_2O_6$) has a relatively high He^3/He^4 ratio. Their results and the analysis of other materials have been tabulated by Faul [9]. Faul also points out that a metric ton of granite containing 2 parts per million of uranium and 10 parts per million of thorium will produce approximately $0.22 + 0.29 = 0.51$ ml of helium per million years (at standard temperature and pressure). Such concentrations can be analyzed even for very small samples.

Data such as the above permit analyses to be made on the age of minerals, as helium is continuously being generated in the earth's crust. Both gross helium content and isotopic ratio measurements allow important interpretations to be made of nuclear and geologic processes. In some minerals the retention of helium is almost complete (e.g., in crystals of magnetite and zircon, diffusion coefficients are so small that there is a negligible helium loss even in a period of a billion years). But in other instances helium is lost by diffusion owing to radiation damage, to general migration to more stable crustal regions, or because the original helium content is contaminated from some other localized source. Hence in practice the basic data must be combined with other information to be of any significance.

In the case of the exceedingly high helium content of Rattlesnake

Table 11.2

Helium Content of the Atmosphere, Natural Gas, and Minerals [9]

Location	*Material*	*He content*	*$He^3/He^4 (\times 10^{-7})$*
Stamford, Conn.	Air	0.004%	12.0
Rattlesnake Field, Shiprock, N.M.	Natural gas	7.7%	0.5
Cliffside Field, Amarillo, Tex.	Natural gas	1.8%	1.5
Great Bear Lake, Canada	Pitchblende	—	0.3
Keystone, S.D.	Beryl	0.022 (cc/gm)	1.2
Jokkmokk, Lapland (Sweden)	Beryl	0.023 (cc/gm)	1.8
Cat Lake, Manitoba, Canada	Spodumene	0.01 (cc/gm)	24
Edisonmine, S.D.	Spodumene	0.01 (cc/gm)	120

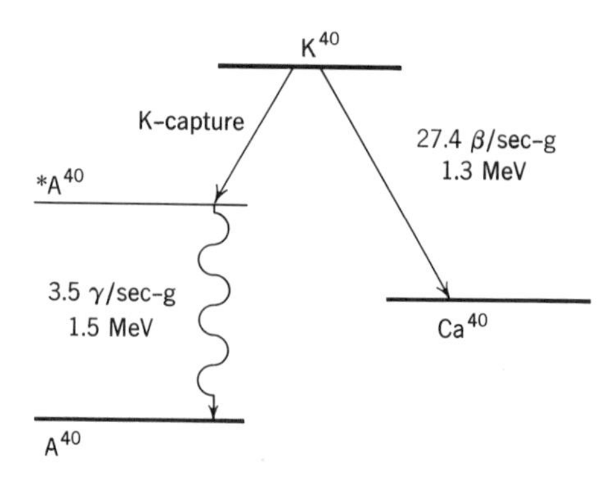

Fig. 11.3 Decay scheme of K^{40}.

Field, N. Mex. (~8%), geologists have had difficulty ascribing this very rich helium source to radioactive products, for no significant radioactivity has been observed in this region. Consequently, this large concentration has been tentatively ascribed to a migration from a reasonably remote source, the helium finally becoming trapped, as in the case of natural gas or petroleum reservoirs. A partial summary of such findings appears in Table 11.2.

Important mass spectrometric measurements have also been made for argon. The abundance of argon in the atmosphere is much higher than that of helium, being slightly less than 1%. In natural gases it is present to only about 0.1%. The isotopic abundance of argon in the air, as reported by Nier [10], is A^{40}: 99.6%; A^{36}: 0.337%; and A^{38}: 0.063%. A^{40} is important in geologic age determinations because it is the radiogenic product of K^{40}—according to the decay scheme indicated in Figure 11.3.

This decay suggests why the A^{40} isotope occurs in relatively high

Table 11.3

Argon Content of Potassium-Bearing Minerals [11]

Mineral	*Argon* (10^{-3} *cc*/gm)	*Potassium* (%)	*Radiogenic argon* (%)	*Est. age* (10^6 *yr*)
Orthoclase, $KAlSi_3O_8$	0.69	14	79.7	1400
Microline, $KAlSi_3O_8$	0.22	16	35.5	350
Sylvite, KCl	0.32	54	13.8	200
Langbeinite, $K_2Mg_2(SO_4)_3$	0.16	20	68.8	200

abundance; it also indicates that the argon content of potassium-bearing minerals should permit the determination of mineral age. A specific method for age determination is presented in a subsequent section, but representative measurements of Aldrich and Nier [11] are listed in Table 11.3.

Variations in the isotopic abundances of krypton and xenon are of interest because they can be related to the buildup of fission-product gases in uranium- or thorium-bearing minerals. These fission products are generated by (a) neutron-induced fission and (b) uranium that un-

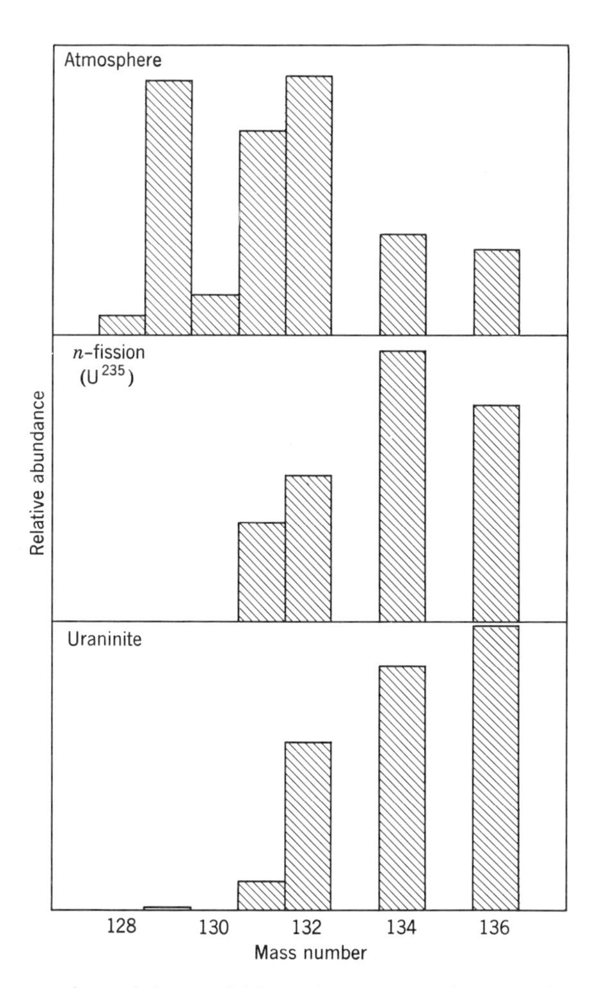

Fig. 11.4 A comparison of the partial isotopic spectrum of xenon with spectra from neutron-induced fission of U^{235} and xenon extracted from uraninite [13, adapted].

dergoes fission spontaneously. Spontaneous fission half-lives are very long, even compared with the age of the earth. Half-lives for Th^{232}, U^{235}, and U^{238} are 1.35×10^{18}, 1.87×10^{17}, and 8.07×10^{15} yr, respectively [12]. The similarities in the xenon spectra of the neutron-induced fission of U^{235} and uranite were first reported by Macnamara and Thode [13]. A comparison of these spectra with the partial atmospheric isotopic spectrum of xenon is shown in Figure 11.4.

The availability of such data has allowed important interpretations to be made with respect to mineral age and the relative concentrations of fissionable nuclides.

ISOTOPIC VARIATIONS INDUCED BY FRACTIONATION

Whereas nuclear transmutations may lead to reasonably large changes in the isotopic abundances of an element, other physical-chemical phenomena will usually result in only minor variations. Some effects will be so small as to be unobservable in any mass spectrometer. If the changes are 0.01% or greater, however, there exists a possibility of detecting fractionation—a process important in geology, hydrology, and meteorology. Isotopic fractionation arises from the fact that the thermodynamic properties of a system of molecules are mass dependent. As would be expected, isotopic fractionations are in evidence primarily when large percentage mass differences exist. Thus one might expect to observe fractionation if isotopic substitutions of low atomic weight elements were made in low molecular weight compounds.

The thermodynamic properties of isotopic substances has been analyzed by Urey [14] with respect to exchange equilibrium. If Q_1 and Q_2 are the partition functions of isotopic molecules having molecular weights M_1 and M_2, and energy states E_1 and E_2, a partition function ratio can be expressed by

$$\frac{Q_2}{Q_1} = \left(\frac{M_2}{M_1}\right)^{3/2} \frac{\sigma_1}{\sigma_2} \frac{\Sigma e^{-\frac{E_2}{kT}}}{\Sigma e^{-\frac{E_1}{kT}}}, \tag{11.3}$$

where σ_1 and σ_2 are symmetry constants and the sum is taken over all the particular energy states. One prerequisite for the validity of this expression is that the temperature of the system kT is large compared with level spacing of rotational energy levels.

Many investigators have reported several geochemical environments wherein this mass-dependent exchange equilibrium may be found. An effect has been noted in the equilibrium distribution of the oxygen isotopes in regions of mineralization or hydrothermal change. Thus mineral species will indicate an enrichment of the O^{18} isotope, the percentage enrichment being a function of the ambient temperature at which equilibrium was established. In one instance the O^{18} enrichment (about 3%) in the $CaCO_3$ of shells of marine organisms was measured over the oxygen content in ocean water [15]. By combining these data with an empirically determined temperature coefficient of fractionation,

$$\alpha = \frac{(O^{18}/O^{16})CO_2 \text{ liberated}}{(O^{18}/O^{16}) \text{ carbonate}}, \tag{11.4}$$

it has been possible to measure paleotemperatures. A recent review of the oxygen isotope method of paleotemperature analysis has been given by Emiliani [16]. Temperature difference of $\pm 1°C$ may be determined if the isotopic abundances of the ocean water is known, although variations in the oxygen isotope composition of marine surface waters may vary corresponding to $\pm 2.5°C$ [17].

It has also been suggested that similar fractionation phenomena might be expected to take place in mineral formation (i.e., between a melt and the minerals crystallizing out of it upon cooling). Thus the progressive cooling of a batholith from the outside toward the core provides the opportunity for unfolding a geological time scale. In this connection the sulphur isotopic fractionation in a Fe-S-O_2 system has been studied [18].

Vapor pressure is another phenomenon that contributes to fractionation. The preferential evaporation of the lighter isotopes of hydrogen and oxygen over ocean waters tends to leave the surface layers enriched in the heavier species. An accompanying result of this effect is that the final vapor precipitate of snow and ice in the polar region is a "light" snow. A few studies have also been made with respect to the enrichment of H^2 and O^{18} caused by evaporation from fresh-water bodies.

The variations in the N^{15}/N^{14} ratio in natural gas fields and associated crude oils are also attributed in part to a multistage enrichment brought about by a complex effect of molecular flow through porous rock and surface diffusion [19].

Fractionation is also a cause for the observed variations in the isotopic abundances of sulphur [20].

GEOCHEMISTRY OF CARBON AND SULPHUR

Extensive mass spectrometric work has been undertaken to measure isotopic abundance variations of carbon and sulphur. Studies of the former have included specimens of carbonates, igneous rocks, diamonds, petroleum, natural gases, fossil woods, marine plants, marine invertebrates, and sediments. The isotopic studies of petroleums and their relation to other natural carbonaceous substances provide an insight into questions on petroleum origin and evolution. Studies with respect to sulphur have been of particular interest because of that element's wide distribution and abundance in the earth's crust and in meteorites.

Nier and Gulbransen [21] were the first to report that C^{13}/C^{12} ratios of sedimentary rock carbonates are 2–3% higher than those of natural organic matter. A recent review by Silverman [22] reports the deviations, δ (in parts per thousand), of C^{13}/C^{12} ratios of natural carbonaceous materials (see Figure 11.5). The high C^{13} content of carbonates of limestone is definitely attributed to a carbon isotopic exchange reaction. In marine and land plants, however, there is a preferential assimilation of $C^{12}O_2$ over $C^{13}O_2$ during photosynthesis, and there are other kinetic effects that are presumed to concentrate C^{12} with respect to atmospheric carbon dioxide. There exists a further line of demarcation between marine or land plants in that CO_2 is provided and transported across cell walls in a different mode. In the former case, CO_2 is supplied from a solution whereas in the latter case carbon dioxide diffuses through stomata along

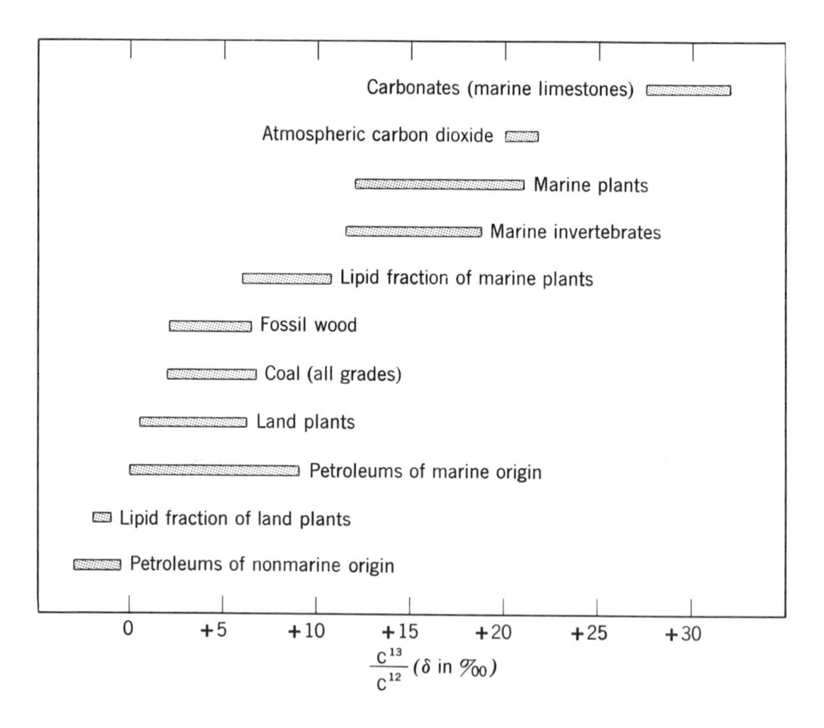

Fig. 11.5 C^{13}/C^{12} δ-value ranges in natural carbonaceous materials [22, adapted].

a concentration gradient. McMullen and Thode [23] have concluded that a clear distinction can be made among limestone carbon, organic carbon of marine origin, and carbon of land-plant origin.

Investigations relating to petroleum and oil-formation products have been extensive. Chromatographic fractions of petroleum and of soluble and insoluble organic constituents of shale have been analyzed to detect isotopic changes that might be attributed to chemical composition. Such changes, if they occur, are exceedingly small. However, significant differences in the C^{13}/C^{12} ratio have been noted between hydrocarbon gas evolving from oil-bearing deposits and liquid petroleum. Fractionation results in a depletion of the C^{13} in the gas phase. Petroleums of marine and nonmarine origin have lower C^{13}/C^{12} ratios than some of the corresponding organic source materials. McMullen and Thode [24] suggest that this would indicate that not all biogenic carbon is converted into petroleum.

Isotopic differences in sulphur provide considerable insight into effects induced by bacterial action, in addition to fractionation arising from purely physical phenomena. Analyses have been made of meteorites; sulfides of igneous origin; gases from volcanoes; sea-water sulfates of the Atlantic, Pacific, and Arctic oceans; free sulphur produced in lake bottoms; petroleum oils; and the vast sulphur wells of Texas and Louisiana. The range of S^{32}/S^{34} ratios of sulfates, sulphur, and sulfides from these latter two sources are shown in Figure 11.6, together with other data on naturally occurring sources.

The S^{32}/S^{34} ratio in meteoritic samples is remarkably invariant and it is assumed that this represents the ratio for terrestrial sulphur during the time the world was formed. For this reason it is sometimes used as a standard. There is considerable evidence to indicate that the Texas and Louisiana sulphur deposits reflect the effects of living organisms. Bacterial reduction of sulfates are known to result in large fractionation effects, and similar enrichments have been reported in the sulphur-producing lakes of Africa, where bacterial reduction of the sulfate has been analyzed [25].

Thode et al. [26] have indicated that we can always expect a kinetic isotopic effect in the reduction of sulfate by bacterial action, with S^{34} enrichment in the sulfate. This effect will occur in sediments of lakes and shallow ocean waters where bacterial action is considerable. A biological process is also ascribed to the fractionation of sulphur isotopes in sedimentary rocks, and the extent of this fractionation has suggested that isotopic ratios might reveal the biological activity at the time of deposition. There is some indication that such isotopic data can yield at least a rough correlation of bacterial action with geological period.

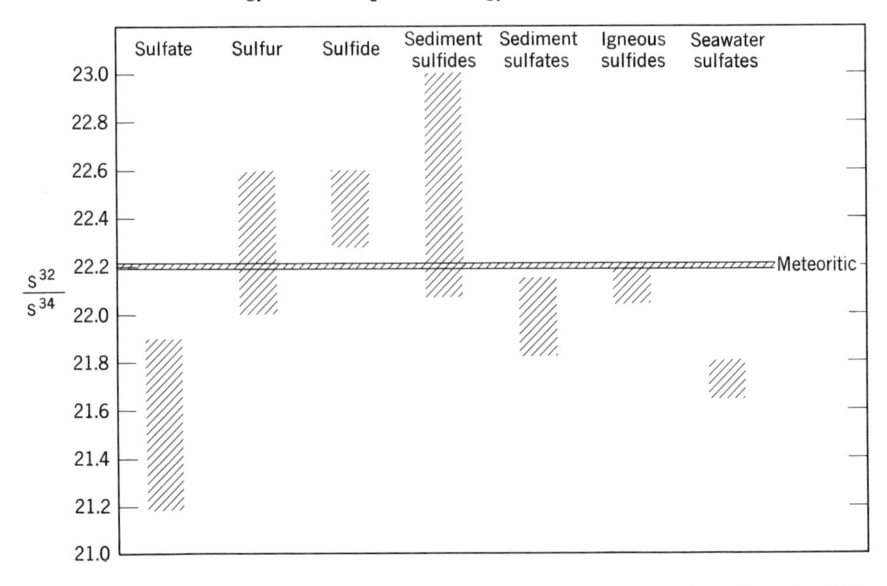

Fig. 11.6 Variations in the S^{32}/S^{34} ratios of naturally occurring deposits [23, adapted].

Isotopic ratios of sulfur have also been measured in asphalts extracted from rocks, as denoting a function of age and depth of deposit. A biological theory of S^{34} enrichment postulates that $(S^{34}O_4)^=$ and $(S^{32}O_4)^=$ are reduced at different rates and that the H_2S produced is depleted in S^{34}. The residual sulfate thus becomes enriched in the heavier isotope.

RUBIDIUM AND POTASSIUM DATING

The age of a mineral may be calculated if upon its formation the mineral included a radioactive nuclide of suitable half-life among the several isotopes of one of its chemical elements. The isotopic dilution method (see Chapter 13) can be employed to measure the present concentration of radioactive atoms. If the decay constant is known, the age may then be calculated. Two cases of considerable interest are the weakly radioactive isotopes of rubidium and potassium.

Rubidium is a relatively rare element with an abundance of about 0.035% by weight in igneous rocks of the earth's crust. The element has two naturally occurring isotopes, Rb^{85} (72.15%) and Rb^{87} (27.85%). This latter isotope decays to Sr^{87} by beta emission, with a half-life of about 5×10^{10} years. Thus if a mineral was found which contained the single Sr^{87} isotope (rather than the four naturally occurring nuclides), the isotopic dilution method could be employed—using one of the other strontium isotopes as the tracer. Knowing the half-life of Rb^{87} and the number of Sr^{87} atoms contained in a specific mineral, an estimate of the mineral's age is possible. That the age of the mineral can be computed directly is seen from the radioactive decay law:

$$t = \frac{1}{\lambda(Rb^{87})} \cdot ln \frac{N_0(Rb^{87})}{N(Rb^{87})}$$

$$= \frac{1}{\lambda(Rb^{87})} \cdot ln \frac{N(Rb^{87}) + \Delta N(Sr^{87})}{N(Rb^{87})}$$

$$= \frac{1}{\lambda(Rb^{87})} \cdot ln \left[1 + \frac{\Delta N(Sr^{87})}{N(Rb^{87})}\right]. \tag{11.5}$$

Here $\lambda(Rb^{87})$ is the decay constant of the Rb^{87} isotope, $N(Rb^{87})$ is the number of rubidium atoms currently measured in the mineral, and $\Delta N(Sr^{87})$ is the number of radiogenic strontium atoms. (The decay constant, $\lambda(Rb^{87})$, is 1.39×10^{11}/yr.)

Actually, the elemental abundance of nonradiogenic strontium in igneous rocks is only slightly less than that of rubidium. Nevertheless, by making careful mass spectral corrections, a clear differentiation can be made of the strontium resulting from rubidium decay and primeval strontium. Fortunately, also, a few minerals contain rubidium with much lower natural strontium concentrations than average—lepidolite, muscovite, and biotite.

Ages of the order of 10^9 years have been determined for various minerals and meterorites. The concentrations of radiogenic strontium are usually small—even for very old minerals—compared with the amount of lead in uranium-bearing minerals; for this reason the U-Pb age method is sometimes preferred.

The decay of K^{40} to A^{40} (by K-electron capture) and to Ca^{40} (by beta emission) also allows numerous dating problems to be solved. The number of radiogenic argon atoms formed after a time, t, is given by

$$A^{40} = \frac{\epsilon}{1 + \epsilon} K^{40}(e^{\lambda t} - 1), \tag{11.6}$$

where ϵ is the branching ratio (see Chapter 5) and K^{40} is the number of atoms of that nuclide remaining after time, t.

Thus the determination of geological age in the case of the potassium-argon system is dependent on both half-life and the branching ratio $(K^{40} \rightarrow A^{40})/(K^{40} \rightarrow Ca^{40})$, which is ~ 0.1. The potassium-argon age method is useful because of the widespread distribution of potassium.

However, the possible escape of radiogenic argon from the mineral by diffusion poses a special problem. The measurement of the small amounts of argon extracted from potassium minerals is further complicated by possible contamination from atmospheric argon. It is therefore necessary to determine the normal argon concentrations of the sample. Mass spectrometric measurements of A^{36}/A^{40}, and A^{38}/A^{40} are always required to differentiate between atmospheric contamination and the primordial argon gas.

GEOLOGICAL AGE FROM LEAD

The lead present at the time of the earth's formation is believed to have contained Pb^{204}, Pb^{206}, Pb^{207}, and Pb^{208}. Geological dating by the lead method presumes the following decay modes:

$$U^{238} \rightarrow 8\ \alpha \text{ particles} \rightarrow Pb^{206},$$

$$U^{235} \rightarrow 7\ \alpha \text{ particles} \rightarrow Pb^{207},$$

$$Th^{232} \rightarrow 6\ \alpha \text{ particles} \rightarrow Pb^{208}. \tag{11.7}$$

Pb^{204} is not of radiogenic origin.

At one time it was supposed that common lead had an invariant isotopic composition throughout the earth's crust. If this were the case, the relative amounts of common and radiogenic lead in a mixture could be computed from the Pb^{204} content. However, in 1938 Nier [27] reported large isotopic variations in common lead samples. These variations have been attributed to the addition of small amounts of radiogenic lead to common lead throughout geologic time. Thus it is postulated that at an early stage—when the earth was fluid—primordial lead, uranium, and thorium were uniformly distributed throughout the planet's mantle. Upon formation of the earth's crust, primordial lead was frozen into rocks together with various amounts of thorium and uranium. The U-Th-Pb composition of each locality was thus fixed. Subsequent thorium and uranium decay then continuously generated radiogenic lead. Therefore it is presumed that if we knew the age and isotopic composition of a number of lead ores, we might compute the time when mixing ceased and the earth's crust became a solid.

It will be noted that isotopic measurements of lead are especially valuable in that several ages may be computed corresponding to the several decay modes. Ores and minerals that have been analyzed include pitchblende, uraninite, monazite and zircon. The ages reported for a zircon from Ceylon are interesting because they indicate exceptionally good agreement for a single specimen. Results are shown in Table 11.4.

Table 11.4

Age Determination of Zircon By Several Methods [28]

Measurement	*Calculated age (yr)*
U^{238}/Pb^{206}	5.40×10^8
U^{235}/Pb^{207}	5.44×10^8
Th^{232}/Pb^{208}	5.38×10^8
Pb^{206}/Pb^{207}	5.55×10^8

According to Nier [29], the measurement of the Pb^{206}/Pb^{207} ratio generally provides a more accurate age than uranium-lead measurements. The half-lives of the U^{238} and U^{235} leading to Pb^{206} and Pb^{207} are quite different (4.5×10^9 and 7.1×10^8 yr, respectively), so that the lead abundance ratio directly yields the age of the mineral. Hence in lead-lead measurements it is unnecessary to analyze for total amounts of uranium or lead in the sample.

ANALYSIS OF METEORITES

Mass spectrometric analyses of meteorites furnish a wealth of information with respect to both chemical composition and isotopic abundances. Similarities or differences in the isotopic composition of elements found in terrestrial minerals and in meteorites permit conclusions to be made with respect to the history of mineral formation as well as the determination of age. Further, isotopic measurements provide the best clues concerning the nuclear reactions that were associated with the formation of our solar system. Meteoritic data are also the best source for estimating past and present intensities of cosmic radiations in interplanetary space. In the near future, other extraterrestrial samples will undoubtedly be procured so that "isotope cosmology" may actually represent a major discipline.

Mention has already been made of the fact that the isotopic constitution of meteoritic sulphur appears to be constant and is close to estimated terrestrial averages for this element and to sulphur of igneous origin. Isotopic-dilution techniques have also been used to assay the potassium and radiogenic argon of stony meteorites, which yielded meteoritic ages in the range of 4.6×10^9 to 4.8×10^9 yr [30]. Duckworth [31] suggests that this value, which may be close to a lower limit for the age of the universe, is in good agreement with values calculated from the isotopic composition of terrestrial lead and meteoritic lead. Age measurements have also been made of meteorites, using uranium-lead and rubidium-strontium methods. Hintenberger [32] reports average ages of 4.5×10^9 to 5×10^9 yr. This is the approximate time when present-day meteorites solidified from the liquid state. The indicated K-Ar age may be considerably less than the U-Pb or Rb-Sr ages. The explanation lies in the fact that even after solidification some diffusion of the radiogenic argon took place at elevated temperatures. K-Ar ages have given values ranging from 5×10^8 to 5×10^9 yr. These dates then correspond not to solidification times per se but rather to the times that the temperature fell to a level below which substantially all the argon formed by radioactive decay was retained within the mineral.

The composition of some meteorites poses an especially intriguing challenge to mass spectrometry. If a meteorite has undergone extensive exposure to cosmic radiations, trace quantities of elements can be expected to have substantially different isotopic composition than terrestrial material. Cosmic rays consist primarily of very high energy protons (GeV) and about 15% alpha particles. Upon striking a meteorite this radiation produces other protons, deuterons, helium nuclei and other light atoms, and nuclei in highly excited states. The ultimate products of these spallation reactions then accumulate in the meteorite and permit some conclusions to be drawn with respect to the irradiation duration, its intensity, and variations in cosmic radiation levels in interplanetary space.

The isotopic abundance of the spallation products are markedly different from the terrestrially observed normal abundances. However the number of spallation atoms formed is often so small ($\sim 10^{-10}$ gm/gm of meteorite) that quite sophisticated methods of mass spectrometry or neutron activation are required. Stable spallation products of helium, neon, potassium, and argon have been observed mass spectrometrically. He^3 has been found in meteorites with an *isotopic* abundance as high as 25%—even though the atmospheric helium fraction is only 1.3×10^{-4}%. The concentration of a spallation product will also be dependent upon the penetration depth of the cosmic radiation into the meteorite. Thus one can expect to find contrast of surfaces of equal concentration of spallation products. For this reason it is possible to reconstruct the original size and shape of a meteorite in space, even though it may have been considerably altered or broken up upon entering the earth's atmosphere. Very large isotopic anomalies that have been observed for potassium by Hintenberger [33] are shown in Figure 11.7. Highly interesting isotopic anomalies have also been recently reported for lithium in the Holbrook meteorite [34]. Li^7/Li^6 ratios varied in specific locations from 9.5 to 7.5. The elemental concentration was also

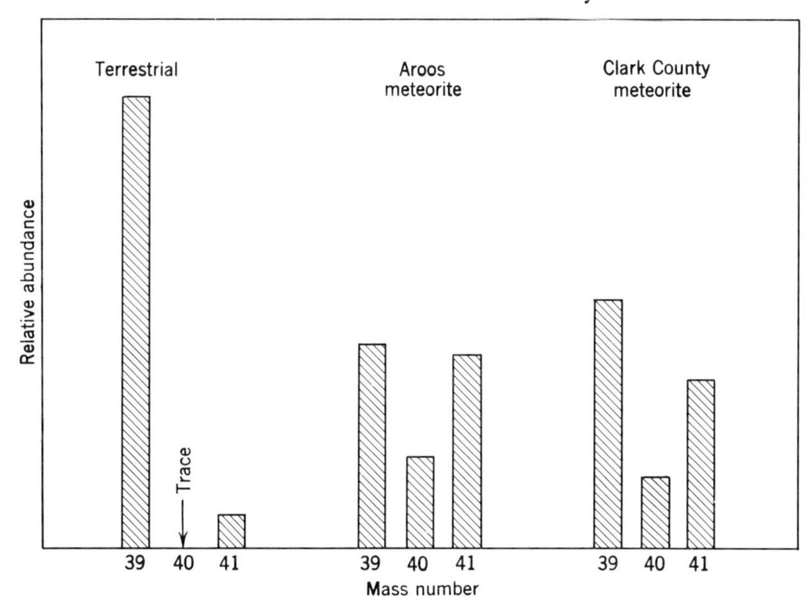

Fig. 11.7 Comparison of potassium spectra for terrestrial and meteoritic material [33, adapted].

exceedingly inhomogeneous, varying from about 1 to 100 ppm. This work also has considerable instrumental significance inasmuch as the analysis was accomplished by means of an ion-microprobe spectrometer that permitted a meteoritic surface assay without chemical preparation. This method of ion-beam sputtering not only permits highly localized regions to be assayed but it also greatly reduces the amount of an element that can be detected and measured without appreciable contamination.

Clarke and Thode [35] have also reported isotopic ratio anomalies for xenon gas extracted from stone meteorites. A large excess of Xe^{129} is tentatively attributed to the decay of I^{129} ($T_{1/2} = 1.7 \times 10^7$ yr) during the early history of the solar system. Deviations of some other isotopes from "normal" are thought to be affected by spallation reactions. The samples of meteorite (10–30 gm) were melted by rf induction heating in a high-vacuum system and fed to a spectrometer operated in a "static" mode. This reported detection limit of the instrument is approximately 10^5 atoms of any isotope of xenon.

A final note with respect to age determination relates to a basic assumption of radioactive decay. Geological dating presumes that radioactive nuclei decay at a constant rate and that conditions of pressure, temperature, and chemical state in no way affect the rate of nuclear disintegrations. This assumption is not completely valid, even though the influence of such factors is exceedingly small. The largest observed difference has been reported for the decay rate of different chemical compounds of technetium (about 0.27%) by Bainbridge et al. [36]. The basic explanation is that one form of radioactive decay (electron or K-capture) can be slightly influenced by the electron density around the nucleus. No conceivable conditions of temperature or pressure in the interior of the earth, however, would compare to the extremes of chemical state for which this phenomenon has been detected. For this reason, radioactive decay is assumed to be invariant. As one geologist has remarked, "The smallness in the change of the radioactive decay constant in an exceedingly sensitive system is more reassuring than the negative results of earlier measurements of considerably less precision."

REFERENCES

[1] A. O. C. Nier, *Phys. Rev.*, **48**, 283 (1935).
[2] A. O. C. Nier, *Phys. Rev.*, **55**, 153 (1939); also **60**, 112 (1941).
[3] R. Brown and W. A. Wolstenholme, *Nature,* **201**, 598 (1964).
[4] L. H. Ahrens, *Distribution of Elements in Our Planet,* McGraw-Hill, New York, 1965, p. 20.
[5] J. M. Robson, *Phys. Rev.*, **100**, 933 (1955).
[6] R. D. Hill, *Phys. Rev.*, **59**, 103 (1941).
[7] W. F. Libby, *Phys. Rev.*, **69**, 671 (1946).
[8] L. T. Aldrich and A. O. C. Nier. *Phys. Rev.*, **74**, 1590 (1948).
[9] H. Faul, *Nuclear Geology,* Wiley, New York, 1954, p. 135.
[10] A. O. Nier, *Phys. Rev.*, **77**, 789 (1950).
[11] L. T. Aldrich and A. O. C. Nier, *Phys. Rev.*, **74**, 876 (1948).
[12] E. Segre, *Phys. Rev.*, **86**, 21 (1952).
[13] J. Macnamara and H. G. Thode, *Phys. Rev.*, **80**, 471 (1950); also H. G. Thode, *Trans. Roy. Soc. Can.*, **45**, 1 (1951).
[14] H. C. Urey, *J. Chem. Soc.*, 562 (1947).
[15] H. C. Urey, et al., *Bull. Geol. Soc. Amer.*, **62**, 399 (1951).
[16] C. Emiliani, *Science,* **154**, 851 (1966).
[17] K. I. Mayne, in *Methods in Geochemistry,* ed. by A. A. Smales and L. Wager, Interscience, New York, 1960, p. 153.
[18] *Ibid.*
[19] T. C. Hoering and H. E. Moore, *Geochim. Cosmochin. Acta,* **13**, 225 (1958).
[20] H. G. Thode, et al., *Bull. Amer. Assoc. Petrol. Geol.*, **42**, 2619 (1958).
[21] A. O. Nier and E. A. Gulbransen, *J. Amer. Chem. Soc.*, **61**, 697 (1939).
[22] S. R. Silverman, in *Isotopic and Cosmic Chemistry,* ed. by H. Craig, S. L. Miller, and G. J. Wasserburg, North Holland, Amsterdam, 1964, p. 95.
[23] C. C. McMullen and H. G. Thode, in *Mass Spectrometry,* ed. by C. A. McDowell, McGraw-Hill, New York, 1963, p. 422.
[24] *Ibid.*, p. 424.
[25] N. B. Slater, *Trans. Roy. Soc.* (*London*), **A246**, 57 (1953).

[26] H. G. Thode, J. Macnamara, and W. H. Fleming, *Geochim. Cosmochim. Acta,* **3,** 253 (1953).

[27] A. O. C. Nier, *J. Amer. Chem. Soc.,* **60,** 1571 (1938).

[28] G. R. Tilton, et al., *Trans. Am. Geophys. Un.,* **38,** 360 (1957).

[29] A. O. C. Nier, *Amer. Scientist,* **54,** 368 (1966).

[30] G. J. Wasserburg and R. J. Hayden, *Phys. Rev.,* **97,** 86 (1955).

[31] H. E. Duckworth, *Mass Spectroscopy,* Cambridge, London, 1960, p. 160.

[32] H. Hintenberger, *Naturwissenschaften,* **51,** 473 (1964).

[33] *Ibid.,* p. 497.

[34] W. P. Poschenrieder, R. F. Herzog, and A. E. Barrington, *Geochim. Cosmochim. Acta,* **29,** 1193 (1965).

[35] W. B. Clarke and H. G. Thode, in *Isotopic and Cosmic Chemistry,* ed. by H. Craig, S. L. Miller, and G. J. Wassenburg, North Holland, Amsterdam, 1964, p. 471.

[36] K. T. Bainbridge, M. Goldkeber, and E. Wilson, *Phys. Rev.,* **90,** 430 (1953).

Chapter 12

Ecology and Environmental Science

Ecology was introduced into the English vocabulary in 1869. It encompasses the complex interrelationship of man to his total environment. Ecology admits to neither artificial barriers in the natural sciences, nor to any real division between the natural and social sciences. Science and society are properly envisioned as a continuum, and man is faced with the fact that his surroundings will ultimately determine his very existence. Only within the past two decades, however, has there developed a general recognition of the great need for environmental engineers and scientists. Since World War II a sizable corps of health physicists have been trained to deal with the potential hazards related to a man-made radiation environment. But the broader aspects of environmental control have not, as yet, been granted the priorities that they merit.

Indeed, the inclusion of this brief chapter in this book has a twofold purpose. One is to call attention to mass spectrometry as an important tool in analytical environmental studies. A second objective is to emphasize the complete interdisciplinary nature of environmental science and to point out that almost any mass spectrometric group, regardless of its primary role, has the opportunity to participate in some facet of ecological monitoring or research. Further, at least a few mass spectrometrists might seriously address themselves to the general question: What contributions can spectrometry make in *ecosystem studies?*

It is disturbing, philosophically, to consider what may happen when large numbers of sociologists, economists, engineers, chemists, physicists, and biologists enter the field of ecology. We will witness a proliferation of research and obtain results from many detailed studies. These detailed studies are admittedly important. But there exists a danger that there may be insufficient collaboration among groups engaged in an effort comparable to that of the United States space program. Even among today's "environmental engineers" we find an artificial compartmentalization of responsibility that should be critically examined. We have health physicists concerned only with nuclear radiations, biologists who are focusing their attention solely on pesticides, engineers who are worried about air pollution but not about a pure water supply, and agronomists who are concerned with soil conservation in a fairly restricted sense. The dilemma is aptly reviewed by Egler [1], who states that although ecology may well be one of the most important of the sciences it is among the least understood and the whole is considerably greater than the sum of the parts.

It would seem that, to a greater degree than most specialists, the mass spectrometrist can afford to be objective. Spectrometry applies to an unusually large variety of analyses and the spectrometrist should be challenged to provide data in every possible relevant area. He is in a position to trace both the short- and long-term imbalances in the earth's biotic constitution or distribution of matter and together with other scientists he should be acutely concerned with the failure of a technological society to appreciate the total scope of the problem.

ECOLOGY AND TECHNOLOGY

The specific successes of science and engineering have perhaps blinded man to the possibility that relatively "minor" changes from the natural environment could have serious consequences. Of course, certain changes appear beneficial and they are prerequisite to the attainment of many large-scale engineering goals. The Panama Canal is an example. Yet other enterprises envisioned as having short-term goals have been clearly wasteful. The Dust Bowl of the 1930's and the soil exhaustion of some 100 million acres is the result of "overuse." But a more subtle and irreversible change may be taking place that is directly attributable to "an advancing technology." Wastes and residues are being broadcast throughout the environment, many of which substances were not previously present in either their chemical or physical form. More than a few of the 10,000 chemical compounds produced per year are also known to be highly toxic, and some are persistent pollutants. Initially they may be widely dispersed and diluted by rain and run-off water. But at a subsequent date such toxic compounds can be reconcentrated in organisms and animal life in a fashion that may result in a genuine hazard.

The more obvious deleterious effects of air and water pollution are now receiving publicity, but progress is far from encouraging. Over 130 million tons of "aerial garbage" are being discharged into the United States atmosphere each year. The fall of combined soot and particulate matter on a single square foot in a city may easily exceed a pound per annum. In metropolitan areas the principal contributors are motor vehicles, thermoelectric power stations, assorted industries, and households. A further compounding of the general problem of pollution is expected from increasingly dense population concentrations. A substantial increase in the global carbon dioxide concentration has already taken place—about 5%, although the exact figure is in dispute. Some predict by the year 2000 an increase in atmospheric carbon dioxide of about 25% over present levels. Whereas a small increase in the carbon dioxide level has little known effect on living organisms, carbon dioxide is directly involved in mechanisms that determine the over-all temperature of the earth. An increase in the latter would cause a melting of a portion of the polar ice caps, a rise in sea level, and accompanying changes of considerable significance. Such changes are obviously of more than academic importance.

Twentieth-century technology is also accompanied by "accidents" that are completely unforeseen. In March of 1967 the 61,000-ton American tanker, *Torrey Canyon,* flying the Liberian flag, broke up on a reef seven miles off the southwestern tip of England. With an oil cargo of 36 million gallons, the loss was reported as the most costly in merchant marine history. But the incident also posed a pollution threat of major proportions.The oil from the vessel's ruptured tanks overran 120 miles of Cornwall's beaches. England's finest vacationland coast was stained, untold wildlife was killed, and oyster beds and inshore fisheries were seriously jeopardized. A similar event occurred on March 3, 1968 when the 12,000-ton oil tanker, *Ocean Eagle,* broke in two at the mouth of the harbor in San Juan, Puerto Rico.

AIR POLLUTION

The more dramatic and sometimes lethal effects of air pollution have, of course, pointed up the seriousness of the problem. In 1930 in the Meuse Valley of Belgium, 100 persons became ill and 63 died in an air-pollution incident. In 1948 a somewhat analogous situation occurred near the town of Donora, Pennsylvania. Fog and a low-level temperature inversion covered the horseshoe-shaped valley of the Monongahela River, resulting in sickness for half the population and 20 deaths. In 1952 in London 4000 deaths above normal were registered during a two-week period. Figure 12.1 shows the dramatic record of the fatality rate [2]. Measurements of the amount of suspended smoke in the London disaster reached 4.46 mg/m^3 of smoke, and 1.34 ppm of sulphur dioxide [3].

The effects of pollutants in terms of material damage and crop spoilage are also costly. Damage to structural metals, surface coatings, fabrics, and other articles of commerce and trade, plus incidental costs required

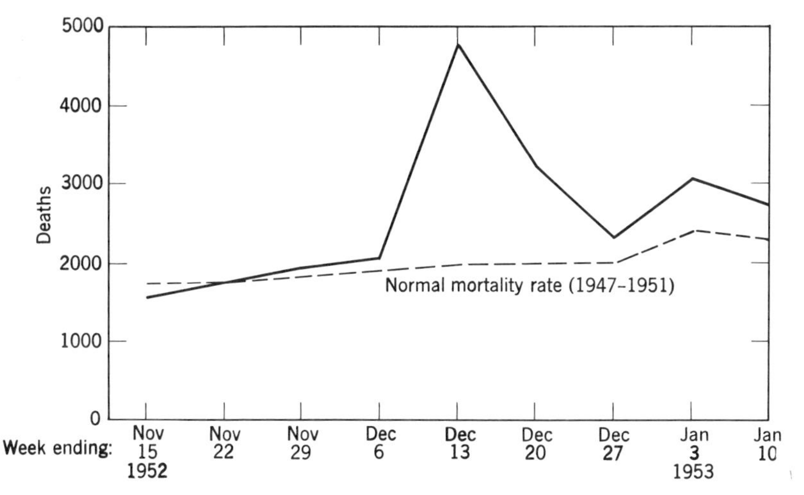

Fig. 12.1 Recorded deaths in the Greater London fog of December 1952 [2, adapted].

for cleansing and protective activities is estimated at several billions of dollars in the United States. A sizable number of agricultural crops is adversely affected. Leaf damage, stunting of plant growth, and the decreased size and yield of fruit are but a few of the more obvious effects from specific pollutants. The state of California has estimated agricultural damage at $6,000,000 per annum—and this figure is probably a conservative one [4].

Air polluting substances are often categorized as primary and secondary pollutants. Primary contaminants consist of the relatively stable pollutants generated by industrial, commercial, and domestic sources, and can be readily traced to their point of origin. Secondary pollutants represent a more intractable class, as they are produced by photochemical or physicochemical interactions between primary pollutants within the atmosphere. Pollutants of this type include oxidation products, often produced by ozone (resulting from photochemical reactions between organic matter and oxides of nitrogen). The most publicized example of this secondary pollution on a large scale is that of the Los Angeles smog, which results from photochemical reactions of unburned fuel. Some estimates report that 3 million cars emit 8000 tons of carbon monoxide and almost 1650 tons of hydrocarbons per day. This direct emission from automobile exhausts to the atmosphere is complemented by the ablation of 50 tons of rubber per day from tires.

Specific gaseous pollutants that have been the object of considerable study include carbon monoxide, sulphur dioxide, the oxides of nitrogen, ozone, fluorides, and many organic vapors. Carbon monoxide is hazardous to humans at about 100 ppm for prolonged exposures and sulphur dioxide is irritating to the upper respiratory tract at levels of only a few parts per million. Sulphur dioxide and sulphur trioxide also attack building materials such as limestone, marble, and mortar—which contain carbonates that are converted to soluble sulphates that can be washed away by rainwater. One of the simplest organic substances, ethylene, a participant in "smog" reactions, is damaging to plant life at the parts per billion level [5]. Superposed on such man-made pollutants are the so-called "natural" pollutants arising from volcanic eruptions, hurricanes, cyclones, and meteoric dusts. Estimates of the extent of such pollution have been as high as 1000 to 2000 tons per day over the entire earth, and meteroic dust particles form a significant source of condensation nuclei.

There are several reasons for anticipating a greatly increased use of mass spectrometry in air pollution studies; these include:

1. The sensitivity for detecting trace elements at the parts-per-million level.
2. Ability to monitor both primary pollutants and complex secondary pollutants resulting from photochemical reactions.
3. Potential advantages of automation in a worldwide analyzing network.

This last reason may not seem important at this time, but it should be pointed out that air pollution is less amenable to control than water pollution. As one environmental scientist has stated: "It is easier for man to control hydrological events than meteorological events; air is not delivered to users in pipes, we simply live in it." Therefore we can ultimately expect a reasonably large-scale effort for providing spatial- and time-dependent data to monitor atmospheric contaminants. Technologically, at least, atmospheric mass spectrometric stations could be partially automated to provide a wealth of data on the contaminants that exist in the gaseous phase.

Many mass spectrometric investigations have already been made in laboratories, even though mass spectrometers have not been used generally at air monitoring sites. Analyses of solvent vapors in air of the order of parts per million have been reported [6] and smog samples have also been analyzed at exceedingly low concentrations (for hydrocarbons and hydrocarbon derivatives) [7]. In the pioneering work of Shepherd et al. [8] analyses of gaseous pollutants were made in connection with studies of the Los Angeles smog. The gaseous phase of the smog was found to be of the order of 0.5 ppm of the air, and about 60 chemical compounds or families of compounds were tentatively identified by mass spectrometric assay. It was shown that the gaseous phase of the smog was primarily a mixture of hydrocarbons and hydrocarbons combined with nitrogen, oxygen, and chlorine. It was further shown that some of these hydrocarbons, when oxidized with ozone and nitrogen dioxide (in the presence of ultraviolet light), produced substances that constituted an appreciable portion of the smog concentrates. The specific steps reported by Shepherd and co-workers are the following:

1. Isolation of the gaseous and vaporous pollutants from the air on a filter maintained at the temperature of liquid oxygen.
2. Without prior evaporation, separation of the isolated concentrate by isothermal distillation or sublimation at low temperatures and pressures.
3. Subsequent analysis of the distillates, which were delivered directly to the mass spectrometer.

This method provided an efficient means for pollutant collections and it circumvented reactions between the vaporized pollutants prior to a mass spectral assay.

Current smog research is now attracting the attention of automotive engineers, as well as scientists in the environmental and medical fields. In fact two of the major smog chambers are almost exclusively devoted

Table 12.1

Range of Metal Concentrations in the Atmospheres of U.S. Cities [9]

Metal	*Concentration* ($\mu gm/m^3$)
Zinc	0.4–49.0
Iron	0.23–30.0
Copper	0.05–30.0
Lead	0.33–17.0
Manganese	0.01–3.0
Barium	0.005–1.5
Tin	0.004–0.80
Vanadium	0.002–0.60
Titanium	0.01–0.24
Nickel	0.005–0.20
Chromium	0.002–0.12
Cadmium	0.002–0.10
Bismuth	0.002–0.03

to the analysis of automobile exhaust fumes. One is operated by the state of California and the other is located at the General Motors Research Center near Detroit. In these chambers, tests can be made of eye irritations, ozone formation, aerosol creation, and the effects on growing plants.

Another aspect of air-pollution research relates to the large numbers of metals that are found in particulate form in the environment of industrial communities. Table 12.1 is a list of metals the concentrations of which have been measured in the air of 20 communities throughout the United States [9]. It should be noted that although those concentrations were determined by techniques other than mass spectrometry, present mass spectrometric methods are well within the sensitivity range that is required.

STANDARDS OF WATER QUALITY

Standards of water quality are now being established to meet all needs—domestic, agricultural, industrial, and recreational. In the United States alone approximately 400 billions of gallons are used daily, with predictions that this figure will rise to 600 billion by 1980. Mass spectrometry is now joining x-ray fluorescence [10] and other analytical methods that furnish quantitative data on water pollutants and concentration levels of specific metallic impurities. Also, because large quantities of water will have to be reused, there is little doubt that a vastly increased analytical effort for maintaining water quality will be required. Water, like steel, will be sold on the basis of specification.

Acceptable standards for drinking water have been established by the United States Public Health Service [11], which has indicated upper limits for many elements and compounds. Table 12.2 gives a partial list.

The establishment of such limits, however, has not automatically resulted in their attainment by all communities. In general, tolerable levels of impurities also vary depending on the forms of life that are sustained by rivers and streams. Even low concentrations of nearly all salts are toxic to certain aquatic species. Chlorides are toxic to fish at 400-ppm concentrations. Copper is toxic to certain bacteria and microorganisms at 0.1 to 0.5 ppm—although, for setting, oyster larvae require a level of about 0.05 ppm.

Until recently there have been few detailed analyses of trace metals in water supplies. Results of one extensive survey [12] by the United States Public Health Service, however, is of considerable interest. Table 12.3 is a compilation indicating the range of certain metallic concentrations in water supplies of major United States cities.

Table 12.2

Upper Limits on Elements or Compounds for Drinking Water Supplies [11]

Substance	*Parts per million*
Lead	0.1
Fluoride	1.5
Arsenic	0.05
Selenium	0.05
Chromium (hexavalent)	0.05
Copper	3.0
Iron and manganese	0.3
Magnesium	125
Zinc	15
Chloride	250
Sulfate	250
Phenolic compounds	0.001
Total solids (desirable limit)	500

Table 12.3

Concentration of Trace Elements in Public Water Supplies of the 100 Largest Cities in the United States [12]

	Concentration ($\mu g/l$)		
Element	*Maximum*	*Median*	*Minimum*
Ag	7.0	0.23	ND*
Al	1500	54	3.3
B	590	31	2.5
Ba	380	43	1.7
Cr	35	0.43	ND
Cu	250	8.3	0.61
Fe	1700	43	1.9
Li	170	2.0	ND
Mn	1100	5.0	ND
Mo	68	1.4	ND
Ni	34	2.7	ND
Pb	62	3.7	ND
Rb	67	1.05	ND
Sr	1200	110	2.2
Ti	49	1.5	ND
V	70	4.3	ND

* ND: not detected.

The general problem of water pollutants is both analogous to and distinctive from that of the primary and secondary pollutants in the atmosphere. Water pollutants may be (a) degradable, (b) nondegradable, and (c) "persistent." The most common degradable pollutant is domestic sewage and organic wastes generated from the food, pulp, and chemical industries. When discharged into clean streams, bacteria feed on these wastes and break them down into their inorganic components—carbon, nitrogen, and phosphorous—which are basic plant foods. If the waste load is too great, however, the quantity of dissolved oxygen will drop below that needed for the support of aquatic life (5 ppm), excessive algae growth will result, and the water will fail to recover to its original state of purity.

Nondegradable pollutants are not attacked by stream biota, and for this reason they may represent a more formidable problem. Salts, heavy metals, and inorganic colloids are in this category. These substances are partially responsible for the deleterious effects to industrial equipment (boilers, pipes, rollers in steel mills, etc.), as they contribute to corrosion, scaling, and adverse surface reactions.

The third class of "persistent" pollutants are a recent product of our modern chemical and nuclear industries. Of the 500,000 organic chemicals currently known, most are synthetically produced. Many of these synthetics have complex molecular chains that cannot be degraded by stream bacteria, at least on a short-time scale. Typical industrial and agricultural effluent in this class includes certain detergents and pesticides. Long-lived radionuclides from the wastes of atomic power plants are also species that cannot be "treated."

It would be a gross exaggeration to cite mass spectrometry as the most important of present analytical methods in monitoring water quality or for detailed studies in research. It is not. But nearly every analytical method is useful. C^{14} dating is even valuable for it allows one to distinguish between industrial chemical pollution (from petroleum or coal products the radioactivity of which has substantially decayed) or from "contemporary" carbon (with a disintegration rate of approximately 13.5 atoms/gm of carbon). However, when one considers that nearly every major industrial processing plant—chemical, photographic, steel, and so forth—uses mass spectrometric equipment for developing its primary products, it is reasonable to suppose that the same tool (with a high sensitivity for both inorganic and organic species) will be increasingly used for monitoring plant effluent. It appears to be an ideal tool for analyzing the recovery of precious metals (e.g., nickel and silver) and with appropriate auxiliary equipment it could provide an "on-line" assay of the pure water output of water treatment plants.

THE IMPACT OF COMMERCIAL NUCLEAR POWER

It has been suggested earlier that radioactive products (from bombs or controlled nuclear power) represent an important class of persistent pollutants that are being added to the environment. No one can predict the magnitude of the "fall-out" from weapons, but good estimates are now being made of the growth of commercial nuclear power. Some authorities suggest that by A.D. 2000 the electric power generated from nuclear sources will increase by a factor of 1000 from its present levels. In the United States this will mean that about half of all the electrical generating capacity will be derived from nuclear fission. This form of energy is attractive from the standpoint of conserving the nation's fossil fuels, and very recently it has become attractive from purely economic considerations.

The deposits of fissionable material are not infinite, but they are large. The earth's crust contains about 4 ppm of uranium, and thorium has an abundance of about 12 gm per million grams of rock. Roughly then, about 16 ppm of potentially fissionable material might be recovered from the earth. This small percentage, when multiplied by a factor of about 2.5×10^6 greater heat-energy equivalent compared with an equal weight of coal, indicates that 1 ton of the earth's crust has a fissionable energy equivalent to about 40 tons of coal. No one is suggesting that all the fissile material can be extracted, but the richer sources will surely be exploited.

The growing magnitude of the nuclear power industry in the United States alone is shown in Figure 12.2 [13]. Participating utility firms and other agencies include: Boston Edison, Carolina Power and Light, Carolinas Virginia Nuclear Power, Central Vermont Public Service Company, Commonwealth Edison, Connecticut Light and Power, Connecticut Yankee Atomic Power, Consolidated Edison, Consumers Public Power District, Consumers Power, Dairyland Power Cooperative, Delaware Valley Utilities, Duke Power Company, Duquesne Light, Florida Light and Power, Hartford Electric Light, Iowa Gas and Electric, Los Angeles Department of Water and Power, Metropolitan Edison, New Jersey Central Power and Light, Niagara Mohawk Power, Northern States Power, Omaha Public Power District, Philadelphia Electric, Pacific Gas and Electric, Public Service Company of Colorado, Puerto Rican Water Resources Authority, Rochester Gas and Electric, Rural Cooperative Power Association, San Diego Gas and Electric, Saxton Nuclear Experimental, Southern California Edison, Tennessee Valley Authority, Virginia Electric and Power Company, Washington Public Power, Western Massachusetts Electric, Wisconsin-Michigan Power, and Yankee Atomic Electric.

Fig. 12.2 Projected growth of commercial nuclear power in the United States. Locations shown are where reactors are expected to be operational or under test by 1972 [13].

The environmental responsibilities assumed by the federal and state agencies and utility corporations in the implementation of this program will be unparalleled. The monitoring of stack gases, the reprocessing of fuel, and the constant surveillance of rivers and streams will be an important part of the total operations of all large-scale reactors. In some instances there have been special requirements beyond those of personal health. For example, the location of a fuel reprocessing plant in southwestern New York State was chosen only after the potential hazards of minute traces of contamination to photographic emulsions (manufactured by the Eastman Kodak Company, in Rochester, New York) had been evaluated. In such instances even short-lived isotopic contaminants may cause serious difficulties. Health physicists and environmental engineers, however, are especially concerned with radioactive species having intermediate or long half-lives; they are persistent contaminants and their long-term effects may be cumulative.

At reactor facilities such as the Hanford Plant in Richland, Washington, the disposal of radioactive waste water has been rigorously monitored on a routine basis, with monthly checks as far downstream on the Columbia River as Portland, Oregon. Some of the more significant isotopes that are carefully monitored include Cu^{64} (half-life 12.8 hr), Na^{24} (15 hr), As^{74} (26.7 hr), Np^{239} (56 hr), I^{131} (192 hr), and P^{32} (348 hr). Checks for Co^{60}, Sr^{90}, and Cs^{137} have been made in ground waters close to intermediate level waste areas.

The greatly increased resolution of nuclear detectors of alpha, beta, and gamma radiations and fission fragments make such devices invaluable in the quantitative assay of reactor effluent. The advent of reverse-biased *p-n* junctions has been especially important, and there is no debate that such tools will continue to provide the great bulk of analytical data relative to reactor environmental monitoring. But there is also little doubt that the present sensitivity for detecting trace quantities of certain isotopes can be matched or surpassed by mass spectrometric assay. Isotopes such as U^{234} (2.48×10^5 yr), U^{235} (7.13×10^8 yr), U^{238} (4.51×10^9 yr), Np^{237} (2.14×10^6 yr), Pu^{239} (24,360 yr) and, Pu^{240} (6760 yr) are among the longer-lived nuclides, and the isotopic dilution method can be employed to provide highly quantitative environmental data on these species. The basic method is the same as was earlier reported for the Idaho Chemical Processing Plant (Idaho Falls, Idaho) in connection with assaying the U^{235} content of spent reactor fuel elements [14]. In this instance U^{233}, available from Oak Ridge, was used as the "spike" isotope and the total uranium content was determined to levels as low as 0.001 mg/ml. Much lower concentrations can now be assayed.

Another environmental problem relates to the natural uranium concentrations in the lakes and potable waters of the United States. To date, man-made contaminations (of uranium) from weapons or nuclear power has been negligible. Consequently, present concentrations of uranium reflect only the long-term geological or other effects. Yet we actually do not possess accurate data of uranium concentrations in any but a few of the main bodies of water in the United States. For this reason a limited number of lakes is currently being analyzed for natural uranium content in the Stable Isotope and Ion Physics Laboratory at the Rensselaer Polytechnic Institute. Aside from providing information of geological interest, these data should provide a natural uranium baseline against which any long-term increases in uranium levels (i.e., over a 50-year period) can be compared. The excellent "memory" of lakes will

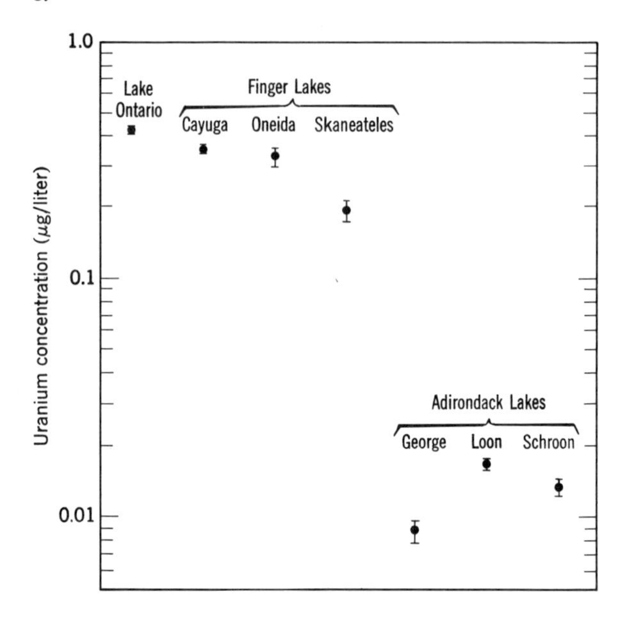

Fig. 12.3 Uranium concentrations in potable waters of New York State [15].

then provide a cumulative measurement of the environmental changes caused by activities of the nuclear industry. Recent measurements by Levin [15], which show significant differences in the present uranium concentrations of the Adirondack lakes as compared with Lake Ontario and several lakes in central New York State, are shown in Figure 12.3.

URANIUM CONTENT OF SEA WATER

Many studies have already been made to determine the uranium content of ocean water. Most workers agree that the mass spectrometric method (using the isotopic dilution method) yields more precise values than polarographic or fluorimetric methods. Some of the results of studies by Rona, Gilpatrick, and Jeffrey [16] and Wilson et al. [17] are presented in Table 12.4.

The reader is referred to the literature for details indicating that significant differences can be detected, depending on depth of sampling and the like. In all the above determinations an enriched U^{235} tracer was added to the sea water prior to the mass spectrometric assay. Much larger variations in uranium concentrations have been reported in freshwater rivers and streams. For example, the average uranium content of the Mississippi and Missouri Rivers at St. Louis have been assayed

Table 12.4

Uranium Concentrations in Ocean Waters [16], [17]

Body of water	*Latitude, longitude*	*Uranium content* ($\mu gm/l$)
English Channel	—	3.31
Bay of Biscay	—	3.31
Gulf of Mexico	26° 52′ N, 95° 04′ W	3.59
Gulf of Mexico	28° 20′ N, 88° 21′ W	3.56
Straits of Florida (Miami)	25° 49′ N, 79° 55′ W	3.53
North Atlantic (Beaufort)	34° 21′ N, 76° 34′ W	3.33
North Atlantic (Woods Hole)	37° 04′ N, 66° 54′ W	3.38
North Pacific	29° 41′ N, 120° 48′ W	3.28

to be 1.83 and 2.39 μgm/l respectively, and concentrations of 0.4 to 1.2 μgm/l have been reported in river waters in Japan [18].

TRANSLOCATION OF FISSION PRODUCTS

Considerable concern has been expressed about the possible contamination of agricultural land from fission products. This concern arises from the sharp increase in the use of nuclear energy as a source of power as well as from the uncontrolled contamination that accompanies the detonation of bombs. Weapons testing has already resulted in a worldwide distribution of Sr^{90} and other fission products.

Sr^{90} and Cs^{137} are important "fallout" products because of their relatively long half-lives (28 and 30 yr, respectively). Sr^{89} and I^{131} are also important, although they have shorter half-lives. I^{131} accumulates in the thyroid gland and produces local radiation damage; strontium rapidly enters into bone and may introduce bone cancer or leukemia. Long-lived Cs^{137} (30 yr) will disperse itself generally in the body and may cause genetic changes. In biological processes strontium behaves similarly to calcium, and cesium is somewhat analogous to potassium. Thus the translocation of these fission products is of great interest with respect to their role in the food chain. Other radioisotopes (e.g., the rare earths) are also produced in fission but are absorbed and translocated in plants very slowly. Further, they are not readily absorbed in the blood stream of animals.

From an environmental standpoint, fallout fission products may be retained by lakes, the aerial parts of plants, or pass on into the soil after successive rainfalls. One danger of strontium is that it is absorbed from the soil into plant life at a rate greater than any other fission product. Cs^{137} tends to become more "fixed" in soils, although it is very mobile throughout a plant once it has entered. The exceedingly high sensitivity of a mass spectrometer (using a thermal ionization source) for cesium will inevitably result in additional analytical studies that employ this tool. At present the sensitivity of the spectrometer is comparable to counting methods of Cs^{137}, although the samples are more difficult to prepare. The fact that both Cs^{135} (3×10^6 yr) and Cs^{137} (30 yr) are nearly equally abundant fission products also makes the mass spectrometric approach an attractive one.

Studies of both environmental and biological significance of fission products include:

1. Mobility of Cs^{137} in plants.
2. Absorption and exchange rates.
3. Passage of fission products into milk.
4. Detailed ecological studies of fallout.

Rubidium, another important fission product, also can be detected in a mass spectrometer with a very high sensitivity. This element plays a somewhat analogous role to that of potassium in metabolism; it is hence of comparable analytical importance.

ENVIRONMENTAL STUDIES OF CALCIUM ISOTOPES

In the previous chapter it has been noted that small variations in the naturally occurring isotopic abundances of carbon, oxygen, and sulphur have been observed, and that these are important in interpreting geochemical and other processes. A few analyses have also been made of calcium samples collected from diverse oceanographic, geologic, and biological environments to ascertain whether isotopic fractionation has occurred for this element. Calcium has six stable isotopes ranging from Ca^{40} to Ca^{48}. The percentage mass difference between the lightest and heaviest nuclide is thus large, so that some fractionation might be expected to take place in the natural environment.

Corless and Winchester [19] report that the isotopes of calcium do indeed undergo significant fractionation. In this study the Ca^{48} content of the samples was determined by a precision neutron activation analysis, using Eu^{151} as a spike that served as an internal standard. The method essentially consisted of comparing all sample Ca^{48}/total-Ca ratios to that of a reagent-grade $CaCO_3$ reference standard. The data of Table 12.5 include only a portion of a wide variety of samples, in which the

Table 12.5

Comparison of Samples to Standard [19]

Sample and location	δ	*Estimated error*
Snail shell, Gay Head, Martha's Vineyard, Mass.	−5.5	7.0
Snail shell, Atlantic City, N.J.	−2.5	4.5
Oyster shell, Atlantic City, N.J.	6.5	7.0
Fibrous apatite, Oka Complex, Oka, Quebec, Can.	21.0	3.0
Sea water, Open ocean off Bahamas, 25° 20/N, 77° 50′ W	25.0	5.0

quantity δ is defined as

$$\delta = \left(\frac{R_s}{R_r} - 1\right) 10^3. \tag{12.1}$$

R_s is the Ca^{48}/total-Ca ratio of a sample and R_r is that of the reference standard; thus δ is the deviation (per mil) of the C^{48}/total-Ca ratio of the sample from an arbitrary standard.

Such work is suggestive of the many isotopic anomalies that probably exist in the natural environment but which require isotopic assays of considerable precision.

THE ROLE OF TRACE NUTRIENTS

One of the more obvious contributions that mass spectrometry can make to environmental science is in the quantitative assay of trace metals. The data that we now possess are exceedingly meager, and nothing approaching a worldwide effort has even begun. Yet isolated studies have pointed up the radical changes in trace-metal concentrations that are accompanying our present civilization. These studies have also indicated that trace concentrations may have either a toxic or beneficial effect—depending on the quantity and the particular system in which the metal is found.

Perhaps the most frequently cited example in the human system is copper. Essential for the complex mechanism of blood formation, approximately one-tenth of a gram is essential to life. Excessive amounts, however, are poisonous. The observation has also been made that trace metals are of greater intrinsic importance to human nutrition than are vitamins. Plants can manufacture all the vitamins that are needed to support human life, but no form of plant or animal life can produce metals unless the environment contains them. Nor does the human organism automatically excrete all metals above the threshold where such elements are toxic. For this reason, interest in trace metals is being stimulated by research in both biology and medicine (Chapter 14) and in agriculture.

Recent reviews include specific examples of the vital role of trace metals in plants, animals, and man. A western peach orchard contained numerous stunted trees and fruit below standards both in quality and quantity. Soon after the installation of some galvanized fencing on the acreage, however, the orchard yielded a much improved output. The explanation concerned a soil deficiency in zinc, which was ultimately supplied by the rain's washing minute amounts of this metal from the galvanized steel fence into the soil [20]. Boron and cobalt are other nutrients that must exist in the environment or be added to achieve "normality." An ounce of boron per acre is sufficient to prevent one type of decay in apples, and a like quantity of cobalt when added to salt licks is sufficient to keep 100 sheep healthy.

Cadmium and chromium are also metals that have attracted the attention of metallurgists and biologists alike. Trace quantities of cadmium enter the body from drinking water that passes through galvanized iron pipes; the metal is also injected indirectly through fertilizers that contain small quantities of it. Presumably it is one of the metals whose concentrations in the liver and kidney accumulate with age. One investigation reported that in a comparison between African and American groups, cadmium concentrations in kidneys of Americans showed five times the amount of those in the Africans. Other autopsies have indicated even greater differences between American and Japanese samplings. Another object of current testing is to identify causes for the wide discrepancies in heart and arterial diseases that exist between national groups.

Similar quantitative assays are being made to determine the role of chromium and lead [21]. Tests with rats and mice have demonstrated an interesting correlation between chromium content and the incidence of diabetes. When chromium was completely eliminated from a diet of a group of experimental rats, 80% showed diabetic symptoms. But the addition of only a 5 ppm chromium content to their drinking water apparently was sufficient to ward off such an abnormality. Lead is another metal under scrutiny as a trace metal, in part because of the large-percentage increase with which this metal is now found in our environment. Assays of tree rings show a lead content greater than 20 times that of a century ago, which can be attributed to leaded paints, gasoline, and so on that have become a part of a technological age. Controlled experiments on lower forms of animal life show definite effects on health and life span even when amounts approximate the relative levels currently found in human organisms. Minute traces of lead that get into the bloodstream through the lungs are also known to deposit in the membrane covering the brain. Some suggestions have been made that such deposits could result in minor brain damage and a potential loss of intelligence.

As our knowledge of trace-metal nutrients increases, some experts predict that society may be taking two new types of pills. One type will be available, as "vitamins," to correct for metal deficiencies. A second could contain certain chelating agents the role of which would be to combine with metal excesses to produce compounds that would render the metals harmless or make them excretable. If such is true, mass spectrometry and neutron-activation analysis will be among the most sensitive methods that will support statistical and chemical studies in this new area of modern positive toxicology.

REFERENCES

[1] F. E. Egler, *Am. Scientist,* **52,** 110 (1964).
[2] Ministry of Health; Reports on Public Health and Medical Subjects, No. 95, Her Majesty's Stationery Office, London, 1954.
[3] J. R. Goldsmith, in *Air Pollution* (vol. 1), ed. by A. C. Stern, Academic, New York, 1962, ch. 10, p. 343.
[4] L. A. Chambers, in *Air Pollution* (vol. 1), ed. by A. C. Stern, Academic, New York, 1962, ch. 1, p. 19.
[5] E. R. Weaver, *J. Res. Natl. Bur. Standards,* **59,** 383 (1957).
[6] G. P. Happ, D. W. Stewart, and H. F. Brockmyre, *Anal. Chem.,* **22,** 1224 (1950).
[7] E. R. Quiram, S. J. Metro, and J. B. Lewis, *Anal. Chem.,* **26,** 352 (1954).
[8] M. Shepherd, S. M. Rock, R. Howard, and J. Stormes, *Anal. Chem.,* **23,** 1431 (1951).
[9] E. C. Tabor and W. V. Warren, *A.M.A. Arch. Ind. Health,* **17,** 145 (1958).
[10] F. J. Marcie, *Environ. Sci. Tech.,* **17,** 164 (1967).
[11] U.S. Public Health Service; Public Health Reports, 61, No. 11, 371–384, March 1946.
[12] D. M. Hadjimarkes, *Arch. Environ. Health,* **13,** 102 (1966).
[13] *Nucleonics,* August, 1966, pp. 94–100.
[14] P. Goris, W. E. Duffy, and F. H. Tingey, *Anal. Chem.,* **29,** 1590 (1957).
[15] A. A. Levin, unpublished data.
[16] E. Rona, L. O. Gilpatrick, and L. M. Jeffrey, *Trans. Am. Geophys. Union,* **37,** 697 (1956).
[17] J. D. Wilson, R. K. Webster, E. W. C. Milner, G. A. Barnett, and A. A. Smales, *Anal. Chim. Acta,* **23,** 505 (1960).
[18] Y. Miyake, Y. Sugimura, and H. Tsubota, "Content of Uranium, Radium, and Thorium in River Waters in Japan," in *The Natural Radiation Environment,* ed. by J. A. S. Adams and W. M. Lowder, University of Chicago Press (1964), p. 221.
[19] J. T. Corless and J. W. Winchester, in *Proceedings of the Symposium on Isotope Mass Effects in Chemistry and Biology* (Vienna, 1963), Butterworths, London, 1964, p. 317.
[20] J. D. Ratcliff, in *Today's Health,* March, 1966, p. 36.
[21] *Ibid.*

Chapter 13

Chemistry by Mass Spectrometry

The mass spectrometer is one of the most important tools in modern chemistry. In basic research it is the prerequisite instrument for gaining insight into molecular structure, ionization phenomena, and chemical kinetics. In industry the mass spectrometer is also meeting the growing demands placed on it with respect to product monitoring and quality control. Industrial product monitoring in the United States dates back as far as 1875, when the Pennsylvania Railroad adopted the practice of purchasing steel by specification; this was the same year as the founding of the Corning Glass Works [1]. Today spectrometers will be found in the laboratories of the latter firm and the nation's leading steel producers. Mass spectrometers are also among the most important laboratory investments in the petroleum industry. In this case consideration of quality control and economics complement its research potential. As one research director has stated: "It is equally important to maintain an octane rating for gasolines which is neither above nor below our specifications." The mass spectrometer also certifies the bottled gases that are delivered to hospitals for anesthesia, and to industry for commercial use. Mass spectra can also be identified with food flavors; one recent publication has reported in some detail on the volatile oils from citrus trees, grapefruit, and tangerine and orange blossoms, in order to determine the origin of citrus flavor components [2].

Species of very high molecular weight are also amenable to analysis, and I am not aware that any upper limit has been established for analyzing them. The electromagnets displayed in the frontispiece can, for example, focus singly charged ions of mass 10,000 at energies of approximately 4,000 eV. There are also a number of potentially important applications for using mass spectrometry in a chemical processing feedback loop. For example, in a gas stream a chemical species can be identified in 10 μsec, the nominal transit time for an atom or molecule to pass from an ion source to an electronic detector.

These few pages will not attempt to document the prolific applications of mass spectrometry to chemistry for this has been done by several authors. Many chemical measurements are presented in books by Kiser [3] and McDowell [4] and special attention has been focused on organic chemistry in the comprehensive works of Beynon [5], Biemann [6], and McLafferty [7]. Nevertheless, a few selected topics will be outlined to indicate the diversity of chemical problems to which the mass spectrometer may be applied.

THE ISOTOPIC DILUTION METHOD

The isotopic dilution method has been mentioned in a number of previous chapters. It is an important technique for the determination of geological age and it is among the most powerful general methods available to the analytical chemist. Fairly extensive literature is available on the subject and a recent paper has given consideration to the attainable precisions that may be expected in specific cases [8]. The method applies to all chemical elements that have two or more naturally occurring isotopes. In the event that an element is monoisotopic, the method may still be considered if a long-lived radioisotope of the element is made available.

The principle of the method is to mix or blend the unknown quantity of an element (sample) with a spike (or tracer) that has an isotopic composition that is substantially different from that of the sample. The isotopic composition of the sample must be known or measured, and both the isotopic composition and the amount of the spike must be known before blending. The method can be applied for gaseous mixtures and samples in solution; the element should be present in the same chemical form in the sample and in the spike. For an element of two isotopes, the following equation yields the unknown weight, W_x, of the sample:

$$W_x = W_t \frac{(R_t - R_m)(R_s M_1 + M_2)}{(R_m - R_s)(R_t M_1 + M_2)}, \tag{13.1}$$

where W_t is the known weight of the spike (tracer), R_s is the isotopic ratio of isotope 1 to isotope 2 of the sample element, R_t is the isotopic ratio of isotope 1 to isotope 2 of the spike, R_m is the isotopic ratio of isotope 1 to isotope 2 of the mixture, and M_1 and M_2 are the atomic weights of isotope 1 and 2, respectively. In (13.1) R_s is usually the natural occurring ratio and is known; R_t is also known or measurable and R_m is the ratio of the mixture determined by mass spectrometry. With a known spike weight, W_t, the unknown weight of the sample element can be calculated.

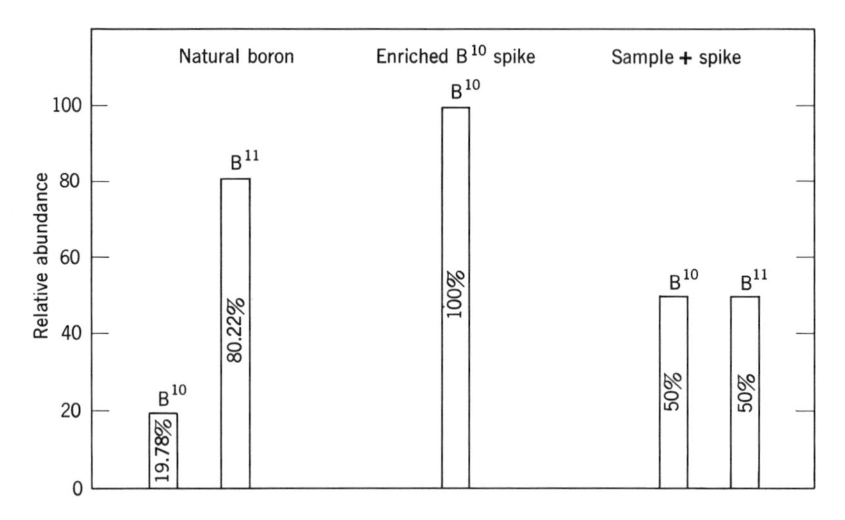

Fig. 13.1 Application of the isotopic-dilution method to boron, using pure B^{10} as the tracer.

Boron is an important element having two isotopes, and the application of the isotopic dilution method for its assay is illustrated in Figure 13.1. In this example, boron, whose isotopic abundance is B^{10} (19.78%), B^{11} (80.22%), can be spiked with pure B^{10}, which is the minor isotope. Table 13.1 is presented simply to indicate the approximate mass spectrometrically observed B^{10}/B^{11} ratios of the mixture, R_m, that would correspond to the ratios of the weight of the sample, W_x, to the spike weight, W_t.

Table 13.1

Isotopic Ratios for Boron Corresponding to Sample-Spike Weights, With a Pure B^{10} Tracer (*1 ppm B^{11}*)

$R_m(B^{10}/B^{11})$	*Sample-to-spike wt.* (W_x/W_t)
1	1.78
10	1.38×10^{-1}
50	2.71×10^{-2}
100	1.35×10^{-2}
1000	1.34×10^{-3}

Such a tabulation suggests that if a 1×10^{-6} gm spike of B^{10} leads to a mass spectrometrically observed B^{10}/B^{11} ratio of 1000 for the blend or mixture, the unknown weight of the boron is approximately

$$1.3 \times 10^{-9} \text{ gm}$$

It is also assumed that (a) the mass spectrometer is capable of measuring such ratios at high sensitivity and (b) that reagent or other impurities will not lead to errors in the analysis. Such assumptions are not always justified. For the example cited, it is often desirable to measure B^{10}/B^{11} abundances not at mass numbers 10 and 11, but at a higher mass number using compound ions (e.g., $Na_2BO_2^+$ at mass numbers 88 and 89) in order to minimize discrimination in the ion source.

For samples comprising elements of more than two isotopes, (13.1) may still be used if a normalization is made to account for the other isotopes; generalized analytical expressions can also be employed [9]. The availability of a large number of separated stable isotopes from the Oak Ridge National Laboratory has made the isotopic dilution method a feasible one for about 80% of the elements. The method has the great advantages of eliminating the necessity for quantitative recovery and, in favorable instances, deleting the need for any prior chemistry. The mass spectrometer itself will often discriminate against other chemical species. Even if impurity mass peaks overlap the mass spectrum of the sample, it is sometimes possible to reduce spurious mass peaks to very low values. For example, large differences between the sample and impurities with respect to volatility or the probability of ion formation (in the case of thermal ionization sources) permit a differentiation of several orders of magnitude.

In the general case, however, and when exceedingly small samples are analyzed, extreme care must be exercised to eliminate contamination from reagents, glassware, and even from laboratory dust. It is because of contamination problems, in fact, that small samples can sometimes be more conveniently analyzed by neutron-activation analysis than by mass spectrometry.

DETERMINATION OF MOLECULAR WEIGHTS

The ability of a mass spectrometer to assign a specific weight to a molecule is unique. If a molecule is ionized and accelerated in an ion source and subsequently analyzed, a mass assignment can be made—provided that the original molecule does not fragment between the ion source and the ion collector. A heavy ion may fragment because of some colli-

Table 13.2

The Mass Spectra of a Few Simple Alcohols [10]

Mass No.	*Ion formula*	*Ethanol*	*1-Propanol*	*2-Propanol*
13	CH	0.54	0.04	0.04
14	CH_2	1.9	0.26	0.35
15	CH_3	4.1	1.2	1.5
16	CH_4	0.15	0.05	0.06
17	HO	0.22	0.63	0.11
18	H_2O	1.6	3.5	0.56
19	H_3O	2.9	0.66	3.5
26	C_2H_2	3.8	1.6	0.36
27	C_2H_3	14.8	8.7	100.0
28	C_2H_4	3.2	3.0	2.2i
29	CHO	9.7	3.6	2.0
	C_2H_5	3.9	5.6	0.29
30	CH_2O	4.6	1.5	0.10
	$C^{12}C^{13}H_5$	0.02	0.16	—
31	CH_3O	100.0	100.0	2.6
32	CH_4O	1.3i	2.0i	0.17
33	CH_5O	0.20i	0.92	—
38	C_3H_2	—	0.47	0.34
39	C_3H_3	—	2.3	1.6
40	C_3H_4	—	0.58	0.28
41	C_3H_5	—	5.1	2.7
42	C_2H_2O	2.5	0.35	0.65
	C_3H_6	—	8.8	0.72
43	C_2H_3O	8.0	1.7	4.2
	C_3H_7	—	0.85	2.5
44	C_2H_4O	1.1	0.52	2.6
	$C_2{}^{12}C^{13}H_7$	—	—	—
45	C_2H_5O	46.6	4.0	54.3
46	C_2H_6O	25.2p	0.10i	1.2i
47	C_2H_7O	1.4	—	0.12i
53	C_4H_5	—	—	—
54	C_4H_6	—	—	—
55	C_3H_3O	—	0.30	0.09
	C_4H_7	—	—	—
56	C_3H_4O	—	0.11	0.18
	C_4H_8	—	—	—
57	C_3H_5O	—	0.90	0.18
	C_4H_9	—	—	—
58	C_3H_6O	—	0.15	0.07
	$C_3{}^{12}C^{13}H_9$	—	—	—
59	C_3H_7O	—	12.5	2.3
60	C_3H_8O	—	11.6p	0.40p
61	C_3H_9O	—	0.44i	0.09
62	$C_3H_8O^{18}$	—	0.03	—

p = parent ion.
i = ions in which isotopic contributions form the major part.

sion phenomenon, but it may also decompose into daughter molecules because of its inherent instability. For this and other reasons an examination must often be made of the entire mass spectrum of a compound.

The spectral intensities of various ion species for some simple alcohols, as reported by Saunders and Williams [10], are listed in Table 13.2.

Clearly, a mass spectrometer of high resolution is desirable if an accurate mass assignment is to be made, but considerable information about a compound can be obtained with fairly simple instrumentation. In the general case in which a molecular formula is to be determined, the multiplicity of isotopes of some elements is also useful. Thus when it is difficult to distinguish between alternative formulae on the basis of a precision mass measurement, differences in the isotopic composition of several compounds (which are reflected in varying peak heights in the mass spectrum) may provide a simple interpretation [11].

The importance of molecular weight determinations can hardly be overestimated. In this connection the mass spectrometer is used for analyzing petroleum products, silicon compounds, and the effluent of gas chromatographic columns. It is also used in the identification of the small quantities of compounds that are responsible for imparting characteristic odors and flavors to foodstuffs. In some cases this identification demands a sizable laboratory effort. For example, in order to identify the volatile aromatic constituents of a particular variety of grape, 5 tons of grapes were needed [12]. Nevertheless, the synthesis of natural flavors and odors can be expected to become an increasingly active field of research.

MEASUREMENT OF DISSOCIATION ENERGIES

In general the dissociation of molecular ions by collision reactions leads to fragment ions that have a spread in kinetic energy and produce relatively broad ion peaks at the detector of a mass spectrometer. However, an interesting phenomenon is observed for the case when the primary molecule and the dissociation products are highly collimated. In particular, consider the case when primary molecular ions are accelerated through a tandem magnet system and the ions are dissociated in a reaction chamber located between the two magnets. A schematic diagram of the analyzing system, with a gaseous reaction chamber at an intermediate focal point, is shown in Figure 13.2. If the primary molecules from S_1 are dissociated in the gaseous reaction chamber at S_2 and only ions are collected that are collinear with the primary ion trajectory, multiple peaks can be observed at S_3. Such a multiple peak is shown in Figure 13.3, where a "triple peak" for O^+ ions results from the dissocia-

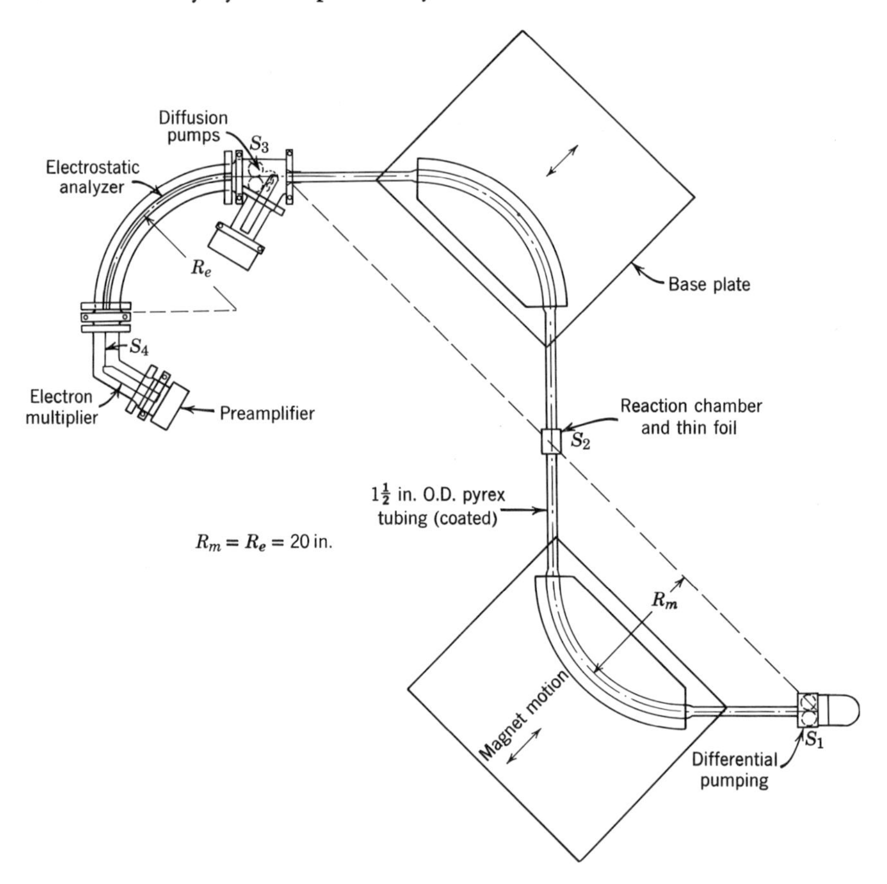

Fig. 13.2 Experimental arrangement for measuring the dissociation energy of a molecule. The primary ion is dissociated at S_2 by interacting with a gaseous reaction chamber [13].

tion of CO^{++} by helium. Rourke, Sheffield, Davis, and White [13] have shown that this type spectrum can yield directly the dissociation energy of a molecule, even though the spectrum is produced from ions of very high kinetic energies.

The mechanism by which the dissociation energy gives rise to these "side peaks" and the method for calculating the dissociation energy from the energy separation of these peaks are briefly outlined. Consider a diatomic ion at rest, and let m_1 and m_2 be the masses of the two atoms. Also let T_1 and T_2 be the kinetic energies given to each by the dissociation process, and v_1 and v_2 the velocities imparted to each atom respectively. Then

$$T = T_1 + T_2 = \tfrac{1}{2}m_1v_1^2 + \tfrac{1}{2}m_2v_2^2 \tag{13.2}$$

and

$$m_1v_1 = m_2v_2. \tag{13.3}$$

In a highly collimated spectrometer only those dissociated ions are collected that do not have an appreciable component of velocity perpendicular to the original direction (i.e., dissociation occurs in line with the trajectory of the original beam). Then a parent ion with velocity v_0 will give rise to atoms of velocity $v_0 \pm v_1$ and $v_0 \pm v_2$. The resultant total kinetic energy of particle 1 would be

$$\tfrac{1}{2}m_1(v_0 \pm v_1)^2 = \tfrac{1}{2}m_1v_0^2 \pm m_1v_1v_0 + \tfrac{1}{2}m_1v_1^2. \tag{13.4}$$

The last term is sufficiently small compared with the other terms that it may be neglected; thus the two side peaks corresponding to a finite value for v_1 are symmetrically displaced in energy from the center peak where $v_1 = 0$. It is also apparent that the energy displacement of the side peaks, $m_1v_1v_0$ will be proportional to v_0, as observed. Setting

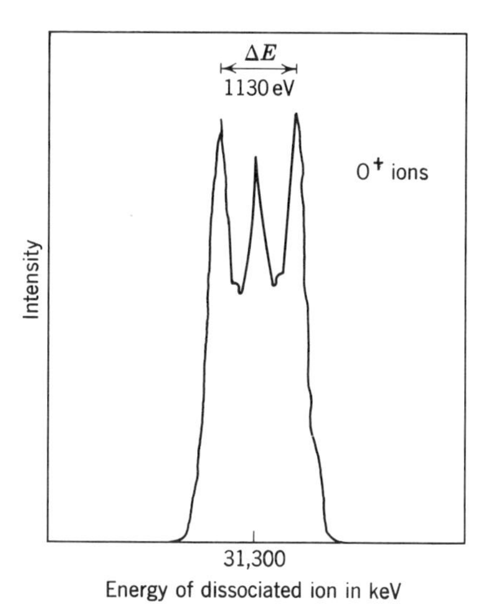

Fig. 13.3 Spectrum of O^+ ions obtained from the dissociation of CO^{++} by helium. Energy of the primary CO^{++} molecules is 54.8 keV [13].

$\Delta E_1 = m_1 v_1 v_0$ and solving for v_1:

$$v_1 = \frac{\Delta E_1}{(2E_0)^{1/2}} \cdot \frac{(m_1 + m_2)^{1/2}}{m_1} \tag{13.5}$$

and thus

$$T_1 = \frac{(\Delta E_1)^2}{4E_0} \cdot \frac{(m_1 + m_2)}{m_1}, \tag{13.6}$$

where $E_0 = \frac{1}{2}(m_1 + m_2)v_0^2$ is the kinetic energy of the undissociated ion. As $\Delta E_1 = \Delta E_2$, the total kinetic energy from dissociation is:

$$T = \frac{(\Delta E_1)^2}{4E_0} \cdot \frac{(m_1 + m_2)^2}{m_1 m_2}. \tag{13.7}$$

In Figure 13.3 the O^+ ion is seen to have a central peak that occurs at the energy expected (i.e., $\frac{16}{28}$ of the original molecular ion energy) and the measured value of T for the side peaks of the O^+ from the dissociation of CO^+ is 4.3 V. Other dissociation energies measured by this method have been molecules of H_2^+, NO^+, N_2^+ and $C_3H_7^+$, and $C_3H_{10}^+$. In the general case, this type of spectral analysis also allows some interpretation to be made relative to the dissociation mode.

A number of other tandem mass spectrometers has also been reported for investigating ion-neutral interactions involving momentum transfers to the product ions. The tandem spectrometer reported by Futrell and Miller [14] consists of a double-focusing spectrometer that serves as an ion source for a collision chamber; a Consolidated Model 21-110 is used as a second stage. One unique aspect of this apparatus is a deceleration lens (located between the first and second stage) that retards the primary ion beam of 100–200 eV to ~1–2 eV, without serious defocusing. Using these low-energy ions, preliminary studies have been made of ion-molecule reactions with an unambiguous specification of reaction products.

Dissociation phenomena have also been reported by Hunt and Huffman [15], using the drift tube of a time-of-flight mass spectrometer as a reaction-collision chamber. Basically, their method was to measure the difference between the flight time of the fragment and that of the parent specie. Observations were made for the 58^+, 43^+, 42^+ dissociations in n-butane, and results were in qualitative agreement with previous dissociation spectra found in magnetic spectrometers.

CHARGE PERMUTATION

Charge permutation as well as dissociation phenomena can be examined by the mass spectrometric arrangement of Figure 13.2. In this case it is convenient to utilize both magnetic and electrostatic analyzers; a thin foil may also replace the gaseous reaction chamber at S_2. A primary ion beam is accelerated from S_1 and ions of a given charge-to-mass ratio are selected by traversing the first magnet. Very thin foils of nickel or vinyl copolymer (250 Å to 500 Å in thickness) are then used to intercept the primary ion beam. Ions will undergo a substantial energy degradation in passing through such foils, and charge exchange will occur. By properly programming the second analyzing magnet and the final electrostatic lens, however, it is possible to remove all ambiguity about the resulting specie and its charge state.

Equilibrium ratios and charge-exchange cross sections are obtained by measuring the counting rate of ions before inserting the films, and observing the counting rate of the new ion species formed by the charge transfer after interaction [13]. For a single charge exchange, the interacting ion beam is only slightly altered in velocity or direction. The equilibrium charge ratios obtained for various ions through the thin nickel films are shown in Table 13.3. The mean ion energy listed is

Table 13.3

Equilibrium Ratios of Charged Ions After Traversing 500-Å Nickel Films [13]

Incident ion	*Mean ion energy (keV)*	*Ratio* (X^{++}/X^{+})
Li^{+}	29	0.0014
Li^{+}	50	0.0035
C^{+}	39	0.050
C^{++}	79	0.13
O^{+}	38	0.036
O^{++}	80	0.16
Ne^{+}	40	0.07
Ne^{++}	87	0.25
He^{+}	41	0.0074
He^{++}	86	0.037
He^{++}	91	0.043
A^{+}	39	0.10
A^{++}	86	0.50

the incident ion energy minus one-half of the total energy loss, which amounted to approximately 5% of the ion energy in most cases.

FREE RADICALS

Stable free radicals are rare chemical species that represent no greater than 0.1% of all known stable chemical compounds [16]. Such species are characterized by having unpaired electrons. An analysis of the velocity of many reactions (e.g., explosions, oxidation, polymerization, and biochemical processes) reveals that unstable free radicals must also exist as chemical intermediates. The lifetimes of some of these intermediates are now known to range from milliseconds to minutes. The detection of such radicals can be observed by means of optical spectroscopy, electron spin resonance, or mass spectrometric techniques.

Early mass spectrometric investigations of radicals were limited to unexcited species, for which information was obtained on ionization potentials, bond-dissociation energies, and reaction kinetics. Quite recently several studies have attempted to extend the scope of mass spectrometric measurements to include vibrationally and electronically excited species—for both free radicals or ordinary molecules. Experimental difficulties in such work include (a) the production of an adequate concentration of excited species to permit detection and (b) the development of short sampling times to minimize losses resulting from molecular collisions, or other thermodynamic equilibrating mechanisms and radiative decay. Both of these difficulties have been substantially circumvented by Foner and Hudson [17], who used a short-duration pulsed electrical discharge for free radical and metastable molecule production and a collision-free molecular beam sampling system. With this apparatus the investigators have reported that an analysis can be made of components with radiative lifetimes as short as 100 μsec. Foner and Hudson [18] have also recently measured the ionization potential of the methylene free radical produced by short-duration pulsed electrical discharges in methane. Their observed ionization potential $I\ (CH_2) = 10.33 \pm 0.1$ eV is in good agreement with the spectroscopically determined value $I\ (CH_2) = 10.396$ eV; their mass spectrometric measurement of $I\ (CH) = 9.86$ eV is also consistent with the spectroscopic determination of 9.843 eV for this latter species.

That the ionization potential of a radical (a fundamental parameter of such a specie) can be measured with such accuracy by mass spectrometry is of considerable significance. As pointed out explicitly by Lossing, Kebarle, and DeSousa [19], a knowledge of the ionization potential of free radicals is of value in three ways. First, the ionization potential together with the appearance potential of the radical ion produced by the dissociative ionization of a derivative permits the calculation of the bond-dissociation energy. Information concerning bond dissociation of simple hydrocarbons and other molecules has thus been obtained. Second, some evidence exists that in certain chemical reactions the ionization potential plays an important role in determining the reaction rate. Finally, it is of considerable importance that accurate experimental values be obtained for the development of theories of molecular structure and of calculational methods for computing ionization potentials of substitutional groups.

The clear advantage of mass spectrometric studies over other techniques is, of course, that the radical may be observed directly. In the past most information on radicals was derived from the over-all kinetics of a chemical reaction, but no detail was available with respect to a particular radical specie. Also, the radical concentrations are often so small that high-sensitivity mass spectrometry is especially advantageous. Further, with respect to the identification of both radical and stable products of catalytic reactions for example, very rapid changes can be recorded.

Most measurements of ionization potentials of radicals have been with electron-impact sources using methods similar to stable mixtures. If the products of the chemical reaction containing the unreacted molecule RX and the radical R are introduced into the mass spectrometer ion source, the bombarding electrons will produce the ionic specie R^+ in the manner [20]

$$RX + e \rightarrow R^+ + X + 2e \tag{13.8}$$

for

$$A_1 \geq I(R) + D(R - X)$$

$$R + e \rightarrow R^+ + 2e, \tag{13.9}$$

for

$$A_2 \geq I(R).$$

In the above, the minimum energy, A_1, which is required for the production of R^+ as stated in (13.8), will be equal to or greater than the sum of the bond dissociation energy, $D(R - X)$, plus the ionization energy of R, $I(R)$.

This latter energy will exceed that required for the production of R^+ species by process (13.9) by an amount approaching the bond-dissociation energy. Hence if the bombarding electron energy, E, is $A_2 < E < A_1$, the R^+ ion will be produced only by the ionization of the free radical (13.9). Accordingly, if the mass spectrometer is sufficiently sensitive, the R^+ ion will be observed at its appropriate mass spectral position. An extensive listing of radicals that have been observed spectrometrically, as well as ionization potentials, are tabulated by Harrison [21].

ANALYSIS BY SPECIALIZED METHODS

It seems appropriate in a monograph of this type to comment on a few of the newer mass spectrometric techniques, even though these methods presently apply to a limited number of analyses. They are (a) photoionization and flash photolysis, (b) photoelectron spectroscopy, (c) the generation of pure atomic beams, (d) analysis by ion impact, and (e) the potential extension of the isotopic dilution method to solids.

Photoionization and Flash Photolysis

Mention has been made in Chapter 3 of photoionization-type sources and the fact that such a source derives its principal advantage from the relative simplicity of the fragmentation patterns that it produces. A disadvantage is the relatively weak radiation intensity that can be realized at the exit slit of a monochromator or from a diffraction grating. Yet it is reasonable to anticipate the availability of more intense light sources and eventually the simultaneous electronic read-out of multiple mass peaks. The collection efficiency of the ions that are produced in the source will also be increased. A significant improvement with respect to this latter parameter has already been reported by Poschenrieder and Warneck [22], who have employed wedge-shaped pole pieces (similar to magnetic analyzers used in β-ray spectrometry) to collect from a greater solid angle. This instrument has been applied to isotopic measurements and to gas analyses generally. Table 13.4 shows some of the data for a test analysis of an air mixture in which the relative photoionization specie abundances were calculated from the photoionization cross section, δ_i (expressed in millibarns), and the known relative concentrations of nitrogen, oxygen, argon, and carbon dioxide in air of standard composition. These data were then compared with the observed ion-beam intensities. A comparison shows a reasonable agreement between calculated and observed values (with the exception of CO_2, for which the observed value is low).

Generally, photoionization mass spectrometry can be expected to be increasingly useful in studies that include pure analytical chemistry, biochemistry, environmental science, and medical research.

Flash photolysis has been used to produce high concentrations of

Table 13.4

Ion Intensities in Air [22]

Mass	*Specie*	*Ion intensity*	*Relative ion intensity*	σ_i (*mb*)	*Relative photoion production*
16	O^+	78	0.037	3.8	0.034
18	H_2O^+	58	0.003	—	—
28	$^{14}N_2$	1620	0.775	23.1	0.780
29	$^{14}N^{15}N^+$	8	0.004	—	—
32	O_2^+	370	0.177	18.9	0.171
40	Ar^+	26	0.013	36.5	0.0146
44	CO_2^+	45	0.002	33.9	0.0006

atomic and free radical species, and this technique has been improved for application to fast gas phase reaction studies. Meyer [23] has recently reported the use of a special flash lamp having a high flux and short light-pulse duration in connection with a time-of-flight mass spectrometer of high sensitivity. His complete apparatus includes a Bendix spectrometer, a capacitor bank, a krypton flash lamp, synchronizing circuits, and an oscillograph drum and camera. The oscilloscope is synchronously triggered at the 20-kHz spectrometer frequency. A display of the flash photolysis of mixed NO_2 and Ar, as reported by Meyer, is shown in Figure 13.4.

The argon serves as an internal standard in the photodecomposition of NO_2 and formation of NO, which is presumed to consist of two steps:

$$NO_2 + h\nu \rightarrow NO + O$$

$$O + NO_2 \rightarrow NO + O_2. \qquad (13.10)$$

The results demonstrate the advantage of time-of-flight spectrometry when high light fluxes can be used with a high-speed and sensitive data-recording system.

Photoelectron Spectroscopy

Within the last few years a new "molecular photoelectron spectroscopy" has been explored. This form of spectroscopy relates to the kinetic energy spectrum of photoelectrons ejected from a molecule under photon impact. The photons, of course, must have an energy in excess of the lowest ionization potential. If the observed photoelectron is emitted in a direction collinear to the electric vector of the photon beam (or perpen-

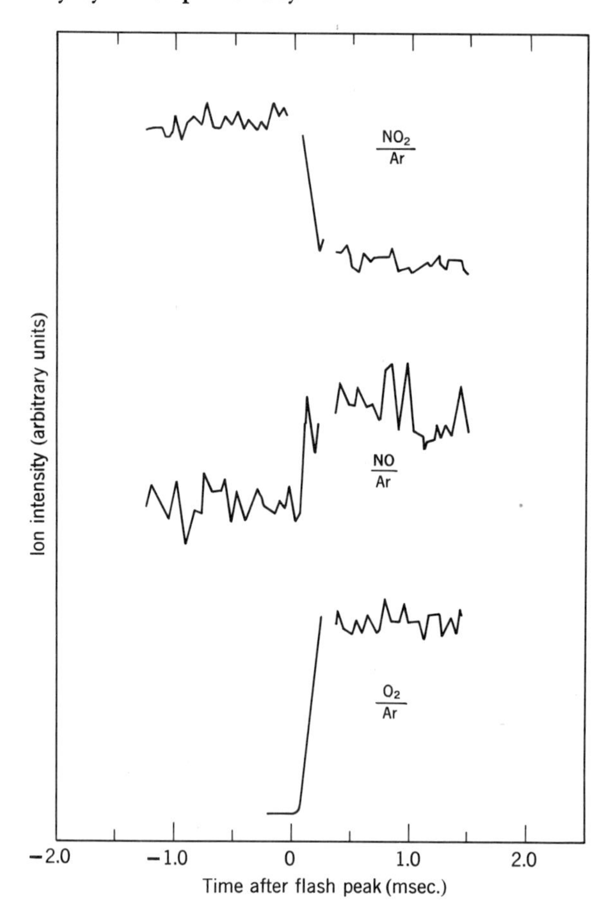

Fig. 13.4 Time-resolved photodecomposition of NO_2 as recorded by Meyer [23, adapted].

dicular to the axis of the light beam), its kinetic energy is given by $(h\nu—I)$, where ν is the photon frequency, and I is the threshold potential for the ionization process. The significance of examining the photoelectron spectrum in detail, however, arises from considerations of conservation of momentum between electron and ion. The energy division will be inversely proportional to the ion-electron masses, so that the error in relating the electron energy to $(h\nu—I)$ is exceedingly small.

Thus Al-Joboury and Turner [24] have pointed out that the photoelectron energy spectrum should consist of a number of discrete energies, corresponding to the orbital energy levels of the ground state of the molecule and also corresponding to the generation of a positive ion—either in its ground state or excited state. In addition, the relative intensities of photoelectrons as a function of energy should reflect the photoionization cross section and the orbital degeneracy of the bound electrons. This method appears to have great potential, for the entire ionization continuum can be examined in great detail without recourse to the usual threshold-ionization technique [25].

Pure Atomic Beams

One of the fundamental limitations in mass spectrometry is that of distinguishing between molecular and atomic ions having the same approximate charge-to-mass ratio. There are no practical ion sources that do not yield both molecular and atomic species, so some mass spectral lines are inevitably superposed. If the resolution of an instrument is exceedingly high, the problem is substantially solved. But in an instrument of only modest resolution it is exceedingly difficult to differentiate completely between molecules, multiatom groups, and single atoms. Furthermore, the detector has not been invented that has a sizable differential response between ions on the basis of their molecular structure. There is a simple and rather interesting technique, however, by which one can insure that only species in the atomic state arrive at a final detector [26]. The method consists of accelerating the primary ion group to a high energy and focusing these ions through a film located between two analyzing magnets (see Figure 13.2). Consider the case in which a thin metallic foil of only a few hundred Angstroms is placed at the detector focal plane of the first magnet. If the incident ions have sufficient kinetic energy, they will penetrate the foil. But in traversing this medium, *all* molecular ions will be dissociated and only atomic species will emerge from the far side of the foil. The foil (or crystal lattice) acts as a strong coulomb force field at a discrete point in the beam trajectory, and the mechanism of interaction is such that uncorrelated forces act on the constituent atoms of a single molecule. Thus if a molecule undergoes an appreciable energy degradation (i.e., >100 eV) it is highly improbable that any two atoms in a molecule will experience exactly the same individual coulomb interactions. Hence all molecular bonds will be broken and only an atomic beam remains. This atomic beam can then be examined by a second analyzing magnet. Figure 13.5 shows the spectrum obtained when a CO_2^+ molecule is incident upon such a foil and the second magnet "scans" the mass spectrum for the atomic constituents. The peak at mass position 44 was obtained by withdrawing the film from the ion beam path (at S_2, Figure 13.2). The dissociation of CO_2^+ was noted to be

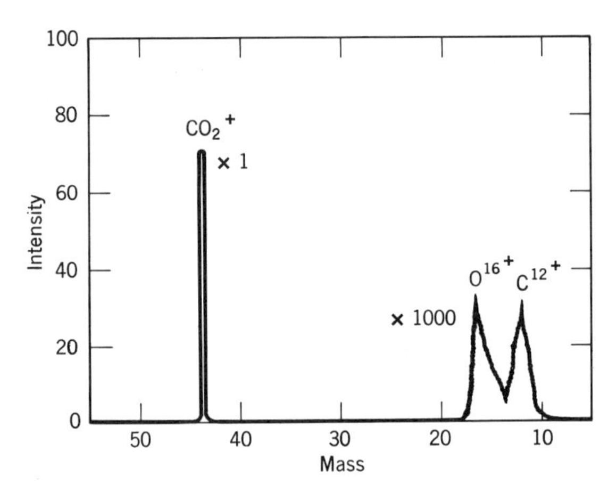

Fig. 13.5 Dissociation of CO_2 and identification of fragments after passage through a thin foil [23].

complete, and the energy of the C^{12+} and O^{16+} atoms was approximately 5 keV for an incident CO_2^+ particle energy of 45 keV and a film of a few hundred Angstroms.

There is a substantial loss in ion intensity because many atoms emerge from the foil as neutrals. But the technique has more than academic interest, as it appears to be a general method by which one can guarantee that only atoms—rather than a heterogeneous beam of atoms, molecules, and multiple atomic groups—are brought to a final focus. If mass spectrometry proceeds toward higher accelerating potentials, the technique may be of some practical consequence in the identification of unknown organic species or in differentiating between ions having a common charge-to-mass ratio (e.g., U^{234} and ${}_6K^{39}$) in an instrument of low resolving power.

Another possible approach that seems attractive for obtaining pure atomic beams of the metallic elements is to make use of a diffusion phenomenon. This method is currently being explored in the author's laboratory. The basic concept presents an alternative to the usual surface-ionization filament, where material is coated on a ribbon filament and vaporized at a high temperature. Consider the case when the sample is totally enclosed by a folded ribbon, a thin-walled, small-diameter tubing or is "clad" by a thin evaporated film. Under these conditions material is forced to diffuse from the inner to outer surface before being ionized. In principle such an approach may lead to a method that yields a pure atomic spectrum, as all molecules will be dissociated in the diffusion mechanism. It is also possible that in some instances a higher degree of ionization can be attained because the atoms are ionized on a clean, high-temperature surface, the work function of which is not decreased by the sample coating.

Analysis by Ion Impact

Mention has already been made of the fact that an ion-probe source has special advantages in the analysis of surface impurities. In this regard it is a preferred technique to that of x-ray fluorescence and other methods because the ion penetration depth is only a few hundred Angstroms. It is also appropriate to consider the potential application of the ion bombardment method with respect to sensitivity for a large number of elements. Such a comparison has been reported by Barrington, Herzog, and Poschenrieder [27] for a "pure" platinum sample. The analysis is interesting because it displays the sensitivity of the method, even though there is admittedly a difficulty in relating the secondary sputtered ion beam to an actual impurity concentration. Table 13.5 is an abbreviated summary of the analysis of a platinum single crystal subjected to (a) neutron activation, (b) spark source mass spectrometry, and (c) analysis by sputtering.

The second footnote to Table 13.5 shows the reported neutron activation limit of sensitivity for the several impurities in platinum rather than an observed quantitative value. Under the particular irradiation conditions for the sample, neutron activation actually failed to yield quantitative data on any impurities. A spark source provided data on the long list of indicated elements, with the assumption of equal ion production for all elements in the source. The last column represents the actual uncorrected electron multiplier output current ratio of the impurities to 0.001 of the sputtered platinum ion current. The spark and sputter currents show good qualitative but poor quantitative agreement. However, it will be noted that the sensitivity for a few elements—Al, Ca, and Zr—is exceedingly high. Clearly, sputter-type mass spectrometry is in the early stages of investigation. But the highly desirable features of low sputtered ion energy spread, high sensitivity, and the possible elimination of any chemical preparation for the sample will certainly cause this type of spectrometry to come into widespread use in the future.

Isotopic Dilution Method for Solids

Few chemical techniques possess the sensitivity of the isotopic dilution method for the quantitative analysis of gases and liquids. The question therefore arises as to the procedure's potential applicability to solids. It is my opinion that some type of isotopic dilution technique will indeed

Table 13.5

Analysis of a Platinum Single Crystal [27]

Element	*Spark source mass spectrograph (ppm atomic)*	*Relative sputter source mass spectrometer current**
B	present	1
C	present	0.8
N	—	0.1
O	300, not uniform	0.5
Na	35	unobserved
Mg	—	0.1
Al†	40	113
Si	present	43
P	present	unobserved
Cl	5	unobserved
K	20	0.05
Ca†	1300	340
Ti	unobserved	14.7
V	—	0.1
Cr	10	2.7
Mn	8	0.2
Fe	150	1.9
Ni†	400	0.1
Cu	10	unobserved
Zn	10	unobserved
Ga†	5	2.6
Ge	5	unobserved
As	2	0.05
Zr†	15	210
Nb	unobserved	2
Rh†	100	1
Pd	40	—
Ag	20	unobserved
In	30	—
Sb	15	—
Te	30	—
Hf†	plate fogged	2.3

* Electron multiplier current ratio of the impurities to platinum, multiplied by 10^3.
† Detection limits by neutron activation (ppm atomic) were as follows: Al, 14; Ca, 49; Ni, 40; Ga, 0.14; Zr, 1900; Rh, 114; and Hf, 0.9.

be developed. For years solids (in the form of metallic foils or ribbons) have been used as "targets" in isotope separators to collect other metals or quantities of gases. More recently, exceedingly small quantities of various elements have been collected in metal ribbons for studies relating to nuclear physics. Hence there is nothing new to the concept of having a crystal lattice serve as a host matrix for one or more impurities. Furthermore, within limits, it is possible quantitatively to dope or ion-implant a solid.

But how will complete mixing be achieved, as in the case of gases and liquids? The most likely answer is by means of diffusion phenomena. Diffusion is an important method for obtaining single crystals of exceedingly high purity, and this same mechanism can probably be ingeniously employed to "mix" dopants (implanted by ion beams) homogeneously with existing trace elements in solids. If such a technique is achieved, it might be used to assay meteorites, semiconductors, or solids generally—with a sensitivity and precision that exceed present methods. In such a case the spike or tracer isotope could be enriched to a $10^6/1$ ratio. Further, the problem of reagent impurities (a serious concern in liquid systems) would be virtually eliminated.

REFERENCES

[1] F. A. White, "American Industrial Research Laboratories," Public Affairs Press, Washington, D.C., 1961, p. 13.

[2] J. A. Attaway, A. P. Pieringer, and L. J. Barabas, *Phytochemistry,* 1273 (1966).

[3] R. W. Kiser, *Introduction to Mass Spectrometry and Its Applications,* Prentice-Hall, Englewood Cliffs, N.J., 1965.

[4] C. A. McDowell (ed.), *Mass Spectrometry,* McGraw-Hill, New York, 1963.

[5] J. H. Beynon, *Mass Spectrometry and Its Applications to Organic Chemistry,* Elsevier, Amsterdam, 1960.

[6] K. Biemann, *Mass Spectrometry: Organic Chemical Applications,* McGraw-Hill, New York, 1962.

[7] F. W. McLafferty (ed.), *Mass Spectrometry of Organic Ions,* Academic, New York, 1963.

[8] P. J. DeBievre and G. H. Debus, *Nucl. Instr. Methods,* **32,** 224 (1965).

[9] R. K. Webster, "Isotope Dilution Analysis," in *Advances in Mass Spectrometry* (vol. 1), ed. by J. D. Waldron, Macmillan, New York, 1959, p. 103.

[10] R. A. Saunders and A. E. Williams, in *Mass Spectrometry of Organic Ions,* ed. by F. W. McLafferty, Academic, New York, 1963, p. 364.

[11] J. H. Beynon, *Op. cit.,* p. 305.

[12] A. D. Webb and R. E. Kepner, *Food Res.,* **22,** 384 (1957).

[13] F. M. Rourke, J. C. Sheffield, W. D. Davis, and F. A. White, *J. Chem. Phys.,* **31,** 193 (1959).

[14] J. H. Futrell and C. D. Miller, *Rev. Sci. Instr.,* **37,** 1521 (1966).

[15] W. W. Hunt and R. E. Huffman, *Rev. Sci. Instr.,* **35,** 82 (1964).

[16] J. Turkevich, *Physics Today,* **18** (1965), p. 26.
[17] S. N. Foner and R. L. Hudson, *J. Chem. Phys.,* **45,** 40 (1966).
[18] *Ibid.,* p. 49.
[19] F. P. Lossing, P. Kebarle, and J. B. DeSousa, "Ionization Potentials of Alkyl and Halogenated Alkyl Free Radicals," in *Advances in Mass Spectrometry,* ed. by J. D. Waldron, Macmillan, New York, 1959, p. 431.
[20] A. G. Harrison, "Mass Spectrometry of Organic Radicals," in *Mass Spectrometry of Organic Ions,* ed. by F. W. McLafferty, Academic, New York, 1963, p. 209.
[21] *Ibid.,* pp. 218–221; 240.
[22] W. Poschenrieder and P. Warneck, *J. Appl. Phys.,* **37,** 2812, (1966).
[23] R. T. Meyer, *J. Sci. Instr.,* **44,** 422 (1967).
[24] M. I. Al-Joboury and D. W. Turner, *J. Chem. Phys.,* **37,** 3007 (1962).
[25] J. E. Collin, in *Mass Spectrometry,* ed. by R. I. Reed, Academic, New York, 1965, pp. 183–199.
[26] F. A. White, F. M. Rourke, and J. C. Sheffield, *Rev. Sci. Instr.,* **29,** 182 (1958).
[27] A. E. Barrington, R. F. K. Herzog, and W. Poschenrieder, in *Progress in Nuclear Energy; Analytical Chemistry* (vol. 7), ed. by H. A. Elion and D. C. Stewart, Pergamon, New York, 1966, p. 267.

Chapter 14

Potential Applications in Biology and Medicine

"What really leads us forward is a few scientific discoveries and their application." LOUIS PASTEUR (*1822–1895*)

Isotopic labeling in biology and medicine is one of the most important developments of our time. On August 2, 1946, the Oak Ridge National Laboratory shipped a few millicuries of C^{14} to the Barnard Free Skin and Cancer Hospital in St. Louis. This was the first delivery of a radioactive isotope by the newly organized Isotope Division of the United States Atomic Energy Commission. Today tagging with radiotracers is a standard technique and there is little doubt that the extension of this method will be invaluable to many unprobed areas in medical science. Nevertheless, there are limits to the use of radiotracers in biological systems, and in this final chapter I should like to confront the reader with an alternative. The alternative is the use of stable isotopes.

THE CASE FOR STABLE ISOTOPES

Indeed, the application of stable isotopes and mass spectrometry to medical research may reveal data that can be obtained in no other manner. I admit to both bias and optimism in making this statement, but I am supported by a number of eminent biologists, physicists, and chemists who are confident that substantial contributions can be made by mass spectrometrists to the life sciences.

What is the basis for this assertion—especially in view of the sensitivity and simplicity of radiotracer instrumentation? Several factors can be cited specifically:

1. With stable isotopes the biological system will not be damaged. Use of a radiotracer may cause damage from the primary ionizing radiation, the recoil kinetic energy of the parent nuclei, or both. Nuclear

radiations are known to disrupt chromosomes and create harmful free radicals.

2. Stable tracers introduce no new chemical species such as are introduced by radionuclides in their decay. Some daughter decay products are potentially toxic.
3. Certain radiopharmaceuticals and compounds are subject to radiolytic decomposition. Such compounds therefore have a definite "shelf life."
4. Stable nuclei have an infinite lifetime. They appear to be a favored candidate for studies extending over periods of many years, as there is no loss in analytical sensitivity as a function of time.
5. In continuous infusion experiments a multiplicity of isotopes (if available in a single chemical element) can provide time-dependent data on turnover rates.
6. Stable isotopes appear useful in trace nutrient studies and in investigations of metal toxicity, where the availability of radionuclides with convenient half-lives is somewhat restricted.
7. It is unlikely that radioisotopes will be extensively used for statistical studies on human subjects, even if radiation effects are deemed negligible.
8. In favorable cases the sensitivity of stable isotopes and mass spectrometry already exceeds radiotracer detection.

There is, of course, little doubt that the radiotracer method will continue to be the favored choice in a vast majority of biological studies. But the time has arrived when instrumental methods must be reviewed with respect to their future rather than their present potential. Today's diagnostic procedures have reached an unparalleled degree of sophistication and there is reason to believe that the entire periodic chart can serve as an information source. Too, a rebuttal must be made to writers who casually state that "stable isotopes are used only when no radioactive ones are available."

Clearly, sampling and instrumental problems will offer difficulty. The reader is reminded, however, of the somewhat analogous situation in optical and electron microscopy. In America commercial electron microscopes were nonexistent until about 1940. The early versions possessed only modest resolving power, difficulties were encountered in sample mounting, and the microscopes were available only at a relatively high cost. By 1953 the Radio Corporation of America alone had marketed nearly 500 instruments to industry, hospitals, and university laboratories throughout the nation. Today, electron microscopes number in the thousands—they make possible otherwise *impossible* measurements. Perhaps the same will be stated of mass spectrometric studies in biology. At present the cost of stable isotopes may seem exorbitant, but the price is artificial. If borderline experiments can be performed by turning to a complimentary method, means will be found to enhance and improve that method. Thus this chapter is concerned with a survey of *possible* applications for mass spectrometry—with the understanding that only the specialist can fairly assess their difficulty and potential merit. A few isolated spectrometric applications will also be cited, but these are believed to represent a small fraction of the research or clinical studies to which mass spectrometry can ultimately contribute.

ISOTOPES OF OXYGEN AND NITROGEN

The rare stable isotopes O^{17} and O^{18} become especially important as no radioactive oxygen isotope is sufficiently long-lived to find general application in tracer work. O^{14} and O^{15} are positron emitters of only 76 and 118 sec, respectively. O^{19} is a beta emitter with a half-life of only 29 sec. Molecular ions are observed in the mass spectrum at mass positions 32, 33, and 34—corresponding to the ions $(O^{16}O^{16})^+$, $(O^{16}O^{17})^+$, and $(O^{16}O^{18})^+$. Compounds of oxygen will be observed throughout most of the mass spectrum. Atomic ions of oxygen are of lesser value because of the usual water-vapor spectra at the lower mass numbers.

O^{18} is the traditional tracer, its natural occurrence being only 0.204%. O^{17} is even more rare, and hence more costly as a tracer, having an abundance of only 0.037%.

Tracer studies of oxygen are of biological importance comparable to carbon and hydrogen. In molecular states, in water, and in a multiplicity of organic compounds, oxygen tracers have been used in metabolism studies, research on the effect of light on plant respiration, and for labeling phosphates. Some of the most important photosynthesis studies to date have depended upon O^{16}/O^{18} mass spectrometric measurements. Cohn [1] has provided an excellent review of the use of O^{18}, including (a) the direct utilization of atmospheric oxygen in biological oxidations, (b) enzymic hydrolyses, and (c) phosphate transfer reactions.

As in the case of oxygen, radioisotopes of nitrogen can be formed, but their lifetimes are too short to have them seriously considered for most applications. N^{12}, N^{13}, N^{16}, and N^{17} have lifetimes of 0.0125 sec, 9.93 min, 7.35 sec, and 4.14 sec, respectively. N^{13} has been used, however; it is formed by a (d,n) reaction on C^{12}. Stable isotopes are N^{14} and N^{15}, the latter (of 0.365% natural abundance) being employed as the tracer [2]. In this study glycine labeled with N^{15} was incorporated in nearly synthesized hemoglobin in human subjects. This tracer was fed by mouth over a three-day period. A series of mass spectral analyses revealed that the life span of the newly formed red cells labeled with N^{15} glycine was 127 days. This work established a new technique for labeling red cells.

The atomic percent of N^{15} in a mass spectral assay will be found by examining the peak heights at mass positions 28, 29, and 30—corresponding to $(N^{14}N^{14})^+$, $(N^{14}N^{15})^+$, and $(N^{15}N^{15})^+$. Precautions must be taken in all measurements of molecular mass spectra. If a mass spectrometer has very high resolving power, it is possible to mass-resolve the small weight differences of hydrocarbons and the like. But for an instrument with a small radius of curvature, $(C^{12}O^{16})^+$ and $(C^{13}O^{16})^+$ will occur also at masses 28 and 29 and lead to gross errors. A peak at mass 40 (argon) sometimes indicates the presence of air, suggesting that other gaseous elements may also interfere with an accurate assay.

A number of N^{15}-labeled compounds are commercially available. Amino acids and N^{15}-labeled ammonium salts have been effectively employed in investigations of nitrogen metabolism.

SPECTROMETRY OF BLOOD GASES AND PLASMA

Two important types of mass spectrometric measurements have been reported for the analysis of blood. One is the in-vivo recording of partial pressures of gases in respiratory research. The second is the analysis of dried blood plasma.

The first is a fairly recent study by Woldring, Owens, and Woolford [3], in which gases were sampled directly from circulating blood through a gas-permeable membrane at the tip of an intravascular cannula that was connected directly to the source section of a mass spectrometer. The partial gas pressures and the membrane permeability determined the gas flow into the instrument. A natural latex rubber was shown to have a good permeability to carbon dioxide and oxygen, and polyethylene tubing conducted these gases to the spectrometer. With the use of this technique, arterial carbon dioxide and oxygen pressures were simultaneously recorded in an anesthetized cat that was subjected to respiratory tests. Figure 14.1*a* shows the response of the spectrometer resulting from the closing of the trachea, with the gradual depletion of oxygen and the gradual buildup of carbon dioxide. Figure 14.1*b* shows the result of the animal's temporary asphyxiation, caused by 100% nitrogen inhalation. The work amply demonstrates that a continuous analysis of gases

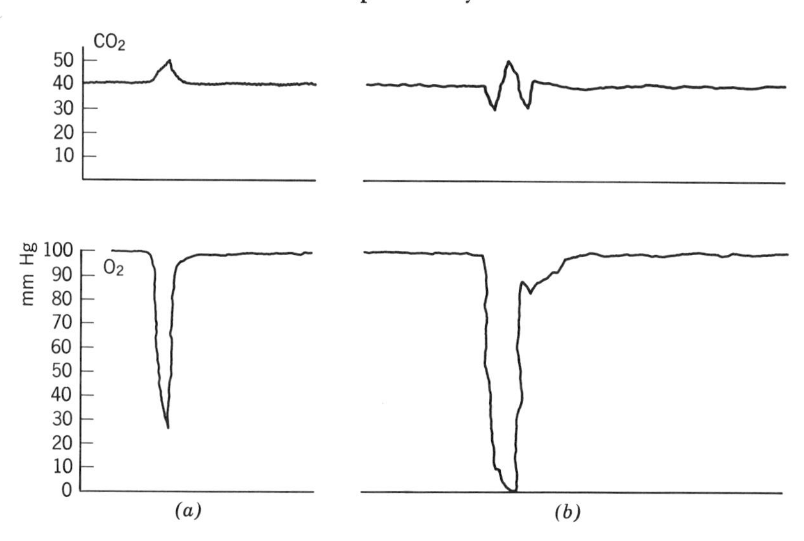

Fig. 14.1 Continuous recording of arterial carbon dioxide in a cat. (*a*) Response to closing of the trachea; (*b*) asphyxiation by nitrogen inhalation [3, adapted].

can be made in circulating blood; the same technique is presumably possible for studies in tissue also.

The analysis of dried blood plasma has been achieved using spark source mass spectrometry. This type of investigation is of interest to the biochemist in determining whether certain trace elements are essential to life. In addition, it is possible that an eventual correlation can be made between pathological conditions and variations in the abundance of specific elements. Wolstenholme [4] has reported the use of a technique for plasma analysis in which the sample was mixed with pure graphite and the graphite-support matrix was subjected to an independent assay. His results, together with several "known" average values for plasma are given in Table 14.1.

In addition to the data of Table 14.1, upper limits for the abundance of more than 40 elements were also reported. Although the agreement between results of the mass spectrometric and other techniques is not too good, the work was essentially exploratory and an accounting could be made for several discrepancies. There is little doubt, however, that with proper calibrations this type of analysis could be exceedingly useful in biology. Further, if a blood "fingerprint" of trace elements ever became highly accurate, this type of analysis might excite considerable interest in forensic circles.

Table 14.1

Analysis of Dried Blood Plasma [4]: Elements Detected (ppm weight)

Element	*Mass spectrometry*	*Typical av. values [5]*
Thorium*	0.4†	
Lead*	0.2	0.7
Barium	0.8	
Cesium*	0.05	
Iodine	1.0	1.4
Antimony*	0.9	
Cadmium*	2.0†	
Silver*	0.2	
Strontium*	0.7	0.56
Rubidium	14.0	34.0
Bromine	50.0	150.0
Zinc*	12.0	31.0
Copper	26.0	19.0
Iron	400.0	20.0
Manganese*	2.0	0.06
Chromium*	5.0	0.4
Potassium	25,000.0	2,740.0
Chlorine	17,000.0	62,000.0
Sulphur	5,400‡	14,600.0
Phosphorus	2,100.0	1,900.0
Silicon	7.0	
Aluminum*	5.0	7.0
Magnesium	500.0	374.0
Sodium	Major component	
Fluorine*	6.0	4.8
Boron	2.0	

* Not detected in unignited sample. Value quoted is for ignited sample.
† Variable results were obtained.
‡ From unignited sample.

IRON INVENTORY AND BLOOD VOLUME

The iron content in hemoglobin is of continuing interest and importance. This substance is the red coloring matter of blood, and each molecule shares an iron atom—thus making iron an indispensable element. Red cells are continually being destroyed and replaced by new ones that are formed in the bone marrow.

The radioisotope Fe^{59} was used in many early investigations to follow hemoglobin cycles and to show that the same iron is reused many times by the blood-forming centers. Today this isotope is used for the in-vivo distribution and movement of iron, for the measurement of plasma and red-cell turnover rates, for the determination of red-cell lifespan, and in many clinical tracer studies. Figure 14.2*a* illustrates the isotope's use in the measurement on newly formed red cells as they appear in the peripheral blood. Graphs show red blood cells for the case of normal patients and those with typically severe anemias [6].

Figure 14.2*b* illustrates the result of a clinical study in which an oral dose of 0.5–10 μc of Fe^{59} was administered to patients. The high retention rates are representative of an iron-deficiency anemia [7]. The radiotracer is also used to measure blood loss, and for various metabolism studies.

These and other uses of Fe^{59} suggest that parallel clinical experiments can be made using stable iron tracers and high-sensitivity mass spectrometry. The adult human body contains about 5 gm of iron; about 3 gm is contained in the hemoglobin of the red cells, 1.5 in the form of storage iron, and the remainder consists of iron being transported in the plasma and iron contained in enzymes [8]. These amounts are sufficiently large as to make feasible the use of a minor isotope of iron

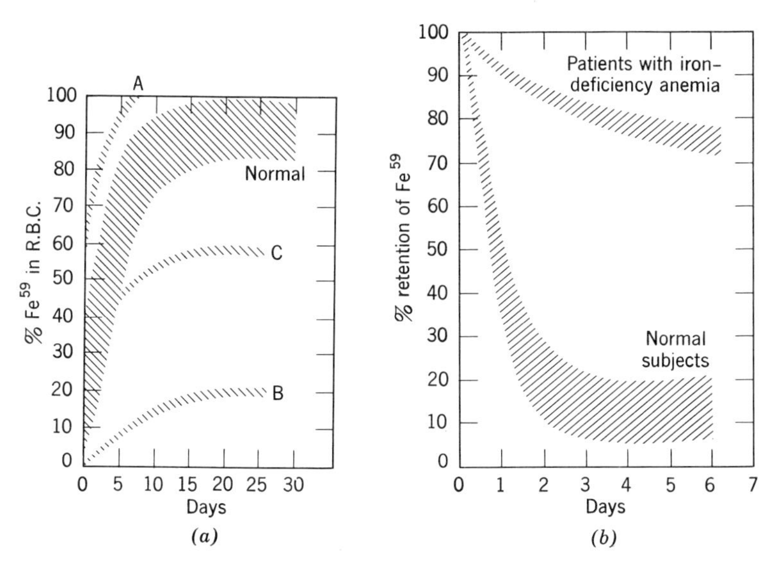

Fig. 14.2 (*a*) Typical Fe^{59} red-blood-cell uptake for normal and iron-deficient conditions. Curves A, B, and C show uptake rates reflecting specific anemias or other disorders. [6, adapted]. (*b*) Percent Fe^{59} whole-body retention in normal subjects and in patients with iron anemia [7, adapted]. (From *Nuclear Medicine,* ed. by W. H. Blahd, *copyright* 1965 by McGraw-Hill; used by permission.)

Table 14.2

Naturally Occurring Isotopes of Iron

Isotope	*Abundance (%)*
54	5.82
56	91.66
57	2.19
58	0.33
59*	†

* Radioactive.
† Half-life, 45 d.

as a stable tracer. Fe^{58} is the least abundant of the naturally occurring isotopes, as indicated in Table 14.2.

Clearly, samples of the stable isotope Fe^{58} could be added in a manner similar to the addition of Fe^{59}. Analysis of the Fe^{58} tracer as a function of time could be monitored by a direct mass spectrometric measurement by means of the isotopic-dilution method. An alternative is to sample the blood, and *after* sampling, employ the neutron-activation method. At least one application of stable Fe^{58} has been reported, with a subsequent neutron activation in a high flux reactor [9]. In this case the mass spectrometer is indispensable in certifying the degree of enrichment of Fe^{58}.

Using isotopic dilution, it should also be feasible to employ Fe^{58} for the measurement of blood volume. Albert [10] has listed seven criteria for a tracer to measure blood volume:

1. The tracer should be of such a nature that it can be easily identified and accurately measured quantitatively in low concentrations.
2. It should neither deteriorate nor break down when mixed with blood.
3. It should have no toxic effect on the organism nor alter the nature of blood.
4. It should remain within the vascular tree throughout the duration of the test.
5. The volume of tracer administered should be small and not affect the over-all volume being measured.
6. The tracer should mix evenly with the constituents of blood within the vascular bed.
7. The tracer should be devoid of antigenic properties.

Also because blood volume is the sum of two components—a solution (plasma) and a cellular suspension that acts as a filler—Albert [11] suggests that a single tracer will not necessarily permit a simple computation of volume. However, it would appear that Fe^{58} could be used in the same role as Fe^{59} for blood volume determinations. Thus Fe^{58} could be injected in a known fluid concentration and a subsequent blood sampling would reveal an anomalously large abundance of this one iron isotope. In other words, if a known volume of known isotopic content is added to an unknown volume having a known isotopic abundance, a mass spectral assay will allow the direct computation of the unknown volume, within the limitations cited above.

There also exists considerable biological interest in blood flow rates. Whether mass spectrometry can provide accurate information of such flow rates is an unanswered question.

SIZE OF BODY POOLS

The total quantity of any body constituent, or the "pool" of that constituent, is always of biological interest. Blood is a single example. Glucose is another, and C^{14}-labeled glucose has been used in connection with isotopic-dilution analysis.

Total body water has also been measured in animals, using isotopic labeling of deuterium or tritium. For rabbits whose water content has been measured by both desiccation and by a tritium tracer in water, there is agreement within a few percent. In fact, for animals whose carcasses are large (e.g. cows; $H_2O \sim 500$ kg) and for cases in which desiccation is difficult, the tritium tracer has been found quite feasible. The interest in the water content of cattle relates to calculating the fat and lean content; a knowledge of total body compositions is desirable to investigators conducting feeding experiments. When the mass spectrometer is used in connection with deuterium for measuring body water, the deuterium is converted to the gaseous phase. Body water has been measured for human subjects by this method, with a reported accuracy within 1 liter for an approximate total body water of 45 liter [12].

For ascertaining the body pool of some important elements, the use of stable tracers and the isotopic-dilution method is, of course, not always applicable. Sodium has a single naturally occurring isotope, Na^{23}. Unfortunately, a very important biological element, phosphorus, is also monoisotopic. Potassium, however, has three isotopes; thus a number of important measurements could be made without recourse to the short-lived K^{42} (12.4 hr). Rb^{86} (19 d) is sometimes used because it exchanges with potassium and its half-life is more suitable. Rb^{85} and Rb^{87} exist as stable isotopes; hence the isotopic-dilution method is applicable, within limits for this element. Rubidium can also be analyzed by mass spectrometry at levels below 10^{-10} gm.

BONE GROWTH

The adult skeleton contains between 1000 and 1500 gm of calcium and is renewed at a rate of about 0.05%/day. In a newborn infant the renewel rate may be 1% or higher [13]. These rates, however, vary considerably in the several parts of the skeleton. Ca^{45} has been an important radiotracer isotope in studying the mechanism of bone growth. It has a convenient half-life of 152 days, with a beta emitter of about 0.25 MeV. It has therefore been ideal for following bone growth by radioautographs (using x-ray film) or by dissection and radioactive counting methods.

For extended investigations, however, the naturally occurring minor stable isotopes of calcium shown in Table 14.3 appear ideal as tracers. One study has already been reported in which Ca^{46} and Ca^{48} were employed [14]. The isotopes were used as tracers in the skeletal metabolism of radiosensitive subjects (young children and pregnant women). The methodology involved the administration of enriched isotopes and the determination of the resulting increased isotopic abundances in blood and excreta by activation analysis. It is again emphasized that if separated isotopes are made available at more modest cost, studies could be safely conducted on specimens with no side effects resulting from radioactivity. There would also be no limit on the lifetime of the study. Presumably intakes of calcium could be programmed in connection with dental studies of children—in whom teeth are easily available specimens over a period of several years. It is also interesting to speculate that, with small animals, it would be possible to supply a completely monoiso-

Table 14.3

Isotopes of Calcium

Mass No.	*Abundance (%)*
40	96.97
42	0.64
43	0.145
44	2.06
46	0.0033
48	0.185

topic diet (e.g., of Ca^{42})—in which case the major isotope, Ca^{40}, could later serve as the "tracer."

Stable strontium has also been reportedly used as a bone tracer [15], but to avoid toxic effects the doses were small. Subsequent analyses by flame photometry failed to yield the desired accuracy. Presumably mass spectral assays could provide an enhanced sensitivity over the results of photometric measurements.

ISOTOPE EFFECTS AND DIFFUSION

In any dynamic system there is a possibility that identical *chemical* species will not undergo identical chemical behavior—an isotope or mass effect will influence a reaction rate. This effect has been established in numerous studies and it has special significance in complex biological systems, where reactions seldom proceed to equilibrium.

Molecules consisting of lighter isotopes are known to react faster than those comprised of the heavier atoms. The effects, of course, are noticeable among the hydrogen isotopes H^1, H^2, and H^3 because of the very large percentage mass difference. Significant effects have also been observed with carbon. Using C^{14} as a tracer, several investigators have shown the isotope effect to be very marked for rapidly growing plants. In reactions of synthesis, in which products of one reaction provide the materials for successive reactions, cumulative differences as high as 24% have been recorded for the C^{12}-C^{14} assimilation from the environment. It is also of interest to note that the very small variations of isotopic abundance (from "normal") that are found in nature are sometimes traced to "enrichments" produced by organisms. Differences in the isotopic abundance of the copper isotopes have been observed in very special situations and are definitely attributed to such an effect.

Diffusion is another kinetic phenomenon that can utilize isotopes as a tool. An example from radiotracer studies serves to illustrate the method. Harrison and Harrison [16] have performed an ingenious experiment to show that vitamin D affects the intestines in a manner that aids in the transport of calcium ions through the intestinal wall. Using an intestinal loop as a diffusion "test tube," these workers arranged to study the passage of calcium through the intestinal wall. A solution of $Ca^{45}Cl_2$ supplied the diffusant material. They conclusively showed that the calcium transfer was exponential—as would be noted in a diffusion-type experiment—and not enzymatic. Comparing healthy and vitamin-D-deficient intestines from animals, Harrison and Harrison observed that the diffusion rates of Ca^{45} ions were reduced by a factor of two for the vitamin-deficient specimens.

This type of diffusion study may also warrant a consideration of stable isotopes and mass spectrometry for use when suitable radiotracers are not feasible.

METABOLISM

In the physical sciences there often exist alternative approaches to measuring a specific phenomenon. In biochemistry, however, labeled compounds appear to be the sole method for answering certain questions relating to "intermediate metabolism." Specifically, the biochemist would like information relating to (a) the history of a particular molecule in a metabolic process, (b) the mode of atomic groupings in molecules as they are mobilized as energy sources in synthesis, and (c) details of cell structure and constituents. Until the advent of tracer techniques, it was virtually impossible to obtain any detailed information of metabolic pathways or to determine the fate of a given atomic grouping subsequent to its disappearance into an organism. In general tracer methodology, isotopes are incorporated into tracer molecules in one of two ways: either as substituents for atoms normally comprising the primary structure of the compound or as radicals attached to the primary structure. Both approaches are widely utilized [17].

Sulfur is one biologically interesting element in metabolism studies, and especially so because of its multiple isotope composition (Table 14.4).

Radiotracers include S^{31}, S^{35} and S^{37}. S^{31} is a positron emitter having a 3.2-sec half-life. The latter two are beta emitters with respective half-lives of 87 days and 5 min. Thus S^{35} is the most useful radioisotope for biochemical research. However all the minor stable isotopes of sulfur could be used as multiple tracers.

Sulfur is important as a label for protein and amino acids. The

Table 14.4

Natural Isotopic Abundances of Sulfur

Sulfur isotope	*Abundance (%)*
32	95.06
33	0.74
34	4.18
36	0.016

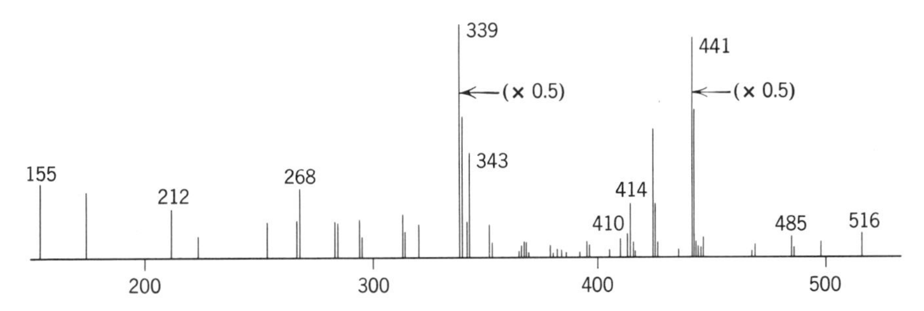

Fig. 14.3 Mass spectrum of peptide [19, adapted].

assimilation of sulfur into tissue as protein has been effected via a theoamino acid that can participate in the metabolic pool reactions [18]. Distribution patterns in animals of tissue protein have been observed using methionine; specific takeup has been noted in the carcass, liver, intestinal tract, kidney, lungs, skin, hair, and other tissues.

It is possible that a multiplicity of stable isotopes and radioisotopes would be an aid in tracing alternative metabolic pathways, as they are available for many metabolites (e.g., from the breakdown of protein or uptake of food). Further, there is rarely a complete transfer of compounds from one site to another and there are often several intermediate forms, each having its own metabolic pool being fed from other sources. Thus it seems clear that a more thorough survey should be made of the possible employment of stable isotopes (or stable isotopes plus radiotracers) that would aid in tracing reactions in living organisms. Mass spectrometric work has been recently reported by Shemyakin et al. [19] on the determination of the amino-acid sequence of peptides. A "typical" mass spectrum of a peptide, as reported by these investigators, is shown in Figure 14.3. Shemyakin et al. suggest that this type of mass spectrometry should make it possible to simplify determinations of the primary structure of proteins.

Perhaps one of the most exciting areas of recent research is that on memory storage processes. Some recent publications suggest that time-dependent changes in memory may be caused by changes in levels of transmitter substances. Rates of learning and RNA levels are reported to be enhanced by certain compounds, although the results are still considered controversial [20]. In any event there is a high probability that mass spectrometry will aid in the development of chemical therapeutics in dealing with memory pathology, as well as in basic research relating to the synthesis of macromolecules.

TOXICOLOGY

Lead has been one of civilization's most useful and most toxic metals. During the period of the Roman Empire it was used in the manufacture of water pipes, cups, sieves, cosmetics, external medicines, and paints. It was a serious toxic element in Roman times, and until very recently lead poisoning was one of the most frequent causes of death by poisoning in children. Lead poisoning has also been cited as probably being the most important toxicological problem in cattle. Stable lead is known to occur in smoke, and investigations have shown a correlation between concentrations of lead in bone and cigarette smoking [21]. There is also current interest in the uptake of lead and other toxic metals because of excessive air pollution.

Aronson and Hammond [22] have summarized many of the uses for the radioactive Pb^{210} isotope (22.5-yr half-life; specific activity $\sim 8.3 \times 10^3$ cpm per μg) and its daughter product Bi^{210}. Using the radioactive tracer, these investigators have reported analyses of bone, feces, urine, soft tissue, the kidney and liver, and the brain. The amount of stable lead per sample analyzed, other than brain samples, ranged from 4 to 12 μg. There can be no doubt of the advantages of radiotracers in some specific instances. But I should like to point out that lead is one of the heavy metals that lends itself to a fairly precise analysis by means of the stable-isotopic-dilution method. Figure 14.4 shows mass spectrum of lead in which the total sample size was 0.3 μg. The spectrum was obtained using a surface-ionization source. There is ample evidence to indicate that much smaller samples can be analyzed by this and other mass spectrometric techniques. It is my estimate that, by using enriched stable Pb^{204} as a spike, quantitative analyses can eventually be achieved below the 10^{-8}-gm level.

Most of the heavy toxic metals also have a multiplicity of naturally occuring isotopes, so the isotopic-dilution method would appear to have fairly general applicability.

Biological interest in cadmium has been stimulated because of the metal's toxic properties in industrial applications and the use of cadmium-plated containers. Levels of 15 ppm in food are reported to produce mild symptoms of poisoning [23]. Cadmium's occurrence in plant and animal tissues is only about 1 ppm, but much higher levels have sometimes been observed in certain body organs. (It is interesting that cadmium was the first common metal in which large nuclear-induced isotopic changes were measured by mass spectrometry.)

Arsenic is widely distributed in nature. Its abundance in soils ranges from 1 to 40 ppm and its abundance in the adult human body is 0.2

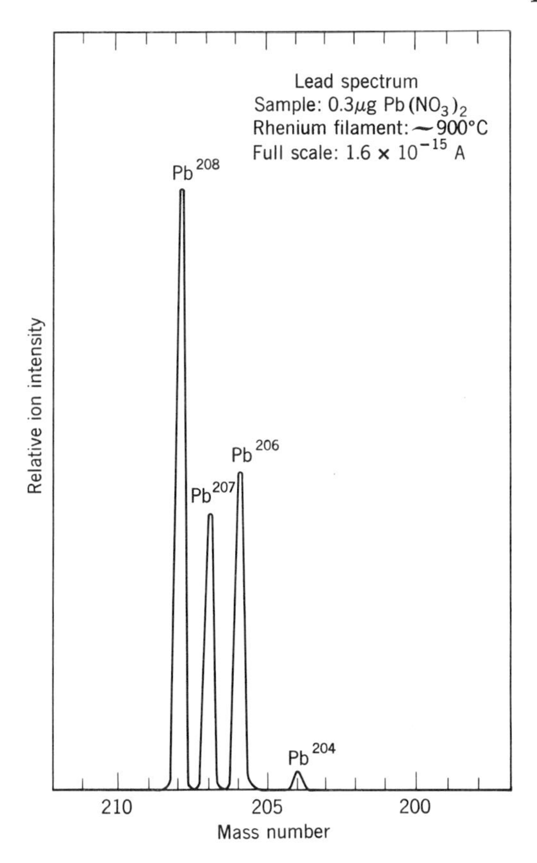

Fig. 14.4 Mass spectrum of a 0.3-μg Pb sample.

to 0.3 ppm [24]. One source of contamination is through the use of arsenical sprays on fruits, trees, and plants. The element can be detected mass spectrometrically at reasonable sensitivity, so its distribution in agricultural applications could be monitored. But the fact that only one isotope, As^{75}, exists in nature, makes spectrometry less attractive for highly quantitative work.

Pathophysiological studies in acute $HgCl_2$ poisoning have been made in an effort to relate the severity of cellular damage in the kidneys of rabbits to changes in renal blood flow [25]. Mercuric chloride was given intravenously in doses of 0.9 to 1.5 mg/kg. The multiplicity of mercury isotopes (seven) that can be made available in enriched form allows the isotopic dilution method to be applied for precision assays if desired.

TRACE ELEMENTS

Brief mention has been made of trace nutrients in Chapter 12, but further comment appears warranted in this section. An analogy might also be made between two completely divergent fields: semiconductors and biology. Electronic engineers will verify the fact that only within the last three decades have serious efforts been made to obtain superpure silicon and germanium. Today semiconductors are fabricated with a high degree of purity because of important technological developments and because even trace impurities would make many devices useless. Also, the role of certain chemicals in a crystal matrix is quite well understood by today's solid-state physicist; thus "impurity levels" can be set for particular elements. The situation in biology has both points of similarity and difference. About 40 years ago the only "trace elements" that were recognized as essential in biological systems were iron and iodine. At present, copper, zinc, manganese, cobalt, molybdenum, and selenium must be added to the "essential" list, and fluorine, bromine, barium, and strontium have been identified as being "probably essential" [26]. Most of the other elements that are found in minute quantities in living organisms are currently presumed to be merely environmental contaminants. However, a few metals such as cadmium, chromium, nickel, vanadium, and rubidium are suspected of having some functional significance, according to Underwood [27].

If the biochemist is required to ascertain the role of trace elements generally, it is imperative that the finest and most sensitive techniques be placed at his disposal. If an objective comparison is made among chemical methods, a variety of spectroscopic or spectrometric methods, and neutron-activation analysis, it is generally agreed that the "optimum analytical method" will depend on the specific element or compound, as well as a number of other factors. However, for the ultimate atomic sensitivity, mass spectrometry can, *in principle,* surpass all other techniques—except for relatively short half-life radionuclides.

There are several metals that exist in humans in the parts-per-million range and that should be amenable to mass spectral assay with some precision. Copper is one such element; it exists at about 1.5 to 2.5 ppm [28] (whole body) and in the enamel of teeth at about 15 to 30 ppm [29]. Blood concentrations are reported at levels of 0.5 to 1.5 mg/l. Metabolism studies of this element have been made, but considerably less is known about its transport and absorption than about iron. Molybdenum exists in quite small concentrations in adult humans. Levels of 3.2, 0.15, 0.14, and 0.14 ppm have been reported for the liver, lung, brain, and muscle [30].

Dietary intakes are known to raise or lower these concentrations significantly. In animals high molybdenum intakes are known to be toxic. Many studies have been made of cobalt because of its association with vitamin B_{12}. Cobalt, however, is monoisotopic and trace analysis is best achieved by techniques other than that of spectrometric assay. Zinc is normally present in man at levels of about 1.4 to 2.3 gm (about 20–30-ppm concentrations) [31]. This is over 10 times the body concentration of copper and less than half that of total body iron. Research on the role of this metal has been extensive—in eye tissues, various organs, blood, and studies of metabolism. Five stable isotopes of zinc are available; probably Zn^{70} would be the best tracer if available in sufficient quantity. A recent study indicates a high retention of zinc in zinc-deficient rabbits to whom the metal was administered orally [32]. Although few if any dietary deficiencies of zinc are known, it has been suggested that an intake of about 0.5 mg of Zn per kilogram of body weight is required to meet the daily needs of children [33].

The discovery of selenium as an essential element has led to many researches; a minimum selenium requirement close to 0.1 ppm has been reported for poultry [34]. Selenium has six naturally occuring isotopes, the abundances of which range from 0.87% for Se^{74} to 49.82% for Se^{80}.

Barium is a trace element in the human body: heart, 0.05; kidney, 0.10; lung, 1.0; muscle, 0.05 ppm [35]. Naturally occurring barium consists of seven isotopes and it can be detected at exceedingly high sensitivity (estimated at 10^{-10}-gm levels) by mass spectrometry with surface-ionization sources.

Boron has two isotopes, B^{10} and B^{11}, and this element can also be detected at high sensitivity by mass spectrometry. The average boron intake from ordinary adult diets is 10 to 20 mg/day [36]. However, while boron is an essential element in certain plants and algae, it is not known to be essential to man.

Both nickel and chromium are trace elements; these elements have five and four stable isotopes, respectively. Nickel is present in normal human blood at a level of about 3 μg/100 ml [37], but little is known of its physiological significance. All plant and animal tissues appear to contain chromium, and Cr^{51} (a radiotracer) has been used for the labeling of red cells [38]. Techniques for the simultaneous determination of red-cell mass and plasma volume have involved the use of I^{131}-labeled albumin and Cr^{51} labeled red cells. A serious difficulty [39] has been that of resolving the major photopeaks of the two radio tracers (I^{131} at 364 keV and Cr^{51} at 320 keV). It is possible that the use of a stable isotopic tracer (e.g., Cr^{54}) might be considered in such cases, together with the isotopic-dilution method for a precise analysis.

DRUG DISTRIBUTION

An increasingly active field of investigation relates to the uptake and distribution of pesticides in agriculture. Likewise the uptake and distribution of drugs in animal organs is a current area of research. The latter has been stimulated in part by the rigorous testing demanded by the Food and Drug Administration. Companies sponsoring the introduction of new drugs, food additives, dyestuffs, and so on, are required to assume burden of proof that exposure to such additives will not be accompanied by undesirable side effects. Chemical testing requiring many years of study can sometimes be circumvented by clearly and quantitatively following the complete patterns of absorption and retention in organs and tissue, as well as excretion rates. Also, the amount of drug that finds its way to the desired organ (i.e., drug efficacy) can also be measured by isotopic labeling. Absorption rates of the active ingredients of ointments and cosmetics also qualify for labeled-isotope techniques. For example, in one instance lanolin was found to favor a faster absorption of radioactive $Na^{24}Cl$ through the skin, as compared with other medications.

While radiotracers continue to be the logical choice in animal studies, stable isotopes and mass spectrometry will be given serious consideration for human subjects. It would be quite feasible, for example, to administer sizable doses of stable labeled compounds and measure excretion rates. Drug-efficacy studies could also be aided when elements of suitable half-life are either unavailable or the associated radioactivity could not be tolerated. The availability of both stable and radioactive tracer isotopes for the same element (e.g., C^{13} and C^{14}, respectively) also affords a unique opportunity for sensing whether a biological system does indeed undergo a radiation effect. Any experimental differences arising from the use of C^{13} and C^{14} labeled compounds might suggest an abnormality in metabolism induced by the radioactive tracer.

PHARMACEUTICAL CERTIFICATION

An important problem in the pharmaceutical industry is that of producing synthetic drugs of required purity. Classical methods of tracing impurities are reported to be sometimes misleading, tedious, or inaccurate [40]. Radioactive isotopes have recently been used in connection with production problems, and neutron-activation analyses have been employed to certify upper limits of elements such as arsenic and lead. In many instances it would appear that mass spectrometric assays might be the preferred technique from the standpoint of ease of analysis and, at the very least, in supplying an upper limit to the impurity concentrations of many elements. Appropriate standard samples would be required.

The important contribution of mass spectrometry in somewhat routine medical applications should also not be overlooked. The preparation and purification of gases for use in anesthesia is an important one. Commercial firms supplying these anesthetic gases to hospitals throughout the world rely on mass spectral assays to certify composition and purity and to monitor the mixing of the constituent gases in the proper proportion.

INDUSTRIAL MEDICINE

Industrial medicine embraces a large area, and one of its many professional concerns relates to the environment of those engaged in quasi-hazardous occupations. Routinely or on special demand, potentially toxic metals must be machined and hazardous chemical compounds processed. Large corporations are continually alert to the hazards of toxic metals and vapors and usually maintain strict controls for safeguarding the health of their employees. Standards have been established throughout various industrial groups and the use of exhaust hoods, filters, and ventilation systems is routinely employed to prevent toxic airborne dust or fumes from being inhaled. Generally speaking, there is little real need for mass spectrometry in this area.

An interesting suggestion, however, was made to the writer by W. J. Jend, Jr. [41], who suggested that results from the high analytical sensitivity of mass spectrometry might be appropriated for use as evidence in legal proceedings.

Consider a potentially hazardous environment in which toxic metal dust might be airborne in minute amounts. A continuous air monitor might consist of a filter paper that would provide for the cumulative collection of particulate matter. This filter paper could be replaced daily or weekly as the need might be. The paper could be dated and filed. There would be no need for incurring the cost of an assay unless some unanticipated legal action were instituted that might place the employer in the role of a defendant. Only at this time would it be necessary quantitatively to analyze the sample in order to establish *positive upper limits* on environmental toxicity. In this case the mass spectrometer would be subpoenaed as an "expert witness."

In view of the increasing sensitivity of the mass spectrometer for detecting potentially toxic metals, this procedure might become quite feasible. For example, recent measurements [42] indicate that a mass spectrometer can detect approximately 10^{-12} gm of beryllium—an amount far below present levels of routine chemical analysis.

CONCLUDING REMARKS

In the preface to this book I cited the risk of predicting the advance of science. Science has undergone a rapid growth and many mutations without revealing a truly predictive pattern, and history contains many pathological examples of scientific prophecy.

Only about a century ago, the eminent philosopher Auguste Comte concluded: "There are some things of which the human race must remain forever in ignorance, for example, the chemical constitution of the heavenly bodies." Today plans are in progress to go to the planets. But long before the advent of this Space Age, the astrophysicist had acquired a considerable knowledge of the distant stars. This knowledge was acquired by means of optical instruments and the spectrograph, which revealed star temperatures, mass, chemical composition, atmospheric structure, motion, and (indirectly) distance and age.

Indeed, the past century has pointed up the consistently important role of advanced instrumentation in all scientific discoveries. The identification of penicillin by Alexander Fleming was achieved through microscopy and the discovery of nuclear fission was made possible only by a new generation of analytical sensors. Instrumentation also seems to function as a catalyst to motivate investigators. The alert scientist is aware that a new tool may yield knowledge that is unsuspected and unsought and that, at the very least, it may allow measurements that were formerly impossible or marginal. Hence one can be realistically optimistic regarding stable isotopes joining radiotracer methods, especially in biological research. The great power of stable isotopes and mass spectrometric analyses, as with radiotracers, lies in the broad scope of application.

A visitor to the Argonne National Laboratory would be intrigued with its "Isotope Farm" where drug-forming plants and algae are nurtured in sealed greenhouses supplied with labeled carbon dioxide. As a radiotracer, C^{14} alone has generated thousands of imaginative studies. But consider the even wider latitude of analogous measurements that could be made if a very large number of stable isotopes could be made in adequate quantity. William Arnold [43] clearly stated the case for stable tracers by suggesting that all *open-field* research (in agriculture) would necessarily favor stable isotopes over radionuclides. Even if the procedure were deemed safe, permission for uncontrolled planting with radiotagging would hardly be granted. Neither would long-term studies in human subjects be allowed.

With stable isotopes, however, studies of trace nutrients could be conducted without licensing and without the restrictions imposed by undesirably short half-lives. Compounds would not "spoil" by "radiation decomposition," nor would radioactivity or toxic daughter decay products obscure the phenomenon under investigation.

Thus a final invitation is extended to the reader to consider the manifold possible applications of stable isotopes, whether presently feasible or not. Perhaps an analogy can be made with respect to electromagnetic radiations. A century ago few layman or scientists could foresee the unlimited exploitation of the electromagnetic spectrum. Today it is not only the primary analytical medium of the scientist but it is also an energy form of such importance that it is literally rationed. The electromagnetic spectrum is, at some frequency, used for every channel of commerce: for transportation, for the protection of life and property, and for education and entertainment—plus its more esoteric uses for telemetering information from outer space. Indeed, the radio spectrum is viewed by the Federal Communications Commission as a natural resource that exists in the public domain—and the agency administers it accordingly.

Stable isotopes also represent a vast natural resource. The almost invariant isotopic ratios in nature permit us to trace the elements that man has extracted from the earth's crust and compounded into countless useful forms. But isotopic spectra are potentially an even greater information source. As an electromagnetic carrier frequency transfers information only when it is modulated, so information retrieval from mass spectral data will depend on the countless variations from the "normal" isotopic abundances that a mass spectrometer can measure. Surely such possibilities should excite both present and future generations of scientists.

REFERENCES

[1] M. Cohn, in *Isotopic and Cosmic Chemistry,* ed. by H. Craig, S. L. Miller, and G. J. Wasserburg, North-Holland, Amsterdam, 1964, Ch. 5, p. 45.

[2] D. Shemin and D. Rittenberg, *J. Biol. Chem.,* **166,** 627 (1946).

[3] S. Woldring, G. Owens, D. C. Woolford, *Science,* **153,** 885 (1966).

[4] W. A. Wolstenholme, *Nature,* **203,** 1284 (1964).

[5] H. J. M. Bowen, United Kingdom Atomic Energy Authority Report No. R-4196.

[6] I. M. Weinstein, R. F. Schilling, and H. N. Wagner, Jr., in *Nuclear Medicine,* ed. by W. H. Blahd, McGraw-Hill, New York, 1965, Ch. 12, p. 314.

[7] W. H. Blahd, in *Nuclear Medicine,* ed. by W. H. Blahd, McGraw-Hill, New York, 1965, Ch. 22, p. 591.

[8] I. M. Weinstein et al., *op. cit.*, p. 308.
[9] J. T. Lowman and W. J. Krivit, *Lab. Clin. Med.*, **61,** 1042 (1963); see also W. Krivit, "Activation Analysis: A New Stable Isotope Method for Human Biological Tracer Studies," United States Atomic Energy Commission, TID-4204, Washington D.C., 1966, p. 286.
[10] S. N. Albert, in *Nuclear Medicine,* ed. by W. H. Blahd, McGraw-Hill, New York, 1965, Ch. 19, p. 530.
[11] *Ibid.*
[12] F. K. Bauer, in *Nuclear Medicine,* ed. by W. H. Blahd, McGraw-Hill, New York, 1965, Ch. 18, p. 517.
[13] G. C. H. Bauer, in *Nuclear Medicine,* ed. by W. H. Blahd, McGraw-Hill, New York, 1965, Ch. 16, p. 470.
[14] *Ibid.*, p. 465.
[15] *Ibid.*, p. 465.
[16] H. E. Harrison and H. C. Harrison, *Am. J. Physiol.,* **199,** 265 (1960).
[17] E. M. Gold, in *Nuclear Medicine,* ed. by W. H. Blahd, McGraw-Hill, New York, 1965, Ch. 21, p. 564.
[18] H. Tarver and L. M. Morse, *J. Biol. Chem.*, **173,** 53 (1958).
[19] M. M. Shemyakin, Yu. A. Ovchinnikov, A. A. Kiryushkin, E. I. Vinogradova, A. I. Miroshnikov, Yu. B. Alakhov, V. M. Lipkin, Yu. B. Shvetsov, N. S. Wulfson, B. V. Rosinov, V. N. Bochkarev, and V. M. Burikov, *Nature,* **211,** 361 (1966).
[20] J. L. McGaugh, *Industr. Res.*, 81 (February 1967).
[21] R. B. Holtzman and F. H. Ilcewicz, *Science,* **153,** 1259 (1966).
[22] A. L. Aronson and P. B. Hammond, *Nucleonics,* **22,** 90 (1964).
[23] S. Frant and J. Kleeman, *J. Am. Med. Assoc.*, **117,** 86 (1941).
[24] E. J. Underwood, *Trace Elements,* Academic, New York, 1962, p. 327.
[25] G. V. Taplin and M. Yamaguchi, United States Atomic Energy Commission Report; UCLA-541, Washington, D.C., 1965.
[26] E. J. Underwood, *Trace Elements,* Academic, New York, 1962, p. 2.
[27] *Ibid.*, p. 3.
[28] T. P. Chou and W. H. Adolph, *Biochem. J.*, **29,** 476 (1935).
[29] F. Brudevold and L. T. Steadman, *J. Dent. Res.*, **34,** 209 (1955).
[30] E. J. Underwood, *op. cit.*, p. 101.
[31] E. M. Widdowson, R. A. McCance, and C. M. Spray, *Clin. Sci.,* **10,** 113 (1951).
[32] E. R. Graham and P. Telle, *Science,* **155,** 691 (1967).
[33] A. Stern, M. Nalder, and I. G. Macy, *J. Nutrition,* **21** (Suppl.), 8 (1941).
[34] E. J. Underwood, *op. cit.*, p. 318.
[35] *Ibid.*, p. 331.
[36] N. L. Kent and R. A. McCance, *Biochem J.*, **35,** 837 and 877 (1941).
[37] E. J. Underwood, *op. cit.*, p. 343.
[38] J. D. Pearson, *Int. J. Appl. Radiat. Isotopes,* **17,** 13 (1966).
[39] B. S. Wood and S. H. Levitt, *J. Nucl. Med.*, **6,** 433 (1965).
[40] A. Alain, Y. M. Dessovky, and R. Shabana, *J. Pharmaceut. Sci.,* **55,** 969 (1966).
[41] W. J. Jend, Jr. (medical director, Michigan Bell Telephone), private communication.
[42] J. A. McHugh and J. C. Sheffield, *Anal. Chem.*, **39,** 377 (1967).
[43] W. A. Arnold (Oak Ridge National Laboratory), private communication.

Appendix 1

Isotopic Abundances of the Elements

Element	*Mass No.*	*Relative abundance*	*Element*	*Mass No.*	*Relative abundance*
H	1	99.985	Ar	36	0.337
	2	0.015		38	0.063
He	3	0.00013		40	99.600
	4	~100.0	K	39	93.08
Li	6	7.42		40	0.0119
	7	92.58		41	6.91
Be	9	100.0	Ca	40	96.97
B	10	19.78		42	0.64
	11	80.22		43	0.145
C	12	98.892		44	2.06
	13	1.108		46	0.0033
N	14	99.63		48	0.185
	15	0.37	Sc	45	100.0
O	16	99.759	Ti	46	7.95
	17	0.0374		47	7.75
	18	0.2039		48	73.45
F	19	100.0		49	5.51
Ne	20	90.92		50	5.34
	21	0.257	V	50	0.24
	22	8.82		51	99.76
Na	23	100.0	Cr	50	4.31
Mg	24	78.60		52	83.76
	25	10.11		53	9.55
	26	11.29		54	2.38
Al	27	100.0	Mn	55	100.0
Si	28	92.27	Fe	54	5.82
	29	4.68		56	91.66
	30	3.05		57	2.19
P	31	100.0		58	0.33
S	32	95.06	Co	59	100.0
	33	0.74	Ni	58	67.76
	34	4.18		60	26.16
	36	0.016		61	1.25
Cl	35	75.4		62	3.66
	37	24.6		64	1.16
			Cu	63	69.1
				65	30.9

Appendix 1 (Continued)

Element	*Mass No.*	*Relative abundance*
Zn	64	48.89
	66	27.82
	67	4.14
	68	18.54
	70	0.617
Ga	69	60.2
	71	39.8
Ge	70	20.55
	72	27.37
	73	7.67
	74	36.74
	76	7.67
As	75	100.0
Se	74	0.87
	76	9.02
	77	7.58
	78	23.52
	80	49.82
	82	9.19
Br	79	50.52
	81	49.48
Kr	78	0.354
	80	2.27
	82	11.56
	83	11.55
	84	56.90
	86	17.37
Rb	85	72.15
	87	27.85
Sr	84	0.56
	86	9.86
	87	7.02
	88	82.56
Y	89	100.0
Zr	90	51.46
	91	11.23
	92	17.11
	94	17.40
	96	2.80
Nb	93	100.0
Mo	92	15.86
	94	9.12
	95	15.70
	96	16.50
	97	9.45
	98	23.75
	100	9.62
Ru	96	5.47
	98	1.84
	99	12.77
	100	12.56
	101	17.10
	102	31.70
	104	18.56
Rh	103	100.0
Pd	102	0.96
	104	10.97
	105	22.23
	106	27.33
	108	26.71
	110	11.81
Ag	107	51.82
	109	48.18
Cd	106	1.22
	108	0.87
	110	12.39
	111	12.75
	112	24.07
	113	12.26
	114	28.86
	116	7.58
In	113	4.23
	115	95.77
Sn	112	0.95
	114	0.65
	115	0.34
	116	14.24
	117	7.57
	118	24.01
	119	8.58
	120	32.97
	122	4.71
	124	5.98

Appendix 1 (Continued)

Element	*Mass No.*	*Relative abundance*
Sb	121	57.25
	123	42.75
Te	120	0.089
	122	2.46
	123	0.87
	124	4.61
	125	6.99
	126	18.71
	128	31.79
	130	34.49
I	127	100.0
Xe	124	0.096
	126	0.090
	128	1.92
	129	26.44
	130	4.08
	131	21.18
	132	26.89
	134	10.44
	136	8.87
Cs	133	100.0
Ba	130	0.101
	132	0.097
	134	2.42
	135	6.59
	136	7.81
	137	11.32
	138	71.66
La	138	0.089
	139	99.911
Ce	136	0.193
	138	0.250
	140	88.48
	142	11.07
Pr	141	100.0
Nd	142	27.13
	143	12.20
	144	23.87
	145	8.30
	146	17.18
	148	5.72
	150	5.62
Sm	144	3.16
	147	15.07
	148	11.27
	149	13.84
	150	7.47
	152	26.63
	154	22.53
Eu	151	47.77
	153	52.23
Gd	152	0.20
	154	2.15
	155	14.73
	156	20.47
	157	15.68
	158	24.87
	160	21.90
Tb	159	100.0
Dy	156	0.052
	158	0.090
	160	2.294
	161	18.88
	162	25.53
	163	24.97
	164	28.18
Ho	165	100.0
Er	162	0.136
	164	1.56
	166	33.41
	167	22.94
	168	27.07
	170	14.88
Tm	169	100.0
Yb	168	0.140
	170	3.03
	171	14.31
	172	21.82
	173	16.13
	174	31.84
	176	12.73
Lu	175	97.40
	176	2.60

Appendix 1 (Continued)

Element	*Mass No.*	*Relative abundance*	*Element*	*Mass No.*	*Relative abundance*
Hf	174	0.18	Pt	190	0.012
	176	5.15		192	0.78
	177	18.39		194	32.8
	178	27.08		195	33.7
	179	13.78		196	25.4
	180	35.44		198	7.21
Ta	180	0.012	Au	197	100.0
	181	99.988	Hg	196	0.15
W	180	0.135		198	10.02
	182	26.4		199	16.84
	183	14.4		200	23.13
	184	30.6		201	13.22
	186	28.4		202	29.80
				204	6.85
Re	185	37.07	Tl	203	29.50
	187	62.93		205	70.50
Os	184	0.018	Pb	204	1.48
	186	1.59		206	23.6
	187	1.64		207	22.6
	188	13.3		208	52.3
	189	16.1	Bi	209	100.0
	190	26.4	Th	232	100.0
	192	41.0	U	234	0.0057
Ir	191	37.3		235	0.72
	193	62.7		238	99.27

Appendix 2

Ionization Potentials, Electron Affinities,* Work Functions, and Melting Points

Element	*IP (eV)*	*EA (eV)*	*Work function (eV)*	*Melting point (°C)*
H	13.60	0.77		−259.14
He	24.58			−272.2†
Li	5.39	0.6	2.4	179
Be	9.32		3.67	1278
B	8.30	0.4	4.5	2300
C	11.26	1.25	4.4	>3550
N	14.53			−209.86
O	13.61	1.47		−218.4
F	17.42	3.45		−219.62•
Ne	21.56			−248.67
Na	5.14	0.3	2.3	97.81
Mg	7.64		3.7	651
Al	5.98	1.0	4.2	660
Si	8.15	1.9	4.7	1410
P	10.48	1.1		44.1‡
S	10.36	2.07		112.8§
Cl	12.96	3.61		−100.98•
Ar	15.76			−189.2•
K	4.34	0.9	2.2	63.65
Ca	6.11		2.7	842
Sc	6.54			1539
Ti	6.82	0.4	3.6–4.3	1675
V	6.74	0.9	4.1	1890
Cr	6.76	1.0	4.6	1890
Mn	7.43		3.8	1244
Fe	7.87	0.6	4.5	1535
Co	7.86	0.9	4.2	1495
Ni	7.63	1.3	5.0	1453
Cu	7.72	1.8	4.4	1083
Zn	9.39		4.4	419.4
Ga	6.00		3.8	29.78
Ge	7.90		4.8	937.4
As	9.81		4.8	817 ¶
Se	9.75		4.4	217 ¶
Br	11.81	3.36		−7.2
Kr	13.99			−156.6
Rb	4.18	0.6	2.1	38.89
Sr	5.69		2.7	769
Y	6.38			1495

Appendix 2 (Continued)

Element	*IP (eV)*	*EA (eV)*	*Work function (eV)*	*Melting point (°C)*
Zr	6.84		4.0	1852
Nb	6.88		4.2	2468
Mo	7.10		4.0–4.3	2610
Tc	7.28			2200
Ru	7.36		4.5	2250
Rh	7.46		4.6	1966
Pd	8.33		4.6	1552
Ag	7.57	(2.0)	4.6	960.8
Cd	8.99		4.1	320.9
In	5.79		4.1	156.6
Sn	7.34		4.4	231.9
Sb	8.64		4.6	630.5
Te	9.01		4.8	449.5
I	10.45	3.06		113.5
Xe	12.13			−111.9
Cs	3.89	0.6	2.1	28.5
Ba	5.21		2.5	725
La	5.58		3.3	920
Ce	5.65		2.8	795
Pr	5.42		2.7	935
Nd	5.49		3.3	1024
Pm	5.55			1035
Sm	5.63		3.2	1072
Eu	5.68			826
Gd	6.16			1312
Tb	5.85			1356
Dy	5.93			1407
Ho	6.02			1461
Er	6.10			1497
Tm	6.18			1545
Yb	6.25			824
Lu	6.15			1652
Hf	6.8		3.5	2150
Ta	7.88		4.3	2996
W	7.98		4.5	3410
Re	7.87		5.0	3180
Os	8.5		4.8	3000
Ir	9.1		5.3	2410
Pt	9.0		5.6–5.8	1769
Au	9.22	(2.8)	4.7	1063
Hg	10.43	1.8	4.5	−38.9
Tl	6.11		3.7	303.5
Pb	7.42		4.0	327.5
Bi	7.29		4.3	271.3
Po	8.43			254

Appendix 2 (Continued)

Element	*IP (eV)*	*EA (eV)*	*Work function (eV)*	*Melting point (°C)*
At	9.5			
Rn	10.75			−71
Fr	4.0			
Ra	5.28			700
Ac	6.9			1050
Th	7.0		3.73	1700
Pa				1230
U	6.11		3.08	1132
Np				640
Pu	5.8			639

* Ionization potentials, electron affinities and work function data were kindly supplied by R. E. Honig.
† 26 atm.
‡ White.
§ Rhombic.
¶ Gray.
• Freezing point.

Author Index

Subject Index